The Culture of Science Education

NEW DIRECTIONS IN MATHEMATICS AND SCIENCE EDUCATION
Volume 3

Scope

Mathematics and science education are in a state of change. Received models of teaching, curriculum, and researching in the two fields are adopting and developing new ways of thinking about how people of all ages know, learn, and develop. The recent literature in both fields includes contributions focusing on issues and using theoretical frames that were unthinkable a decade ago. For example, we see an increase in the use of conceptual and methodological tools from anthropology and semiotics to understand how different forms of knowledge are interconnected, how students learn, how textbooks are written, etcetera. Science and mathematics educators also have turned to issues such as identity and emotion as salient to the way in which people of all ages display and develop knowledge and skills. And they use dialectical or phenomenological approaches to answer ever arising questions about learning and development in science and mathematics.

The purpose of this series is to encourage the publication of books that are close to the cutting edge of both fields. The series aims at becoming a leader in providing refreshing and bold new work—rather than out-of-date reproductions of past states of the art—shaping both fields more than reproducing them, thereby closing the traditional gap that exists between journal articles and books in terms of their salience about what is new. The series is intended not only to foster books concerned with knowing, learning, and teaching in school but also with doing and learning mathematics and science across the whole lifespan (e.g., science in kindergarten; mathematics at work); and it is to be a vehicle for publishing books that fall between the two domains—such as when scientists learn about graphs and graphing as part of their work.

The Culture of Science Education
Its History in Person

Edited by

Kenneth Tobin

The Graduate Center, City University of New York, USA

Wolff-Michael Roth

University of Victoria, Canada

SENSE PUBLISHERS
ROTTERDAM / TAIPEI

A C.I.P. record for this book is available from the Library of Congress.

Paperback ISBN: 90-77874-33-X
Hardback ISBN: 90-77874-35-6

Published by: Sense Publishers,
P.O. Box 21858, 3001 AW
Rotterdam, The Netherlands

Printed on acid-free paper

CONTENTS

CONTENTS

PREFACE

This book about the culture and history of science education told through the auto-biographies of key persons in the field, as any book, is the result of a historical process that we, as any individual, produce and are subjected to. Much in the same way that history is not made by individuals who act independently but by individuals who concretize cultural possibility, this book is not merely the outcome of two author-editors getting together to put our spin on history. Rather, there is a point in the cultural history of a field where realizing such a book that takes the cultural history as its topic becomes a general possibility, which is then realized in concrete form by particular scholars. When we conceived of this book, we set our goal to be an exploration of some issues in science education as they have developed historically and internationally.

In the past there have been several efforts to capture the culture and history of science education, often in ways that we do not especially value or learn from the experiences of those in the field. A clear alternative is to tell the history through the lives of science educators. Accordingly, we sat down and identified 20 science educators other than ourselves whose participation in science education research has been noteworthy, international, and varied in terms of gender, race, focus and place on the career ladder. The approaches they have taken to research differ too. We thought that it would be interesting to read autobiographies of these people, accounts that capture the ways in which they have participated in science, education and science education as teachers and researchers. These accounts would touch on issues considered salient by the authors, but probably would deal with entry, progress and finding ways to succeed in science education. Issues of mentoring would be of interest, as would contradictions experienced in activities such as tenure, promotion, editing, publishing, participating in national meetings, being a consultant, obtaining external support etc.

We asked the invited authors to write a chapter of no more than 6,000 words in which they were to deal with some or all of the following:

1. Your participation in science and science education;
2. Key research foci at different critical points in your career;
3. Significant peers and roles models—the people and their work (including dissertation advisor);
4. Other biographical experiences that shaped your approach to research and teaching in the field;
5. The road ahead (the field and the people);

6. Review of key accomplishments—a review of your research (up to 2,000 of the 6,000 words)—perhaps identifying up to five key papers or books; and

7. Former doctoral students and postdoctoral associates (presented in terms of their work—not a list).

After receiving each contribution, we worked with each author through a series of iterations until his or her chapter was in a form that met both their own and our approval and the needs of producing a volume that cohered rather than constituted a collection of independent essays.

Works such as this book are impossible without the support of those surrounding us. We are grateful to our respective spouses Barbara Tobin and Sylvie Boutonné for their patience, which has allowed us to spend all that time required by the completion of this volume. Inevitably, the writing of this volume produced its tensions between us. However, as scholars and friends and a commitment to difference, we have produced a strong text and maintain a deep respect for one another and the principle of learning from diverse perspectives.

New York, USA
Victoria, Canada
October 2006

HISTORY AND AUTO/BIOGRAPHY

"Great people, leaders with exceptional qualities, make history"—following Winston Churchill, many believe that exceptional people make history and that others are mere foot soldiers following in their path. From this perspective, there are some "great" individuals whose actions causally determine the direction a field takes. This perspective leads to the allure of biographies, which tend to celebrate the greatness of the persons showing how their genius or political savvy affected intellectual fields, world events, and history more generally. Biographies of Galileo Galilei, Albert Einstein, or Marie Curie tend to construct and highlight the genius in these people generally omitting other aspects of their lives. Thus, early biographies of Sir Isaac Newton tended to highlight his contributions to mathematics and physics, celebrating his outstanding qualities and genius. On the other hand, they did not at all concern themselves with the religious, mystical, and obsessive sides of the person. A recent biography, however, shows a different person: he was a twisted, tortured mystic with homosexual tendencies, an ability to hold grudges for decades, an egomaniac, and a very petty individual (White, 1999). Newton spent decades and decades with alchemy and the attempt to decipher Old Testament prophecies, thinking that the design of Solomon's temple was a code for the entirety of recorded human history. The great Sir Isaac Newton was not so great after all. The point is that the biographies of the earlier form are used to construct the image of individuals as shapers of the history of a field; the individuals are portrayed as the causes of history.

There are also individuals believing that—conversely—it is history that makes great people. According to this perspective, there are situations in the history of a country or discipline that provide opportunities for certain individuals to stand out. Thus, to take an example from political history, a rather unassuming and pale George W. Bush, projected to go down as being a president with a transitional role in American history, found his place after the terrorist attacks on the twin towers in New York, an event now referred to as *9/11*. That is, it is this historical event that provided the individual currently holding the presidency to take on a special role, allowing him to set in motion a war machinery that devastated two countries in the Middle East. Bush made history and he made himself a name—but only after history provided an appropriate slot for doing so. In such explanations, history is used as the cause to explain why a certain individual has become a noted and notorious politician who shaped worldly events.

Both forms of explanation fall far short of a good account of the relationship between history and persons, whose lives are accounted for in auto/biographies, because both lead to deterministic ways of understanding the world. A more variegated and sophisticated account of the relationship between individual biography (personal history) and the history of institutions (disciplines, countries) has recently been articulated (Holland & Lave, 2001). Accordingly, historical struggles in person (history in person) and historically institutionalized struggles (enduring struggles) are dialectically related processes that continuously play themselves out in local contentious practice. History thereby turns out to be "history in practice" (p. 6), where practice always means real humans being engaged in realizing whatever cultural-historical form of activities they currently participate. In and through participation, identities are continuously produced and reproduced. Let us concretize this perspective in the following account of an identity- and history-defining event.

For the 1992 annual meeting of the National Association of Research in Science Teaching, we had organized what turned out to be a well-attended symposium focusing on the relationship between beliefs about epistemology, nature of science, and learning as students and teachers articulate them. Michael—a high school teacher at the time—had presented a paper entitled "Physics students' epistemologies and views of knowing and learning." In the paper, he provided an account of his physics course, which predominantly consisted of open-inquiry. To counterbalance the realist presentation of physics in the textbook, Michael had asked students to read and discuss articles and book chapters that took a social constructivist perspective on science and knowledge—including, for example, such authors as David Suzuki (Canadian geneticist and environmentalist) and Gregory Bateson (anthropologist, epistemologist). During the discussion period, a science education historian vehemently attacked Michael for "indoctrinating" students into an epistemology that was outright wrong. Struggling with a response, Michael deferred to Ken, who responded in the former's defense.

This event constitutes a moment of history in the field of science education: a moment of local contentious struggle in which the historically institutionalized struggle between "constructivists" and "realists" was played out and became a defining moment in the historical struggles of personal identities. In attacking, the historian not only articulated Michael as a constructivist demon, denoting him as a member of a particular subculture, but also provided a resource for constructing himself as a realist. That is, even without saying that he is a realist, the nature of the historian's actions constituted a resource for anyone present to construct his identity. Ken, too, in defending what Michael had done as a classroom teacher, provided resources for anyone present to construct an aspect of his identity. Thus, through his actions, Ken not only provided material that allowed others to say that he was a constructivist—even though in his talk he did not say so—but also provided materials to construct his identity as that of a mentor, advocate for junior members in the field, and so on. Michael, too, even without having to say that he was a constructivist, the account he provided of what he had done in the classroom provided sufficient materials for others, including the historian, to attribute to him the identity of a "constructivist."

This analysis shows how the historical debates within the field of science education that confronted "constructivist" and "realist" ideas about the nature of science and epistemology played themselves out in local contentious practice—here the production and reproduction of a symposium at a scholarly conference. In fact, the ad hominem attack turned this symposium into a particular one, distinguishing it from others, though other aspects still reproduced it as a culturally recognizable form. The event also contributed to the construction of identities to the extent that some NARST members who remember the event refer to it in terms of a battle of personalities. Thus, persons and personal identities are closely related to the institutions of which they are a constitutive part—though the relation is not deterministic. But of what sort then is the relation between the two?

The contributions to *Auto/biography and Auto/ethnography* (Roth, 2005a) provide us with a first answer to this question, as they suggest a dialectical relationship between individual and collective. Thus, individual lives are concrete realizations of possible lives, where possibilities always exist at a collective level. More so, biographies and autobiographies never are singularities but both in content and form produce and reproduce culturally available contents and forms. If the content and form of a narrative truly were singular, they would be written in a private language, which constitutes an irresolvable contradiction—a completely personal language would only be understood by the person speaking it and therefore would not constitute a language at all.

To better understand the relationship between auto/biography (person) and history in general and the relationship between the autobiographies of science educators and the history of science education more specifically, we need to understand culture as dialectic. In the following, we first articulate a way of understanding the relationship between individual and collective through the lens of culture as dialectic and, consequently, the relationship between autobiography and history.

CULTURE AS DIALECTIC

The concept of culture has been identified as one of the two or three most difficult concepts in the English language. There are two fundamentally different senses in which the term has been used: (a) as a "theoretically defined category or aspect of social life that must be abstracted out from the complex reality of human existence" and (b) as a "concrete world of beliefs and practices" (Sewell, 1999, p. 39). As the history of the category showed, both senses have their limitations because they exclude important dimensions of culture. Recently, scholars have increasingly turned to framing culture in dialectical ways. Yet the initially proposed dialectical theories have been critiqued, among others, because they lend themselves to deterministic readings, which do not do justice to the indeterminate and emergent nature of culture (Sewell, 1992). Fundamentally, a dialectical approach to culture emphasizes that it is both a system of structurally related symbols and artifacts and a system of patterned actions (i.e., practices). The main question therefore is not whether culture should be conceptualized in terms of symbols/artifacts *or* practices, but how to theorize the articulation of practices and structure. There are different ways of cutting a cake; the following constitutes the way it makes most sense to us.

3

In a dialectical approach, culture is a concrete universal, that is, it constitutes both a generalized set of action possibilities and a set of actions through which some of these possibilities are concretely realized. From a first-person perspective, an action is both mine and not mine: I recognize in my actions the actions of others; and in the actions of others I recognize my own. Culture therefore always exceeds the ensemble of observable actions. The possibilities reside in—but are not fully *determined* by—existing structure consisting in symbols, tools, and artifacts.

Concrete actions realize abstract possibilities; they therefore reproduce culture. But actions have outcomes (results), which add to the existing structures in more or less permanent ways. Concrete actions therefore also produce culture by bringing about change in available resources for subsequent actions. Because an action both reproduces and produces culture, action constitutes a dialectical category. Transformation is built into this concept, both practically (an action may differ from anything else observable in the culture) and theoretically (in dialectical logic, *inner* contradictions such as that of the simultaneous production and reproduction are the engines of change). Culture is therefore continuously reproduced and produced anew, which leads to its constant evolution. These evolutionary changes normally are slower at the level of the individual, but are accentuated when new individuals join a group who may introduce larger variations in the way possible actions are concretely realized. (Adults are much slower to adopt new technology, to change their practices, than are newcomers, young people.)

Culture is a phenomenon of collectives—the term denotes actions and action possibilities that have histories that transcend any individual. The category implies (human, simian) societies in which patterned ways of doing something are reproduced across generations. That is, practices exist at a *collective* level, they are characteristic of society or particular groups; but, of course, societies or groups do not act: *individuals* always realize practices in concrete ways. There is therefore a dialectical relation between individual and collective. The adjective *dialectical* here means that the two opposing terms *presuppose* each other; none can be thought without the other. Because practices and structures (rules, tools, symbols, language) are characteristic of the collective life form, their lifespan is not tied to the individual and they are therefore preserved even though some individuals depart and others arrive.

Language, too, has a dialectical nature, which French philosophers appreciate in the relationship of *langue* and *parole*. *Langue* denotes language as a generalized (abstract) system, universally accepted within a community; *parole* denotes the actual linguistic behavior of people. The relation of *langue* and *parole* is dialectical because every instance of language use (*parole*) constitutes a concrete, particular realization of possibilities that exist at the generalized, universal level (*langue*). The language we use to write these lines, this text, is therefore both ours and not ours—which leads Jacques Derrida (1998, p. 2) to write, "Yes, I only have one language, yet it is not mine." Language therefore constitutes an unlimited resource for discursive actions, especially because each new (poetic, scholarly, philosophical, artistic) creation of language expands what can be expressed verbally. More so, even linguistic innovations, though new and unheard of before, presuppose their own intelligibility. Even the most poetic of poems, innovating the language we use for talking about certain phenomena or bringing new phenomena into discourse, presupposes its own intelligibility, and therefore, the possibility for this language at the collective level (culture).

AUTO/BIOGRAPHY AND NARRATIVE FORM

Narrative forms (genres), too, are resources that can be transformed into new forms at the very moment that they reproduce an aspect of culture, producing and communicating narratives. The simultaneous production and reproduction of culture, genre, and language is exemplified in the following episode and analysis, drawing on an interview conducted by a doctoral student with a world-renowned scientist.

01	Scientist:	I'm doing what I love doing. And I think everybody should do what they love doing.
02	Interviewer:	I'm sure.
03	Scientist:	And I'm fortunate that I, been able to develop a career in something that I cherished from my childhood. So . . .
04	Interviewer:	Um, that's great. Your parents were supportive of you doing science rather than medicine?
05	Scientist:	Um there were a few arguments here and there. But I said I, this is what I will enjoy doing. Becoming a doctor and you know, it's not something that I will enjoy doing it, you know . . .
06	Interviewer:	Were you the oldest son?
07	Scientist:	Nope, nope.
08	Interviewer:	Sometimes there are pressures on the oldest . . .
09	Scientist:	Yes, yes. I think all of my brothers, actually, we all had, we all had the grades, and excellence in school to go into medicine or engineering. All of us chose actually to be in fundamental science.
10	Interviewer:	Oh?
11	Scientist:	Yes.
12	Interviewer:	Then do you think your siblings had some influence on you?

There are several different levels of events that occur in this episode, all of which can be traced back to the dialectical nature of culture. First, the scientist and interviewer, although they do not know each other, produce an interview; and they do so in a way that allows readers to recognize the event *as* an interview. The two participants know this as well as the readers although this particular event, recorded on videotape, is highly singular and occurred only once (in this form). Second, the two participants understand what the respective other is saying, even though they may never have heard a particular question or statement before. Thus, the scientist previously has talked about his parents wanting him to be a medical doctor. In turn 04, the interviewer, following the scientist's statement about having cherished the idea about becoming a scientist, asks whether the parents were supportive of his alternative choice. The scientist has an immediate response, which is concerned with occasional arguments and the explanation he developed for his parents. Third, both interviewer and scientist draw on a particular aspect of telling a biography—influence of parents and peers. The scientist voluntarily provides information about the influence his brother has had or that he has cherished the idea of becoming from early childhood on. Even at the very moment that one of them begins to draw on the family repertoire in auto/biographical accounts, he presupposes the possibility and intelligibility that family members may play a significant role in autobiographical accounts.

At all three levels, the participants realize cultural possibilities for doing interviews and for constructing auto/biographical accounts of their careers. That is, despite the very singularity of *this* interview and *this* scientist's autobiography, we recognize in the event and the narrative produced culturally possible forms of doing interviews and telling auto/biographies. Now the singular nature of the event and autobiography also means that they have not existed before, which means that they are not reproduced but are newly produced forms of interview and auto/biography. Yet the very fact that they are recognizable (*recognizable*, cognizable again)—*do* an interview and *construct* an auto/biography tells us that they *reproduce* a cultural form.

Auto/biographies are narratives about the lives of individuals, viewed by others and themselves. As we noted, the contents of these auto/biographies inherently are intelligible because they realize cultural possibilities for telling the lives of members of the culture. As such, each auto/biography is a concrete realization of culture as such. When we therefore listen to or tell an auto/biography, we simultaneously are confronted with a singular account of a singular individual, on the one hand, and with the possibilities of culture, on the other. That is, in auto/biographical accounts we are confronted with cultural-historical possibilities of being a person in a particular cultural context and therefore, we are confronted with culture more generally.

This means, however, that biography always also is autobiography: we recognize aspects of ourselves in the autobiographical accounts of others, and we narrate biographies in the ways the protagonists recognize themselves. Our autobiographies are fundamentally biographies of others, because "out of the other as compact exteriority I make *my* other, just as *it* makes me *its* other" (Nancy, 2002, p. 58). My self is told in terms of what I perceive the other to be because "[t]he simple position of the I is an abstraction" and because "the concrete awakening of the I is its awakening to the world and by the world—the world of alterity in general" (p. 60).

The upshot of this analysis is that each auto/biography that is contained in this book not only tells us about the science educator who has written it, but also about the culture of science education at the particular point of history featured in the account. Each auto/biography tells us as much about the particular person as it tells us about the culture, as each individual concretely realizes possibilities that existed at the collective (i.e., cultural) level. This book therefore constitutes a science education culture and history in and through science educators: it is a dialectical project where in each narrative we are confronted with history in particular persons, who thereby come to be persons in science education history.

STRUCTURE OF THIS BOOK

A popular adage states that there are many ways in which one can cut a pie; the same is true for ordering narrative accounts of the culture and history of science education through the lives of its members. There are many ways and groupings that one could choose, and even once categories have been constructed, individual contributors could appear in and exemplify one or the other. After repeatedly ordering and reordering the contributors to this volume and articulating possible central themes characteristic of the field of science education and the changes it has undergone over the years, we have decided on six of these: (a) shaping forces across the decades, (b) scientists become science educators, (c) US-trained science educators who make their mark abroad, (d) reduction of gender barriers, (e) science education around the world, and (f) the opening up of conversations across cultures. We then located each contributor in that category that we thought she or he was *most* representative of, while maintaining a balance of contributors within each category. Thus, for example, we included Jane Kahle in part A, because she has been a shaping force in science education across the decades; but we could have equally well included her in part D dealing with the reduction of gender barriers. Also, Elizabeth McKinley could have been included in part D because of her radically feminist approach, but we felt that she was even more characteristic of part F, because she has been an essential force in the opening of the conversation across the cultures. To provide but another example, Michael Roth was trained as a natural scientist obtaining a master's degree in physics from a German university before becoming a science teacher in Canada so that he could have been included in part B, scientists who become science educators; he also completed his doctoral studies in the US before returning to Canada where he has been teaching and researching ever since, making him one of the US-trained science educators who make their mark abroad such as all those whom we included in part C. Our ultimate choice was our sense of the science education community and our knowledge we have acquired in and of the field after having been members in the discipline for a combined 50 years.

We begin each of the six parts of the book with a brief introduction to the particular dimension of science education that is characterized in and through the auto/biographies included. In a synthesis that follows the auto/biographies, we then delve at greater length into the cultural and historical aspects of science education, providing further examples and directions to the general possibilities the discipline offers or has offered as exemplified in the featured narratives.

Part A

SHAPING FORCES ACROSS THE DECADES

PARTICIPATING IN SCIENCE EDUCATION

In this part of the book, we gathered the biographies of five science educators who, for one or another reason, has had significant influence on the field of science education. Here we understand influence in a broad way, as it expresses itself, for example, through the impact a person has on a particular theoretical, methodical, or topical dimension; influence may also come from the number of doctoral students a person trained who became researchers in their own right; and influence may be gauged in terms of the number of publications that an individual produces or the number of times his or her work is cited by others. Although other science educators who contributed to this book could have been categorized in this part—e.g., Peter Hewson because he had coauthored a seminal article on conceptual change that has had tremendous impact on the field—we added them elsewhere in the book for the sake of achieving a balance across the different parts. After many iterations of ordering, seeking commonalities and grouping the different contributors, we included in this first part James J. Gallagher, Jane Butler Kahle, Barry Fraser, and both of us.

All five of the science educators featured in the first section of the book were science teachers, participated in curriculum design and evaluation, and maintained active involvement in science teacher education with new and more experienced science teachers. Like so many science educators of this era, Jim Gallagher came to science education rather than pursuing a career in applied science—in his case medicine. His development as a science educator established roots to which he would return throughout his career and connected him with people around which social networks were developed to support his successes as a science educator. Kahle's biography also shows a strong allegiance to science, making connections that were affordances in establishing collaborative projects supported by massive funding by the National Science Foundation. Fraser and Roth both studied science and then became science teachers before they created alternative routes to research in science education. Fraser got involved as an evaluator on a national science curriculum project in Australia and these experiences provided a segue into his doctoral research and then a career as a researcher of learning environments, especially in science classes. Roth went from teaching into graduate school, flirted with the goal of pursuing a doctorate in science, opted instead to get his degree in science education, and then took a postdoctoral appointment in the United States. In some respects Tobin's career trajectory was different from the others. He came into science as an out-of-field high school teacher who studied physics part time as he taught. Like Roth, he considered research in science, but decided to undertake research on the teaching and learning of science. In the early 1970s Tobin also was involved in curriculum, developing resources to support the teaching and learning of high school science—setting a context for his classroom-based research. Gallagher, Kahle, and Roth also were involved in curriculum design and enactment before research in science education became their primary mission in science education. Gallagher had a chance to work with Jerome Bruner on *Man a Course of Study*, Kahle was on the advisory board for the *Biological Sciences Curriculum Study*, and Roth collaborated with Provincial educators in Canada to develop and evaluate science curricula.

Strong social bonds, with researchers and universities in the United States, Australia and Canada, connect the five researchers included in the first section of the book. The National Association for Research in Science Teaching has been an organization that has fostered relationships among the five scholars. However, several universities also have been involved, hosting large international studies, principally, Curtin, Michigan State, Miami, Penn, Victoria, and Queensland University of Technology. Each of the researchers has nurtured individuals who constitute the next generation of science education research. Like Fraser, all have had a plethora of former doctoral students who have become leading researchers in their own right. However, the five scholars also have been active mentors for emerging scholars in science education, irrespective of where and with whom they studied for their doctorate.

JAMES J. GALLAGHER

A CAREER EVOLVES

Five Decades and Still Engaged

Like many members of our community, I began my career by entering university
as a science major immediately after graduating from high school. And like many
people of my age and background, my parents and older siblings encouraged me to
be a physician. With two uncles in well-established medical practices, the field
held some call until my junior year, after taking on a part-time job as a clinical
laboratory technician at the local hospital and university infirmary. During my two
years in the hospital lab, I found that I did not enjoy the clientele, and was much
more drawn to work with healthy, exuberant youth, than with the infirmed. Therefore, I decided that medical school was not for me and that secondary science
teaching was my career choice.

Upon graduation with a degree in natural sciences from Colgate University in
January of 1954, I continued studies toward a master's degree in education, with
the intention of starting to teach science in a high school the following September.
In the meantime, it was apparent that I could complete the coursework for the degree during Spring and Summer semesters, leaving only the master's thesis to be
completed during my first year or two as a teacher.

Teaching secondary school science was both enjoyable and challenging for me.
During my first two years, I was the entire science department in a small, rural
school in New York State's Catskill Mountains, teaching general science for
grades 7, 8, and 9, biology, and earth science, along with chemistry and physics in
alternate years. I treasured the students, my colleagues, and rustic environment. I
had excellent support from peers and the administration. However, the range of
subjects and ages was somewhat overwhelming, and after two years, I sought more
specialization in my teaching.

The search for a new position was easy in the late 1950s, as there was a shortage
of qualified science teachers. My new venue achieved a desired degree of specialization—physics, earth science and grade nine general science, three other science
teachers who became supportive colleagues, and an excellent environment for
teaching and raising our young family.

During the "Sputnik era," opportunities abounded for science teachers. Summer
study became very accessible through the National Science Foundation's program
of summer institutes for mathematics and science teachers. I entered a summer
institute program at Antioch College in Yellow Spring, Ohio—a four-year, sequen-
tial program that led to a Master of Science Teaching degree. Antioch was an excellent match for my professional needs, as it deepened my understanding of sci-

K. Tobin, W.-M. Roth (Eds.), The Culture of Science Education, 13–23.

ence content and provided excellent pedagogical models. In addition, the financial support added to the quality of life for our family. We were succeeding very well, both in the excellent environment for teaching and the stimulating environment for continued professional growth provided by National Science Foundation at Antioch College. At the end of the four summers, I completed the master's degree with a 4.0 GPA. Life was good and it was about to get better!

Early in 1962, I received an invitation from the Director of Harvard University's NSF-funded Academic Year Institute to apply for this program. A few weeks later, the die was cast, when my application for the program was accepted. After another NSF supported summer institute at New Mexico Highlands University with visiting professors James Bonner from Cal Tech and Guido Pontecorvo from University of Edinborough, I began the program at Harvard with fifty other science and mathematics teachers from across the nation. Fall and spring semesters each involved four graduate level courses, and summer added two more in a combination of advanced science content, history and philosophy of science, and science pedagogy.

A few weeks into the Fall semester, I approached Fletcher Watson about admission to the doctoral program in science education, which he headed. During our first meeting, he described the courses needed for the degree and admission procedures. While he made no guarantees about admission, our encounter appeared promising. During my second semester in the Academic Year Institute at Harvard, two milestones were achieved. I was accepted into the doctoral program and I was offered a second year of financial support in the Academic Year Program. I concluded my doctoral studies within three years. The financial support from the National Science Foundation was a key factor for me, as it was for several of my generation of leaders in science and mathematics education, paying tuition, books, and a living allotment equivalent to my teaching salary, thereby eliminating much of the financial burden that usually accompanies graduate school.

During my second year at Harvard, I enrolled in Jerome Bruner's course in cognitive psychology, and wrote a term paper based on an initial investigation that was the foundation of my dissertation. Bruner read the paper and sent me a letter with praise for my work and requesting I come to his office. In this meeting, he invited me to become part of his Instructional Research Group that was engaged in the development and trial of *Man: A Course of Study,* a middle grades social studies curriculum being developed with colleagues from several universities. Seminars with this group occupied my Friday afternoons for nearly all my remaining time in Cambridge and also provided exciting employment during one summer. The work of our group was a component of Bruner's book *Toward a Theory of Instruction*—a sequel to *The Process of Education.*

Again, life was good and about to get better. As my dissertation neared its completion, I was fortunate to receive two offers for positions—one at Stanford as a post-doctoral fellow with Paul Hurd and a second one as a staff member of the *Science Curriculum Improvement Study* with Robert Karplus at University of California, Berkeley. I accepted the former and in late summer of 1965, our family embarked on a transcontinental move from Massachusetts to California.

The two years with Paul Hurd and his colleagues at Stanford were life-changing. Paul was deeply involved in synthesizing research and development arising from the post-Sputnik science curriculum reforms. He had been an important force in developments by the *Biological Science Curriculum Study* and was on advisory boards for other groups. He was a frequent traveler to Washington with influence on policy and programs at NSF and other agencies. His design for my post-doctoral position was to co-author a book about eight novel elementary science programs being sponsored by NSF at that time (Hurd & Gallagher, 1968). Except for my experiences with *Man: A Course of Study* and my dissertation research, I had virtually no professional experience with elementary schools, children, or teachers. However, the two years resulted in a sharp learning curve for me in understanding elementary school teaching, learning, and curriculum. Most importantly, it added to my vision and knowledge about curriculum and materials design as important factors in science teaching and teacher education.

To complete this developmental sequence, I moved from Stanford to the Educational Research Council of America in 1967, where I was part of a science curriculum development group, with my work focusing on staff development in elementary science, with some additional responsibilities in a newly designed course for low achieving high school students. After two years in this position, the organization came on hard times financially, and I took over as director of a much reduced science group for one year. In 1970, I joined a new university being formed in Illinois: Governors State University. Six years later, I moved to Michigan State University, where I served as Director of the Science and Mathematics Teaching Center for six years and professor of science education for 24 additional years until my retirement on January 1, 2006. I now continue working in a part-time role as Co-director of the Center for Curriculum Materials in Science.

PEOPLE AND EXPERIENCES THAT SHAPED MY CAREER

Given my long career, there are many people who have served as significant peers and role models, and many experiences shaped my research and teaching. I make no attempt at ranking them in importance or influence, as that is both difficult and dangerous! However, these do appear somewhat chronologically. Unfortunately, due to limitations in space established for this book, I am able to describe only a few of the people and events that influenced me. To my regret, many are omitted.

The Early Years–Prior to 1965

Ruth Griffith, my junior high school science teacher, and professors including John Woodruff at Colgate and Oliver Loud at Antioch stand out because of their influence in nurturing my love of science. Oliver Loud added a special dimension, introducing me to the history and philosophy of science, and connecting me with Everett Mendelsohn and Leonard Nash at Harvard, who also enriched that domain of my knowledge. These three significantly influenced one of my most important publications (Gallagher, 1971). Paul Joslin, my closest colleague as a secondary school science teacher, stands out as a mentor and role model. We worked together, sharing laboratories, teaching space, equipment, ideas, and teaching strategies for four years in LeRoy High School. We then followed similar careers in higher education. He taught me so much about teaching, learning, and how to relate effectively with all kinds of people.

The Harvard years were also enriched by my work with Jerome Bruner, Fletcher Watson, and a fellow graduate student, neighbor, and co-commuter to Harvard Square, Andrew (Chick) Ahlgren. Obviously, there were many others during the three years there, including other graduate students, faculty, and visiting scholars who came to work or lecture there.

Support during my dissertation came from three sources: Maurice Belanger, my dissertation director who was restricted in mobility for several months with vertebral surgery. As a result, unlike many graduate students, my dissertation director was readily accessible, and even glad to see me! A student and colleague of Jean Piaget, Maurice was an excellent source of information, ideas, and methods for my research. Another high level of support came from Mary Henle, a visiting professor from the New School for Social Research. Her work related to children's logic, and one important lesson she provided was that children's logic is sound, but it often is based on premises that differ from those held by adults. That principle was important to me as parent, teacher, and researcher. Both Maurice and Mary were very interested in my dissertation topic—children's explanation of science phenomena—and were extremely generous in their time and effort. Last, and certainly not least, Barbara, my wife, not only guided our children and made our home a place for productive writing of the dissertation, but she also typed the many drafts of the dissertation, including the final, letter-perfect one, on an electric typewriter—a technology that preceded word processing computers!

Embarking on My Post-doctoral Career 1965–1976

The experience at Stanford has already been touched on, and one can only imagine the impact of a first post-doctoral position with Paul Hurd, a larger-than-life mentor at a great university, on a person only three years removed from high school teaching. The change was huge, and the support from Paul and my young family was both superb and needed.

After two years at Stanford, I moved to Cleveland, Ohio and became part of the Science Staff at the Educational Research Council, which was headed by Ted Andrews. After three years there, I joined a newly forming faculty at Governors State University, where Andrews had become Dean of the College of Science. Obviously, he had a very strong influence on my professional life. In the former position, we were involved in science curriculum development and guiding teachers with its implementation. In the latter position, we worked together to create a college, literally from the ground up. Governors State was established by the State of Illinois as an upper division university to serve the unmet educational needs of low income and minority students from the Chicago area. It was a laboratory for trying out new approaches for teaching an older, diverse population whose average age was above 30. With Dean Andrews' encouragement, I assumed responsibility for guiding development and enactment of the curriculum for the college's several degree programs: environmental science, urban planning, nursing, health care administration, and science education.

During my years at Governors State University, I also began working with Robert Yager, who influenced me in many positive ways during his term as president of NARST and then later as president of NSTA. We worked closely over these years and he helped me enter the international science education community advance my place in the national community. Those were some of the people who influenced me in the first two decades and more of my career. Now to my thirty years at Michigan State University

New Roles and Responsibilities: Michigan State University 1976–2006

During my initial six years at Michigan State University, I continued in an administrative role as director of the Science and Mathematics Teaching Center. Working in this capacity, the center expanded its capabilities with several large grants, including a biomedical sciences project for low-income and minority high school students and grants dealing with environmental and energy education. Using salary savings from these grants, I was able to sponsor six post-doctoral fellows at the Science and Mathematics Teaching Center. Four of these went on to be key leaders in science education—James Stewart (University of Wisconsin), David Treagust (Curtin University of Technology), Patricia Heller (University of Minnesota), and Charles "Andy" Anderson (Michigan State University). David and Andy continued to be close associates over the years and greatly influenced my thinking and actions.

Michigan State University provided many wonderful opportunities for me to develop in new areas. Three stand out. First, I transformed my research approach from one based in quantitative, experimental method, learned as a graduate student at Harvard, to a qualitative, ethnographic method. With the encouragement of Dean Judith Lanier, I enrolled in Fred Erickson's three-course sequence on educational ethnography during the 1981/1982 academic year, which enabled me to embark on this new research trajectory.

The second opportunity led me into the realm of international science education. Dean Lanier, her successor, Dean Ames, and Jack Schwille, Associate Dean for International Programs, supported me in two efforts that were transforming. They encouraged me to engage in several international ventures that included extended assignments with David Treagust, Ken Tobin, and Barry Fraser at Curtin University, leadership of the NSTA International Committee, and several projects in Latin America, Southeast Asia, Europe, Africa, and the Middle East. The most extensive of these involved over twenty trips to Thailand and Vietnam. In addition, I undertook two substantial projects in Latin America. All of these activities were excellent, enriching opportunities cross-culturally and to expand my understanding of teaching and learning. They also continued a long tradition of international education at Michigan State University.

17

Another part of my international work began in 1983, with the organization of a course plan that was repeated eleven times with an enrollment of over 300 participants, mostly Americans teaching overseas. The course was offered in exotic venues—Hawaii, Europe, Africa, Australia, and Alaska. My key colleague on these endeavors was Arthur Reed, a professor of marine biology at the University of Hawaii. Together, we organized a very popular, yet demanding course that helped teachers increase understanding of the natural environment, people's impact on it, and how to use the local environment as a laboratory to support learning. In each new venue, Art and I had to prepare new readings, learn new environmental connections and examples, and make contact with new local experts about the specific issues we studied.

The third opportunity was also life changing, as it required long-term interactions with practicing teachers. Starting in 1984 with an ethnographic study of secondary science teaching sponsored by the MSU Institute for Research on Teaching, a team of graduate students that included Okhee Lee, I established the groundwork for fifteen years of applied work in a wide array of urban and rural schools at home and abroad. The research with these graduate students, coupled with research with Ken Tobin in schools in Western Australia, began a series of activities that continued through most of the remainder of my career. This long-term work with teachers and administrators created a detailed understanding of their work, their successes, and the obstacles to success. This enabled work to assist them in improving their teaching and their students' learning, which became a hallmark of my career—my research and my professional mission for more than two decades.

Five related enterprises emerged from this research, adding to my experiential base and enriching my understanding of, and ability to apply, contemporary theories to educational improvement. These include:

- A project with the Toledo Public Schools and its teachers' union, the American Federation of Teachers, beginning in 1988, continuing for twelve years. It involved guiding development of a team of teacher-leaders, known as "support teachers" within the district. The effort focused on improving teaching effectiveness and the development of leadership skills where these teachers could help others improve their teaching and students' learning. Extended association with middle school science teachers in Toledo for this long duration was extremely influential in my development of understanding of teaching science for understanding.
- A temporally overlapping research assignment at the Otto Middle School in Lansing, Michigan, as part of MSU's Professional Development School program, provided quite different experiences with teachers. This collaborative work with middle school science teachers and graduate students, on a near-daily basis, complemented my work in Toledo.
- Winning a grant from NSF to develop formative assessments in science and mathematics involving teachers from these two school districts. The research and development for this project extended work with these teachers in partnership with my colleagues, Joyce Parker, Perry Lanier, and Sandy Wilcox. This four-year project had a major impact on my professional growth, and signifi-

cantly increased my humility. The key lesson was about how complex the job of teaching science really is. This work also resulted in a series of booklets for teachers, staff development personnel, and teacher educators to support the use of embedded assessment in teaching (Gallagher & Parker, 1996).

- Involvement in a Local Systemic Initiative provided the context for extended collaboration with Charles (Andy) Anderson and Sarah Lindsay the Science Coordinator for the Public Schools in Midland, Michigan. We planned and learned together as we tried many new approaches to helping this large group of K–8 teachers from this single, well-supplied, "lighthouse" school district develop the skills and knowledge to teach science for understanding.

- Working with two eminent research scientists from MSU's Kellogg Biological Station and with nearly 100 teachers were from relatively small rural schools on an NSF Supported Retention and Renewal project. Because of our success with the Local Systemic Initiative, it was only natural to bring Andy Anderson into this project. His efforts advanced my understanding of how to help teachers develop the knowledge, skills, and vision that are part of teaching science for understanding.

In the final years of my career, three additional projects continued to influence my growth. Robert Floden, Joan Ferrini-Mundy, and I conducted a Study of Leadership Development in Science and Mathematics Education under NSF Sponsorship. I also worked with Robert Yager, and Senta Raisen on the Salish Project. In addition, I was a consultant on the TIMSS-R Video Project, directed by Kathleen Roth, a former graduate student and colleague, from whom I learned so much about teaching and learning over many years.

Beginning in 2002 and continuing to the present, my association with colleagues in the Center for Curriculum Materials in Science constitutes another very important source of learning and influence on my thinking. It renewed a deep working relationship with Ed Smith, with whom I had worked in varied projects over all of my years at MSU and brought me into almost daily contact with colleagues at Project 2061, University of Michigan, and Northwestern University—partners in this project.

Over the years, my work has taken several turns. With two master's degree theses, a doctoral dissertation and a forty-year, post-doctoral career, there has been significant evolution in my research approach. In addition, as a "devout opportunist," I took up many research opportunities in the interest of pursuing potentially promising new lines of inquiry. In the limited space that this chapter allows, I will focus on a few of these, including my dissertation research, my transformation from quantitative to qualitative research, and selected projects that occurred over the years that followed the transformation.

SELECTIONS FROM MY RESEARCH

I already have indicated how my research approach changed over time. In the paragraphs that follow, I provide a few windows into this research beginning with my dissertation, followed by some examples of the changing research scope during the years that ensued.

Dissertation Research

My dissertation was a watershed experience for me. As indicated earlier in this chapter, I had keen support from committee members and my wife in all aspects of it. Unlike what many doctoral students report, I enjoyed all parts of the work, from beginning to end.

The research began with a paper for Jerome Bruner's course in which I explored explanations of students at various grade-levels regarding a small set of exemplars of scientific phenomena. For this term paper, I hoped to determine how students' explanations developed over the years of schooling. Therefore, I interviewed about eight students in the odd-numbered grades (1–11) from a school district in a working class suburb of Boston asking them to explain the causes behind the operation of exemplars of a small set of phenomena exemplified in the Crookes' radiometer, a palm glass, and two bar magnets.

These data, from what was to become a pilot study of my dissertation, showed a slow progression in the use of science concepts and reasoning through the grades. However, with some of the exemplars, the conceptual base was not supported by school science to any significant degree and so progression was limited. For example, students' school learning about magnets went little beyond "like poles repel and unlike poles attract," which most students learned by fifth grade. As a result, students in higher grades gave explanations that differed little from those of fifth graders. The conceptual level of explanations appeared to be limited by the conceptual levels incorporated in the school curriculum, in these years before diverse television programming and the creation of the Internet. Moreover, students' explanations were also limited by a lack of understanding of the logic of explanation.

As a result of this study, my dissertation took on two aspects of this work—an empirical portion involving the chronological development of students' explanations based in a small set of concepts and an experimental portion involving how these explanations can be modified through instruction. In my dissertation study, I omitted magnets from the study because of the conceptual difficulties that surround explanations beyond repetition of the law of magnets. I added explanation of the bi-metallic strip to the study so that the explanations were all heat-related, thus delimiting the instructional aspect of the study. A considerable amount of effort went into the baseline study resulting in a simple scale of explanation that paralleled ideas of children across the school years relating to the three exemplars—the Crookes' radiometer, the palm glass, and the bi-metallic strip. This became basis of the pre-test and post-test for the study.

The experimental part of the study entailed a pre-test, treatment, post-test design, intended to foster modification of explanations by middle school students in grades 7–9. The instructional treatment involved two parts–a rather traditional, short, lesson sequence on concepts related to heat that was directly related to the three exemplars and a second, short lesson sequence based on the "Mousetrap

Game" that was designed to help students understand the logic of explanation, mainly focusing on sequencing of causes in a logical manner. These lessons were presented to students in small interactive groups. The study also involved three treatment configurations and a control—instruction in both concepts and logic of explanation, instruction in concepts, and instruction in logic, with no instruction for the control group.

The results of this study showed that students who had both treatments and those who received the logic treatment showed greater increases on the post-test than those who had the concept treatment only. All did better than the control group (Gallagher, 1965).

The dissertation was given positive praise by many people as high quality work. I presented it at a NARST meeting, in the mid-60s where it received little notice since most people at NARST were thinking about inquiry teaching and Piagetian developmental stages. The lack of interest in this work was a bit discouraging to me, and I was not confident and visionary enough to make the connection between explanation and inquiry. Unfortunately, this connection did not become important to me for years. Consequently, because of my perceptions of the intellectual climate of the science education community at the time, and my lack of foresight, I did not formulate a publication based on my dissertation, a result that I deeply regret.

Early Postdoctoral Research

During my time at the Educational Research Council, I prepared a paper for the journal *Science Education*, titled "A Broader Base for Science Teaching," which dealt with the inclusion in the science curriculum of the nature of science and its applications in technology and society. The paper was published without modification and it received no apparent response from the readership of that journal (Gallagher, 1971). I thought it was of no consequence until several years later I attended a session on Science-Technology-Society at an international meeting run by Avi Hofstein. He began his talk by saying that one of the conceptualizers of S-T-S was sitting in the front row, a reference to my 1971 article. Until then, I had no idea that the article had any impact. The claim was also reiterated in George De-Boer's history of science education (DeBoer, 1991). This may be an interesting issue related to information transfer in our field, or it may only be an artifact of my naiveté!

About this same time, as part of our curriculum development work at the Educational Research Council, I took on the task of studying how an experienced teacher in a local high school was enacting a new curriculum for low achieving students. This was part of the trial of innovative, new materials being developed by my colleagues. This was my first ethnographic study, and although I was conceptually ill-prepared to conduct it, what I was doing "seemed right." I learned so much about the program and its reception by the teacher and his students as a result of watching, listening, and taking careful notes. While this study was in progress, I received a call from Willard Jacobson, then a senior professor at Teachers College. He indicated he was organizing a session for NARST on studies of teacher practice and he had heard I was engaged in this type of work. After hearing about his plan and my co-presenters, I was pleased at the prospect and accepted his offer.

The day of the presentation, I sat at the front of a large ballroom at a plenary session at NARST with about two hundred members present, as Roger Anderson and Arno Bellack from Teachers College presented their quantitative, carefully developed work, and I brought up the rear with my qualitative report based on loosely structured observations in one classroom. As you might expect, I received a negative response for my study from a few of those in the audience, with at least one person saying what I was doing was neither science nor research. This may have marked the beginning of a conflict within the NARST community over qualitative research in science education that went on for over fifteen years.

A Major Transformation in Career Direction and Research Approach

Twelve years later, with my Dean's encouragement, I spent a substantial portion of my time studying ethnographic research under the guidance of Fred Erickson for a year. As part of this work, I engaged in small studies of classrooms to sharpen my skills and deepen my understanding of research methods. Then, in 1984, two events came into play that had a large impact on my research and ultimately impacted my understanding of the field. First, I received support from Michigan State University's Institute for Research on Teaching for a substantial study of middle and high school science teaching. The support enabled me to work with four graduate students as co-investigators over a period of three years. We studied 27 teachers in three school districts, and learned a tremendous amount about the attitudes and beliefs of these teachers. One key finding was that most secondary science teachers believed that learning was dependent on their "coverage" of science content, and it appeared that responsibility to nurture and support students in learning was not part of their responsibility as teachers. Most teachers in our study noted this in statements such as "My job is to teach, while learning is the students' responsibility." Couple this with statements also added by many teachers, "The good students will learn, while the rest won't." This appeared to be an abdication by teachers of responsibility to support learning beyond presentation of content to students. This seemed to be an inappropriate stance, and a serious misunderstanding of the role of science teachers. Moreover, it seemed commonplace, and therefore worrisome.

The second watershed event of 1984 was my first trip to Curtin University, then known as Western Australian University of Technology, where I worked with Ken Tobin on classroom-based studies in high schools, grades 8–12. Ken and I spent much of my two-month stay in one local high school that had seen change in student demography from upper-middle class to lower middle-class. Teachers there were struggling with the new population, while trying to maintain the same approaches that had worked well for them in an earlier era. We observed classes and met with teachers informally almost daily and gathered a huge amount of data. When combined with the studies that had begun at Michigan State University, being carried out by my team there (Okhee Lee and others) we added considerably to our knowledge about teachers in two diverse settings, half a globe from each other. This effort led to several publications including Tobin and Gallagher (1987a, 1987b), Gallagher and Tobin (1987), and Gallagher (1989). It also led to the publication of the NARST monograph on Interpretive Research (Gallagher, 1991), which was widely distributed by NARST that was concurrent with a significant change in the research approach and productivity of members of the organization.

Applications Grounded in Ethnographic Methods at Home and Internationally

This ethnographic work led to three additional areas of work—with teachers, with formative (embedded) assessment, and internationally—as I already have delineated. Much international work emerged as a result of the ethnographic studies between 1984 and 2004 including staff development for researchers in Brazil, Panama, South Africa, and Taiwan and projects in Thailand and Vietnam. In addition, I coordinated two international conferences—a hemispheric conference on science education in the Americas (Gallagher et al., 1985) and a US-Japan Conference on Science Education (Gallagher & Cline, 1988). These were at least loosely tied to my ethnographic work.

One other track of development and application beginning in 2000 was my development and enactment of three on-line courses at the master's degree level as part of the Michigan State University On-line master's degree program. My three courses were *Teaching Science for Understanding*; *Inquiry, Nature of Science and Science Teaching*; and *Action Research in School Subjects*. These courses applied much of my experience with teachers and school students, my ethnographic work, and my work in embedded assessment in science. They also provided a new challenge—establishing a working rapport with teachers whom I never met face-to-face! And finally, these experiences informed two final efforts in the field, my work with the Center for Curriculum Materials in Science and my development of a book for teachers and teacher educators that carries the title *Teaching Science for Understanding* (Gallagher, 2007).

It has been a long, varied and exciting journey. This essay only includes selections from what could be a longer description. On reflection, there is little that occurred that I would change, the exception being to give more attention to publication. And thinking back, I am so pleased that I did not follow a medical career. My life has been so enriched by all of the opportunities that came my way and people encountered as a teacher-researcher and member of the international science education community. Around each bend in this long road, there have been exciting challenges and great opportunities.

James J. Gallagher
Professor Emeritus
Michigan State University

JANE B. KAHLE

THE ROAD TO REFORM: A PERSONAL JOURNEY

Writing this chapter has been a delightful experience, and I hope that it will provide enjoyable reading. Briefly, the chapter is divided into four sections: first, an autobiographical account of my career; second, a brief review of my work that has impacted the field of gender research; third, a mention of a few colleagues who have strongly influenced my research and career; and fourth, a brief look at future directions.

MY CAREER

The Early Years

My journey in science education began as a teacher—first teaching science for fourth through sixth grade students in a small Indiana town. It should be noted that I was totally unprepared to teach those grades. I had graduated from Wellesley College with honors in botany and zoology and had completed high school biology certification requirements at Teachers College, Columbia University (New York City) and Purdue University (Lafayette, Indiana). However, the superintendent stated that I was too young to teach high school students (i.e., boys), so my career began at Washington Elementary School on the banks of the Wabash River. That year, however, taught me a very important lesson: no one works harder than an elementary teacher! The next year I replaced my former biology teacher at the local high school. For ten years, I taught biology in rural high schools, while also demanding to use the new Biological Science Curriculum Study (BSCS) texts and memorizing the *BSCS Handbook for Biology Teachers*. Although Wellesley did not have formal courses in education, all of my science involved a hands-on, inquiry approach (probably due to the lack of teaching assistants)! I was better prepared than my science teacher colleagues to use various media, to culture various organisms, and to dissect various critters. It was a wonderful ten years, interspersed with two maternity leaves.

Very fortunately for me, the first "crisis" in science education occurred during that time when Sputnik circled the Earth. The National Science Foundation (NSF) responded to the outcry that the US was losing the space race by instituting its teacher training institutes. I was accepted to the biology institute at Purdue University. However, I did not receive financial support, because I was married with children and, obviously, would leave the profession. Like many of us who attended those institutes, they started me on a personal journey in science education. One highlight along the road was receiving a Shell Merit Fellowship to Cornell University for two summers. That honor was particularly nice, because the same person

K. Tobin, W.-M. Roth (Eds.), The Culture of Science Education, 25–35.

who earlier had indicated that I would "soon leave the profession" was the program's director, Joe Novak. Joe also became my first mentor and opened many doors through wise advice, manuscript suggestions, and sabbatical experiences.

With a Master of Science degree in microbial genetics from Purdue and with my Cornell experience, I was ready for a full-time doctoral program. I returned to Purdue (only 35 miles from my home) and studied with Sam Postlethwaite in the Department of Biological Sciences. Sam had been very successful teaching large sections of botany to undergraduates through an individualized approach called, the Audio Tutorial System, that placed audiotapes, live materials, and Super 8 millimeter projectors in a carrel. Students progressed through the materials at their own speed and, in what we now call, asynchronous time. Sam and his colleague Floyd Nordland were interested in adapting the system for high school biology, and we all were concerned about underachieving African American students. Floyd and I adapted the materials, a colleague built portable carrels, and we were off to Gary, Indiana, and Chicago, Illinois, to test our theory that self-paced, multi-media instruction could improve achievement for at-risk students. The theoretical underpinning of the materials was Ausubel's theory of advanced organizers, and my dissertation assessed the efficacy of advanced organizers to enhance learning among economically disadvantaged African American high school students.

After finishing my degree, I was employed by Purdue University as a lecturer to teach methods of teaching biology and to supervise biology student teachers. But, again, events unfolded in my favor—the federal government was reviewing major research universities for any discriminatory practices affecting women and minorities. If evidence of discrimination was found, federal research funds were withheld (as was done at the University of Michigan). Without any discussion or notice, I received a letter retroactively changing my appointment from a non-tenure track lecturer to a tenure track assistant professorship! Given my early experiences, part of my personal journey has been to repay those who struggled for equal opportunities before me and to continue to smooth the road for those who follow.

Opening Doors

As an assistant professor, I enjoyed my students, raised my children, and wrote proposals and papers. My proposals focused on economically disadvantaged students and alternative modes of teaching biology, and the research was funded by private groups (Bristol-Myers, Carnegie, Norman, and Squibb Foundations as well as the Lilly Endowment) and by national agencies (NSF, the U.S. Department of Education, and the National Institute of Education [NIE], among others). These early experiences with externally funded research set a career pattern for me, and I have consistently sought external funds and been involved with intervention and research projects throughout my career.

En route to tenure, one annual review provided particularly sage advice—advice that I have shared with many of my students. Bob Kane, a mathematics educator and chair of the Education Department at Purdue University, suggested that I think about where I wanted my career to be in five years and prepare a plan that would get me there. I followed his advice and received tenure and promotion at the end of my five-year plan. In retrospect, I think that most boys (and young men) receive that type of advice from many sources—almost sublimely. Males are raised in our culture to believe that they need to have careers and to be successful. But, it was certainly the first time that anyone had indicated to me that I needed a long-range plan for success. Bob also mentored me in another way; he suggested that I select a focus for my research and stay with that focus, eventually becoming a national and international expert. Again, I have shared that advice with my students and younger colleagues, both female and male!

My decision was to focus my research on the emerging field of gender equity— or girls in science and mathematics. In the early 1980s, differences between girls' and boys' success in mathematics was viewed through what Evelyn Fox Keller called the "characteristics lens" that is, girls' failure to achieve equally with boys on the mathematics components of NAEP, ACT, PSAT, or SAT was attributed to the characteristics of girls. If one interprets girls' underachievement in mathematics and science through the characteristics lens, differences that favor boys are related to girls' maladaptive motivational patterns. Using this theoretical context, the burden of change is placed on the girls.

However, social psychologists such as Alice Eagley and Kay Deaux at Purdue were conducting research that indicated that girls' interests in a particular field was influenced by society's characterization of the field as masculine or feminine. Further, Elizabeth Fennema at the University of Wisconsin and Jacque Eccles at the University of Michigan and their graduate students were beginning to investigate and interpret gender differences in mathematics in a different way. Both investigated the relationships among sociocultural and classroom factors on girls' enrollment, retention, and achievement in mathematics, and both developed theoretical models to support their hypotheses. In 1983, Jacque's academic choice model clearly explicated environmental factors such as causal attribution patterns, opinions of parents and teachers, gender role stereotypes, and perception of task difficulty that affected girls' and boys' choices of future courses. Two years later, Fennema and Peterson's autonomous learning behavior model examined the effect of external or societal influences on the internal motivational beliefs held by girls and boys that affected autonomous learning behaviors contributing to gender differences in high cognitive level skills. The data supporting both models indicated that girls' under-enrollment and underachievement in mathematics were due to a combination of sociocultural and educational factors. Research by Jacque, Eliz, and others at this time clearly indicated that gender differences needed to be interpreted through a "response lens"; that is, they were in response to teaching environments that produced motivational patterns unconducive to learning for girls.[1] Their research opened up the field to investigating multiple influences on girls and boys that contributed to differential enrollments and performances in mathematics and, later, in science.

However, in spite of gender research about girls in mathematics that indicated external, not internal, factors were the root of achievement differences, the most publicized research at the time was Camilla Benbow and Julian Stanley's study of

27

gifted youth. They attributed the gender differences (favoring boys on the PSAT-M) to innate (biological) differences, and their findings were widely published in the popular press.[2] Pallas and Alexander questioned the appropriateness of the statistics used by Benbow and Stanley, but the popular press did not print their refutation. During the 1970s and early 1980s gender differences were viewed by and large through the characteristics not the response lens—with the exception of gender researchers.

Building upon the foundation provided in mathematics, science educators began to explore factors affecting the entrance, retention, and achievement of girls and women in science. For example, Marsha Lakes (now Matyas) and I examined differences in what girls wanted to do in science and what they actually did. Mary Budd Rowe studied dyads of students faced with solving a problem using scientific apparatus. Her research led her to conclude that girls, particularly white girls, needed time to become familiar with the equipment. When such time was provided (called "tinkering time") girls' solved the problem correctly as often as boys did. Other research identified differential treatment of girls and boys in science classrooms, including boys' domination of whole class activities; the selection of more boys, compared to girls, as "target students"; and boys', compared with girls', active involvement with the equipment and apparatus of science.

Issues of test bias were examined in Marcia Linn's and her colleagues' analysis of the "I don't know" response on the National Assessment of Educational Progress' science test. Compared with boys, girls significantly more often selected the "I don't know" response, giving boys, a 20 to 25 percent advantage (depending on whether there were four or five foils) for a correct answer based on guessing. Another of Marcia's studies, which she dubbed, "the coat rack syndrome," analyzed family visits to museums in which the fathers and children used the equipment while mothers held the coats, reminded us that sociocultural factors also needed to be examined. These are but a few of the studies that investigated gender differences in science. As the field expanded, researchers began to establish that girls' under enrollment and underachievement in science were not due to girls' maladaptive behaviors but rather were in response to lack of science-related experiences in-class and out-of-class, differential treatment in class, test construction, and socialization.

During this period, I received a grant from NIE's Minorities and Women's Program that enabled me to combine my concerns about women and African Americans. Bernice Coar Cobb, who was associated with the Alabama Center for Higher Education and a professor at Miles College in Birmingham, Alabama, and I conceived a project focused on enabling women faculty at minority institutes to enhance their research skills and, thus, improve their chances for promotion and tenure. In the summer of 1979, seven women, all of whom were African American, arrived at Purdue's rural campus to begin an intense four-week course, designed to prepare them as future researchers in science education. Nothing had prepared me for the experience. Shortly after I received notice of the award, Purdue's lawyer called to state that Purdue (and I) might be vulnerable to a reverse discrimination suit, because participation was limited to women. Then, when the women arrived on campus—some with several children in tow—Purdue's Building and Grounds department refused to put more beds in the graduate student apartments where the participants were housed. A midnight phone call to a sympathetic dean resolved the bed issue, but the incident highlighted how non-academic factors can affect women's education and careers. The project continued for two more summers, producing many research papers and a book. Further, all of the participants conducted and published research, several completed doctoral programs, all eventually received tenure, and two became academic deans! But, my colleagues at Purdue and I learned more from the participants than they learned from us.

Gender research in science education in the US was recognized first at the 1978 Annual Meeting of the National Association for Research in Science Teaching, held in Toronto, where several papers reported on gender studies. I particularly remember five other women in attendance whose work focused on gender and science. They were Ann Howe, Rita Peterson, Marcia Linn, Jane Bowyer, and Dorothy Gabel; and each influenced my research.

During the 1980s, gender research in science and mathematics grew internationally. Researchers, particularly in the United Kingdom, the Netherlands, Denmark, Norway, and Australia examined factors behind gender differences in science courses and careers. An international professional association began in the Netherlands that provided a forum for such research. It was named, *Girls and Science and Technology* (GASAT).[3] International researchers, such as Jan Harding and Alison Kelly in the UK; Svein Sjøberg, Svein Lie, and Doris Jorde in Norway; Helena Sorenson in Denmark; and Lesley Parker, Gilah Leder, and Léonie Rennie in Australia, contributed to the research concerning factors affecting girls in science.

By the early 1990s there was sufficient research about science and gender that three colleagues, Lesley Parker, Léonie Rennie, Dana Riley (now Black), and I proposed a theoretical model which researchers could use to examine gender differences in science achievement and retention (Kahle, Parker, Rennie, & Riley, 1993). Our model illustrated how classroom factors interacted with both teachers' and students' previous experiences to affect student outcomes (*performance in science, choice of science subjects, choice of science-related careers,* and *extracurricular choices*). It built on Eccles' academic choice model that explicated factors affecting girls in mathematics.

PUTTING IT TOGETHER

Two opportunities clearly influenced my career: first, my selection as part of the biology team for Project Synthesis in the early 1980s. The team, selected and chaired by Paul DeHart Hurd, included Bob Yager, Rodger Bybee, and me. I have often referred to it as my first "big time" consulting job, and it opened doors. The second opportunity was serving on the first of many National Research Council committees. In the late 1980s, the Committee on High School Biology Education was formed to study biology teaching and teacher preparation. The committee was composed of a distinguished group of scientists; two of whom influenced my career, namely, Stephen Berry, a physical chemist at the University of Chicago and

Tim Goldsmith, a biologist at Yale University. Tim provided an excellent example of a how a thoughtful scientist approached science education, while Steve alerted Ken Wilson at The Ohio State University of my move to Miami University in Ohio.[4] Ken, a Nobel Laureate in physics, and I joined forces to write the proposal and to provide leadership for Ohio's Statewide Systemic Initiative (SSI), one of the first ten SSI projects funded by NSF.

I think that it is fair and accurate to say that Ohio was the only SSI that focused on equity—in terms of gender, race, and socio-economic levels. The project targeted middle school science and mathematics teachers (because all students still take those subjects in those grades), economically disadvantaged urban and rural schools, and Ohio's largest minority group, African American students. In selecting participants, priority was given to teachers in the 41 school districts, identified by the Ohio Department of Education as "at risk." Most of those districts were in Ohio's cities. Further, we used research on successful professional development to design six-week, content-based summer institutes that were taught by inquiry and used cooperative learning techniques. Research by various science educators had indicated that those instructional strategies were effective with students who traditionally had lower enrollment and achievement levels in science and mathematics. The institutes were followed by four to six professional development sessions during the academic year. The academic year activities stressed authentic assessment and equity issues. In the third year, we selected an equity team that traveled the state providing guidance and expertise for all instructors and teachers. And, last, when we realized that the leadership teams selected by each regional council in the state[5] did not reflect the mix of students in Ohio, we provided funds for an additional team member as long as the person selected was a member of an underrepresented group in science.

Midway through Ohio's SSI, the Project Director Steve Rogg and I developed a comprehensive research and evaluation plan to assess the progress of the reform. We found evidence that the achievement gap between white and African American students was narrowing in mathematics. And we found that our recruitment plan enabled us to tap into the previously resistant pool of teachers in extreme urban and rural schools. In fact, the majority of SSI teachers were clustered in Ohio's eight major cities. Later, NSF awarded a follow-up grant to Steve, Ken Tobin, and I to extend the evaluation into a full-fledged research study of the five systemic initiatives in Ohio.[6] That project, *Bridging the Gap: Equity in Systemic Reform*, highlighted the importance of equity in systemic reform. *Bridging's* findings suggested policy changes for Ohio and the nation. One analysis showed that 15 percent of the variation in students' science achievement scores was due to teacher differences and that frequency of use of standards-based instruction (inquiry) had a more powerful influence than did teachers' grade level or type of certification. That finding influenced the content and pedagogy standards that Ohio was developing for science. Further, attitudes about length and cost of professional development changed among teachers and administrators, each group indicating that sustained professional development was more efficacious than short workshops. This finding changed the type of professional development funded by the state. Another finding indicated that the gender gap in science achievement between African American girls and boys (favoring girls) was narrowed with increased use of inquiry in science instruction.

The transition into both doing and studying systemic reform was a good one for me. It enabled me to bring the skills of a researcher to practical problems, conducting studies that actually changed policies about science and mathematics education in Ohio. But, my education and experiences as a teacher and as a researcher did not prepare me (or others who directed the SSIs) to oversee massive projects that involved stakeholders from classroom teachers to governors. My leadership of Ohio's systemic initiative was truly a learning experience, and the best decision I made (both for the Ohio project and for myself) was the decision to study the reform while doing it. Responsibility for a multidimensional, multi-year project made me realize limitations in educating the next generation of science educators. For example we educate future science educators to do research and to teach; we do not educate them to think systemically about the educational system nor to work effectively within or with educational bureaucracies. We educate them to focus on specific aspects of research, not to look across a myriad of research components to ascertain how they interact. We educate them to focus on one level of education, not to understand the system from preschool through graduate school. However, graduate programs in science education have been slow to change, and with the end of several NSF programs[7] that provided funding for doctoral students and for graduate program development, the likelihood of change diminishes.

NOTABLE CONTRIBUTIONS

Although I published research on other topics early in my career and more recently I have written extensively about systemic reform, the body of my research has been about gender and science; and I have chosen to focus on that writing in discussing the impact of my research. Reflecting back on a career that has been based on research and writing, a few publications stand out: both in terms of their influence in the field and their contribution to my own learning. Unquestionably, the first of those papers was "The myth of equality in science classrooms" (Kahle & Lakes, 1983). It was one of thirteen papers selected for the *Journal of Research in Science Teaching's* anniversary issue and has been one of the most frequently cited papers in science education (Holliday, 2003). In our secondary analysis of NAEP's items that queried students on what they had done in science and what they would like to do, we found that boys had experienced many more out-of-school—as well as in-school—science activities. The gap in experiences that girls brought to class was not due only to in-class experiences. At that time, girls' lack of experiences was viewed through the characteristics lens and explanations included: girls' "lack of interest," "low risk-taking behaviors," etc. However, our analysis of what girls would like to do indicated that girls wanted to do more science activities—both in and out-of school! This finding along with Alison Kelly's study of gender differences in spatial ability helped move the field away from the "characteristics lens."

Because a model was needed to test whether gender differences in science education were due to girls' characteristics or girls' responses to teaching/learning environments, Lesley Parker, Léonie Rennie, Dana Riley, and I proposed a model that interwove sociocultural influences with those found in classrooms. That paper, "Gender differences in science education: Building a model" (Kahle, Parker, Rennie, & Riley, 1993) provided an impetus for gender research in science education similar to the catalytic effect of Eccles' and Fennema's models in mathematics education.

In 1994, Dorothy Gabel, editor of the National Science Teachers Association's *Handbook of Research on Science Teaching and Learning*, asked me to contribute a chapter in the area of gender and science. Judith Meece and I collaborated to examine educational, social, and cultural factors as well as intervention programs that had been successful in closing retention, achievement, and/or choice gaps between boys and girls (Kahle & Meece, 1994). We were able to firmly conclude that classroom environments could, indeed, affect girls' interests, attitudes, and achievements in science. The "response lens" was gaining credibility in the field of gender research.

A fourth publication was the result of a valuable sabbatical experience as a Fellow at the National Institute for Science Education at the University of Wisconsin. By the time of this sabbatical (1996–1997), I had spent five years as co-principal investigator of Ohio's Statewide Systemic Initiative. The sabbatical allowed me to work with Norm Webb, who challenged me to develop a metric for measuring a reform's progress toward equity. My research into the national databases indicated grade levels or ages when specific educational conditions or challenges had to be met in order to ensure that a student could continue on a science/math pathway. The metric proved valuable in comparing various reforms, including the systemic initiatives. It also was useful in studying progress toward equity in schools and classrooms, as described in papers by Mary Kay Kelly and Peter Hewson. The metric, as well as other important research that studied equity within the context of reform, was published in a special double issue of the *Journal of Women and Minorities in Science and Engineering* (Kahle, 1998). That issue provided a venue for much of the research done in the 1990s and solidified the importance of considering equity issues in systemic reform.

As gender research progressed it became increasingly obvious that "one size did not fit all"; that is, the factors affecting Latinas varied from those affecting Asian American or African American girls. When Judith Meece, Kate Scantlebury, and I analyzed achievement, attitudinal, and behavioral data from Ohio's SSI, we found that attitudes as well as peer participation and home support correlated with science achievement differently for African American girls and boys. Achievement for girls, compared to that of boys, was more strongly correlated with the high ratings for home support and peer participation, while for boys, compared to girls, achievement was correlated with positive attitudes toward science (Kahle, Meece, & Scantlebury, 2000). This particular research allowed me to connect two aspects of my career—my early research with African American students in Gary and Chicago with my later study of systemic reform with African American students in Cincinnati, Dayton, Toledo, Columbus, Cleveland, and Akron.

In 2002, I was privileged to receive the American Education Research Association's (AERA) Willestine Goodsell Award for research about and advocacy of women. My address was presented in 2003 when the theme of the annual meeting was accountability, and my paper focused on that theme. In it, I examined possible test bias and was able to define how content emphasis in standardized tests, for example, NAEP and TIMMS, differentially affected the scores of girls and boys. In addition, gender issues in the *No Child Left Behind* report needed to be aired.[8] My address was published by the *Journal for Research in Science Teaching* as an extended editorial in 2004, and it is my choice as a fifth notable contribution. In this paper, I was able to track the effect of standardized testing on girls' and boys' scores in science on NAEP's first five assessments in science. In addition, I examined how the format and type of questions found on the science assessment of the Programme for International Student Assessment (PISA) resulted in higher scores for girls, essentially eliminating the gender gap favoring boys (Kahle, 2004).

MENTORS AND MENTEES

Mentors have been very important in my career, and both men and women have provided guidance and encouragement. Along the way I have mentioned Joe Novak, Floyd Nordland, Ann Howe, Mary Budd Rowe, Paul DeHart Hurd, Elizabeth Fennema, and Jacque Eccles. The research of each of them provided insights and their friendships opened doors. As my career progressed, a different type of mentorship was needed—one that challenged my views while supporting my work. A research project in Perth, Australia, with Ken Tobin and Barry Fraser required me to learn new skills (qualitative research), and I was challenged to defend the gender differences that I alone observed in the classrooms involved at our weekly research meetings. Ken Wilson and Steve Meiring were mentors during the SSI, and Luther Williams provided support and challenging ideas when I was Division Director for Elementary, Secondary, and Informal Education at the National Science Foundation. From all three, I learned to negotiate bureaucracies, to condense complicated research to brief statements, and to defend every assertion with solid data. The above people are among many who have provided opportunities and aided my career.

Also, along the way, I have been privileged to guide hundreds of master's degree students and a large number of doctoral students. Although I have learned from each of them, and all brought insights to my research agenda, I only have space to mention a few. When I was at Purdue University, Marsha Lakes Matyas was instrumental in conceptualizing our analysis of NAEP data that destroyed the myth of equality in science classrooms. April Gardner helped dispel other gender myths through her study of women's persistence in college majors that required knowledge of science and mathematics but were characterized as "feminine" (nursing), "neutral" (biology), or "masculine" (engineering). Her exit interviews with women and men revealed that women leaving biology and engineering majors had higher grade point averages than men but less confidence in succeeding in the major or in a related career. Steve Rogg's study of small groups suggested that cooperative learning groups were not a panacea to eradicate gender differences in science attitudes and achievement. Kate Scantlebury's study of gender issues that occur during student teaching provided a lens to help prospective and practicing teachers understand classroom interactions and to correct them. Those are but a few of the dissertations that added significantly to the literature about gender and science. At Miami University, all my doctoral and master's degree students focused on systemic reform, but all disaggregated data by race/ethnicity and gender. Nate Carnes, Arta Damnjanovic, and Mary Kay Kelly did rich studies of urban schools, while Dana Riley (now Black) studied inquiry in informal science education.

Now, I find myself in the position where the roles are switching—some of my former students or younger colleagues are in the position to mentor me! I have learned much from Judith Meece and Bill Boone, and I enjoy consulting for Kate and Dana. One of the most enjoyable, and valuable, aspects of our profession may be the intergenerational excitement and support that one experiences throughout one's career. I have been privileged to experience the excitement of research and to receive fiscal support to do research throughout my career. And, Bob Kane's advice—long ago—has held: find an area of research on which to focus. My area has been gender research, followed by research in systemic reform (with an equity emphasis). That research has enhanced my understanding of complex societal, cultural, and educational issues, and I hope that it has contributed—in some small way—to improving science education for many students.

THE ROAD AHEAD

My research, and that of many of my colleagues, did not examine the patriarchal nature of the scientific enterprise. The models developed by Fennema, Eccles, and my colleagues did not examine how the scientific enterprise itself might affect girls' and women's attitudes and careers. Rather, we basically accepted the nature of the scientific enterprise and attempted to understand factors that either aided or impeded girls' and women's opportunities within it. That is, we worked within the paradigm of science. Today, many gender researchers concur that the scientific enterprise itself is flawed. They are developing new models and theoretical frameworks to guide their research. The postmodernist feminists bring a different perspective to the field and will advance our understanding in different ways. The richness of their research will provide other lenses through which to view and to interpret any continuing gender differences. In 1875, Maria Mitchell addressed the Third Congress of Women, stating: "In my younger days, when I was pained by the half-educated, loose, and inaccurate ways which we [women] all had, I used to say 'how much women need exact science,' but, since I have known some workers in science who were not always true to the teachings of nature, who have loved self more than science, I have now said 'how much science needs women'" (Rossiter, 1982, p. 15). Researchers in gender equity will continue to explore both the scientific enterprise and women's roles; and some women scientists, like Barbara McClintock, will change the scientific enterprise. I look forward to continuing to learn from them and about them.

NOTES

1 Interestingly, today there is concern (although little research) that boys' underachievement in school (usually writing and reading are the subjects) is due to teaching environments that are unconducive to boys.

2 Recently, the *New York Times* published an article with the headline, "Numbers Don't Lie: Men Do Better Than Women" with the subheading, "SAT scores accurately reflect male superiority in math."

3 After Americans became more involved, it was renamed *Gender and Science and Technology*.

4 In 1989, I accepted an endowed professorship at Miami University, Oxford, Ohio.

5 The leadership teams were composed of a scientist and a science teacher in physical and life sciences and a mathematician and math teacher. They taught the summer institutes and provided support to teachers in their classrooms throughout the academic year.

6 In addition to the SSI, Ohio's three eligible cities, Cincinnati, Columbus, and Cleveland, had Urban Systemic Initiatives (USIs) and 13 counties in southern Ohio ware part of the Appalachian Rural Systemic Initiative (RSI).

7 Both the Systemic Initiative and the Centers for Learning and Teaching programs that had human resource building capacities have been eliminated.

8 *NCLB* requires that data from the National Assessment of Educational Progress and from the National Center for Educational Statistics be disaggregated by gender only *to the extent feasible*.

Jane B. Kahle
Professor Emeritus
Miami University

BARRY J. FRASER

MENTORING AND THE MULTIPLIER EFFECT

When I was in senior high school in Melbourne in Australia, I had no particular career goals. But I specialized in science and mathematics subjects because this is what you did in those days if you were a reasonably capable student. Actually, English was my favorite subject, but I didn't anticipate then that later I would spend much of my professional life in writing and editing in science education.

SOME EARLY ACCIDENTS IN CAREER DEVELOPMENT

By the time I was ready to enroll in higher education at the University of Melbourne, I still didn't really know what I wanted to do for a career. So, I followed in my older brother's footsteps and started an engineering degree. After a couple of years, it became clear that I didn't have much appetite for this engineering course, except for its physics and mathematics subjects. So, I swapped over to a science degree and majored in physics and mathematics, although I still had little idea of what this might lead to.

When I took on part-time jobs to save up to buy a second-hand car, these included private one-to-one tutoring of secondary students in mathematics and science in their homes. This accident of history was a turning point. I found out that I had a keen interest in mentoring, helping and teaching others so that they could advance themselves and achieve their goals. My next step had become clear: when I completed my science degree, I would enroll in a Diploma of Education so that I could become a secondary science and mathematics teacher with the Victorian Education Department.

At that time, Monash University was still quite young and frequently was in the news because of student activism and protests about the Vietnam War. I decided to try out this new and potentially interesting university for my introductory studies in the field of education as I undertook a one-year preservice teacher education course. This decision to go to Monash was another turning point. Not only did I earn a teaching qualification, but I began to develop a scholarly interest in the field of education. In no small part, this was due to a quite unique team that was assembled at Monash at that time, including Sidney Dunn (sometimes credited for bringing educational measurement to Australia), Peter Fensham (Australia's first professor of science education), Peter Musgrave and Lindsay Mackay (who introduced me to curriculum evaluation through his research on PSSC physics). Last but not least, prior to his invitation to work at Stanford University, Denis Phillips was an inspirational example of a young ex-science teacher who had become an insightful and eloquent scholar of education.

K. Tobin, W.-M. Roth (Eds.), The Culture of Science Education, 37–46.

When I became a full-time schoolteacher the next year, I didn't delay in returning immediately to Monash for part-time studies in the educational field as part of a Bachelor of Education degree. I was especially interested in educational measurement and curriculum evaluation, which were quite new on the Australian scene at the time. Before long, I already was planning to go on to study a Master of Education part-time when my Bachelor of Education was finished.

Although I was happy as a schoolteacher, others had different plans for me. When working in my mother's garden one day, I received a totally unexpected telephone call from the dean of education at Monash to say that he had a scholarship for full-time master's study and it was mine if I wanted it. As money was not of great concern to me at that time, I grabbed this opportunity to spend a stimulating year as a full-time master's degree student. The recognition of being invited to take up this scholarship helped to shape my self-esteem as a researcher.

During that year when I was a full-time master's degree student, I noticed an advertisement in the newspaper for a position called "Research Officer" at the *Australian Science Education Project* (ASEP; Australia's first national curriculum development project) also located conveniently in Melbourne. Interestingly, this position involved writing multiple-choice diagnostic test items and conducting curriculum evaluation (although I had an unclear idea about what the latter meant). Although the position was full-time, I brashly wrote to say that I had studied educational evaluation under Sidney Dunn and Lindsay Mackay and that I would like to work part-time in this position for just one day per week. After telephone calls between ASEP and Monash, I was offered this job for one day a week.

My time at ASEP was another turning point. Gregor Ramsey, Deputy Director of ASEP, had just returned from undertaking his own doctorate in the US and from close involvement in CHEM Study. He was young and inspirational and was trying to apply ideas from the then-new field of curriculum evaluation to evaluating ASEP. My next direction had emerged. When my master's was finished at the end of that year, I would enroll in a doctorate at Monash with my research topic involving an evaluation of ASEP.

My time working at ASEP turned out to be significant in another way. I came into contact with Darrell Fisher, a young science teacher from Tasmania, who had been brought to ASEP as a curriculum writer. This was the start of a lifelong collaboration and friendship that started with Darrell undertaking his doctorate with me and later led to us working together as colleagues at Curtin University of Technology.

Then, at the end of the year as a full-time master's degree student, there was another unexpected telephone call. I was being invited to a full-time junior teaching position at Monash for the next several years. I gladly accepted this as it provided an ideal situation for pursuing my doctorate seriously.

Over the next few years, I worked away on my doctorate entitled "An Evaluation of ASEP Involving Instructional, Aptitudinal and Environmental Variables," which was squarely located in the field of curriculum evaluation. Interestingly, as a somewhat minor afterthought, my doctoral study also involved evaluating the psychosocial learning environment in ASEP classrooms. This and other parts of my doctoral study had been inspired by the pioneering research of Herbert Walberg in relation to Harvard Project Physics. At that time, little did I know that this initial work on classroom learning environments would blossom in the years and decades ahead into probably my most significant contribution to science education.

Under the Australian university system, two world experts examine doctoral theses externally. When completed, my doctoral thesis was sent for evaluation to Gregor Ramsey (a central figure at ASEP) and Herbert Walberg (whose work in curriculum evaluation and learning environments had shaped my own research). I quite was taken back by the strength of these examiners' positive comments about my thesis, especially those of Walberg, who also confirmed his high opinion by inviting me to write a chapter based on my thesis for his book *Educational Environments: Evaluation, Policy and Productivity* that appeared in 1979. Again, this recognition buoyed up my self-esteem as an emerging scholar and began a long-term mentoring and collaborative relationship with Walberg that would influence my career in several salient ways.

SHAPING MY OWN FUTURE

Whereas numerous accidents of fate characterized much of the evolution of my entrée into doctoral work and academia, my own decisions and choices were the main determinants of subsequent directions in my career.

Being prepared for the inconvenience and expense of relocating interstate to take up new jobs was very important. By the time when I was close to completing my doctorate in 1976, already I was relocated in Sydney in my first tenured university position at Macquarie University. Then, about six years later, I moved to Perth to take up a position at the then Western Australian Institute of Technology, which later became Curtin University of Technology (named after Australia's Prime Minister, John Curtin, during the Second World War).

Also, as a young scholar without the luxury of funding support from research grants, I decided to draw heavily on personal funds to travel internationally to expand my knowledge and extend my networks. This included a number of six-month sabbatical leaves (of which the first two were watersheds) and regular participation in overseas conferences.

I survived the "Blizzard of 1979" in Chicago during a sabbatical with Herbert Walberg at the University of Illinois, as well as spending shorter amounts of time at a few other locations. This sabbatical with Walberg not only built on earlier intersections when he was my external doctoral examiner and invited me to write a book chapter, but spawned many years of invaluable collaboration and mentoring which ultimately led, among numerous other things, to a co-authored literature review on psychosocial environment in science classrooms published in 1981 and *Studies in Science Education,* a co-edited book in 1991 entitled *Educational Environments: Evaluation, Antecedents and Consequences* (Pergamon Press) and another co-edited book in 1995 entitled *Improving Science Education* (National Society for Study of Education). It also involved the productive three-way collaboration involving Wayne Welch described later in this section.

This sabbatical leave also enabled me to try my hand at more substantial writing and editing tasks. During this time, I wrote the monograph *Learning Environment in Curriculum Evaluation*, which was published in 1981 as a whole issue of the journal then called *Evaluation in Education* (now called the *International Journal of Educational Research*) and guest-edited a special issue of the journal *Studies in Educational Evaluation* devoted to the topic of "Classroom Learning Environment."

The sabbatical leave in 1979 also involved short amounts of time with Bob Stake at the University of Illinois at Urbana Champaign and with Barry Mac-Donald at the University of East Anglia in England. These visits were focused on the field of curriculum evaluation and ignited my interest in emerging trends in the use of qualitative data-gathering approaches in educational program evaluation.

In another sabbatical leave in 1984, I had the opportunity to spend substantial time at Stanford University working with Denis Phillips (who had impressed me when I was an Education student at Monash) and another great pioneer of human environments research, Rudolf Moos, from the Stanford Medical School. During this large chunk of time at Stanford, I wrote my first and only major single-authored book entitled *Classroom Environment* that was published by Croom Helm in 1986.

Also, during this sabbatical leave, I made a brief but important visit to the University of Minnesota to work with Wayne Welch on secondary analysis of his NAEP (National Assessment of Educational Progress) databases. This formed part of a three-way collaboration with Walberg using his educational productivity model as a guide. This fruitful three-way collaboration on educational productivity and secondary analysis won a NARST Outstanding Paper Award in 1986 and was published as several articles (e.g., *Journal of Research in Science Teaching* in 1986) and as a whole issue of the *International Journal of Educational Research* in 1987 (also with John Hattie).

As well, I invested considerable time and financial resources in attending annual meetings of American Educational Research Association (AERA) and National Association for Research in Science Teaching (NARST), again for the purposes of creating and expanding networks and disseminating research. For example, up until 2006, I have presented a total of 150 papers at AERA and 78 papers at NARST.

Involvements with NARST and AERA have been very important in my career. In numerous ways, these associations provided me with key professional development opportunities and built my self-image as a researcher through recognition in various ways.

Through five awards, NARST acknowledged my accomplishments as a science education researcher at various stages of my career. It was especially encouraging to win the Outstanding Paper Award for the very first paper that I ever presented at a NARST conference. This was with Darrell Fisher in 1984 for a paper on learning environments based on his doctoral research. This Outstanding Paper Award also was given in 1986 for a paper on educational productivity and secondary analysis with Wayne Welch and Herbert Walberg, and again in 1988 for a paper on exemplary teaching with Ken Tobin. As well, NARST awarded me a Practical Implications Award in 1986 (with Darrell Fisher) and its prestigious Outstanding Contributions to Science Education Through Research Award in 2003.

I was elected to NARST's Board of Directors during 1991–1994, when I was able to make a considerable impact on NARST's provisions for overseas members as Chair of the International Committee. Then, I had the privilege to serve as NARST's first President Elect, President and Past President from outside North America during 1994–1997. This included the daunting job of organizing the NARST annual meeting in San Francisco in 1995.

Within AERA, I was invited to serve on the *International Relations Committee* (IRC), first as a member and then as Chair during the early 2000s. This included being program chair for IRC sessions at AERA annual meetings.

But, no doubt, my most significant activity within AERA, with Chad Ellett, was establishing a Special Interest Group (SIG) on Learning Environments. Today, this thriving SIG continues to provide a forum for the dissemination and discussion of learning environment research from around the world.

CONTRIBUTIONS THROUGH DOCTORAL EDUCATION: MULTIPLIER EFFECT

Doctoral education has an enormous multiplier effect. The effort put into helping one person achieve a doctorate not only substantially changes that person's life, but he/she usually makes use of that doctorate in having an impact on the lives of numerous other people. During my career, I chose to devote a lot of effort to achieving this multiplier effect through doctoral education.

At my centre at Curtin University of Technology, I consider flexibility and access in doctoral programs to be of paramount importance. Only a small privileged group can afford the luxury of full-time on-campus doctoral study. Yet so many people could benefit themselves, and use this knowledge to others' benefit, through achieving a doctorate. Consequently, we encourage part-time doctoral study among people who are working full-time and living anywhere in the world. In many locations around the world, Curtin professors "bring the university to the students" by traveling to where groups of doctoral students live to support and help them to achieve this goal.

In Sweden, the word for doctoral advisor or supervisor is "doctor father." This term captures something of the special, significant and enduring qualities of the relationship between a doctoral student and his/her advisor/supervisor. So far, I have been fortunate to be the "doctor father" of the 55 doctoral graduates from 11 countries, including 13 doctoral graduates from Australia, 7 from Singapore and 25 from North America (including 11 from Miami). At present, I am the main advisor/supervisor for dozens more currently enrolled doctoral students.

Many of my doctoral graduates work in positions of influence in universities around the world, with some of them now having their own second generation of currently enrolled or already-graduated doctoral students. For example, Darrell Fisher (my doctoral "child") has now successfully graduated over 50 doctoral students of his own (my doctoral "grandchildren"). In turn, some of Darrell's second-generation doctoral students now have responsibility for third-generation doctoral students (my doctoral "great grandchildren"). The multiplier effect is very pervasive.

Some of my doctoral graduates have gone on to assume educational leadership positions in politics (e.g., one was the Minister of Education in Fiji), in university administration (e.g., two graduates were presidents of large Indonesian universities), as university professors, as school principals and as key figures in educational systems. Still other doctoral graduates play central leadership roles in schools as department heads and mentors to other teachers.

Overall, my work in doctoral education has contributed to capacity building not only in Asian and African countries, but also in urban school districts in the USA (especially Miami).

EDITORIAL WORK

Fairly early in my career, others recognized that I had ability as an editor. During my first tenured position at Macquarie University, I was pleasantly surprised to be invited by a group of distinguished curriculum scholars from the University of Birmingham in England (Philip Taylor and William Reid) to become Australasian Editor for the *Journal of Curriculum Studies* (Taylor & Francis). This was followed later by being Coeditor of the Carfax journal *South Pacific Journal of Teacher Education* and coeditor of Elsevier's *International Journal of Educational Research* (IJER). This work with IJER involved an annual editors' meeting in The Hague in the Netherlands and enabled me to form strong liaisons with significant European researchers such as Neville Postlethwaite (Germany), Bert Creemers (the Netherlands), and Erik DeCorte (Belgium).

At the time of writing this chapter, I am serving officially on the editorial boards of the *American Educational Research Journal, Research in Science Education, International Journal of Science and Mathematics Education, Journal of Research in Childhood Education, South African Journal of Education, Curriculum Perspectives* (Australia), *Science Education*, and *Research in Science and Technological Education*. Previously, at various times, I served officially on the editorial board of journals such as *Journal of Research in Science Teaching, Studies in Educational Evaluation, International Journal of Educational Research, Curriculum Forum* (Hong Kong), *New Directions in Program Evaluation*, and *Asia-Pacific Journal of Education*.

In terms of journal-editing contributions, my proudest accomplishment was the creation in 1998 of *Learning Environments Research: An International Journal* published by Kluwer/Springer. I have been its Editor-in-Chief since this journal's inception, working with regional editors in Europe (Theo Wubbels), North America (Ken Tobin and Hersch Waxman) and Australasia (Darrell Fisher). The existence of this journal has contributed much to the consolidation, expansion and recognition of the field of learning environments research. It also has provided an opportunity for me to "add value" to the work of learning environment scholars and to help them in their career development.

In addition to editorial work involving journals, editing books also has provided a similar opportunity to enhance the research of others by adding value to their work and providing dissemination outlets. As well, the editing of books has afforded me wonderful opportunities to work closely with excellent colleagues.

My first edited book (with Ken Tobin and Jane Kahle) is entitled *Windows Into Science Classrooms: Problems Associated with Higher-Level Cognitive Learning* and was published by Falmer Press in 1990. It was the product of a stimulating and revealing study conducted when Jane Kahle made an extended visit to my University (see Tobin's chapter in this book for further details). With Herbert Walberg, I coedited *Educational Environments: Evaluation, Antecedents and Consequences* for Pergamon Press in 1991 and *Improving Science Education* for the National Society for the Study of Education in 1995.

In 1996, two further co-edited books were published. Teachers College Press published *Improving Teaching and Learning in Science and Mathematics* that I co-edited with Reinders Duit and David Treagust. Kluwer published *Gender, Science and Mathematics—Shortening the Shadow* which was co-edited with Lesley Parker and Léonie Rennie. An interesting feature of both of these two books was my belief that both science and mathematics could be covered within the same book, which is at variance with most other publications that tend to encompass either science education or mathematics education.

One of the largest and most daunting undertakings of my career was my role as coeditor with Ken Tobin of the two-volume, ten-section 1,200-page *International Handbook of Science Education* published by Kluwer in 1998. Not only was the sheer size of this task unwieldy, but also numerous types of problems and delays arose. Many chapters in this handbook have been widely cited. Through this handbook, we made widely available a valuable and comprehensive resource to guide colleagues and doctoral students in their own research.

SOME EXAMPLES OF RESEARCH AND COLLABORATIONS

I am fortunate that my career has involved more studies and more collaborators than can be described adequately in this chapter. So let me sample a few. My early research—during my doctoral studies, my years at Macquarie University and my early years at Curtin University of Technology—focused largely on curriculum evaluation or educational program evaluation. This program of research was supported by around a dozen grants from federal and state governmental agencies in Australia, including the federal Department of Education, Employment and Training, the Commonwealth Schools Commission, the Commonwealth Department of Education and Youth Affairs, the national Curriculum Development Centre, and the Education Department of Western Australia. This work is encapsulated in many journal articles and in my monograph *Case Studies in Curriculum Evaluation* in 1988 (Social Science Education Consortium, Western Australian Institute of Technology).

This period also saw the beginnings of another area of contribution, namely, the careful development and validation of questionnaires that later would be used by hundreds of researchers, thousands of teachers and millions of students. The Australian Council of Educational Research published three of these over a period of around a decade: *Test of Enquiry Skills* (1979), *Test of Science Related Attitudes* (TOSRA, 1981) and *Individualised Classroom Environment Questionnaire* (1990). Of these, TOSRA (or parts of it) are still widely used today and requests for this out-of-print instrument continue. In addition to these published questionnaires, I have developed several other widely used learning environment questionnaires that are mentioned below.

Undoubtedly, my most significant and unique contribution to science education research has been my research on learning environments. Previously in this chapter, I discussed the central importance of Herbert Walberg, who inspired my interest in learning environments and was an examiner for my doctoral thesis. He was a mentor and collaborator who opened doors and shaped my self-confidence as a researcher. Some details of our research and publication activities already have been described earlier.

The significance of the field of learning environments is that, although no-one would dispute the importance of achievement and other learning outcomes, having a positive classroom climate is equally important and should be considered as both a means and an end. Much past research has consistently replicated the advantages of a positive learning environment in terms of promoting improved student outcomes, but a positive classroom environment also is a worthwhile process goal of education.

A striking feature of the field of learning environments is the availability of numerous robust, validated and widely used questionnaires that assess the learning environment through the perceptions of the students and sometimes the teachers in the classrooms. The use of participants' perceptions offers numerous advantages because it provides an assessment method that is economical, practical and nonintrusive. However, although a lot of my research effort (see below) has been invested in developing, validating and using questionnaires, there are considerable merits in combining quantitative and qualitative methods in learning environment research. Essentially a multimethod approach capitalizes on the unique strengths of each method and helps to overcome the weaknesses of each.

Earlier in the chapter, I mentioned my initial encounter with Darrell Fisher when we both worked at the Australian Science Education Project. When I moved to Macquarie University in Sydney and he had moved back to Tasmania (to what is now the University of Tasmania), Darrell undertook his doctorate with me on a part-time basis. During this time, we traveled interstate to visit each other frequently for stimulating discussions about his doctoral research. Through this doctoral study, we made unique contributions to the emerging field of learning environments, including a 1982 article on outcome-environment relationships in the *American Educational Research Journal* and a 1983 article on person-environment fit in the *Journal of Educational Psychology*. Also this work won NARST's Outstanding Paper Award in 1984. Darrell Fisher relocated to Western Australia around 15 years ago to become a colleague at Curtin University of Technology. Our research and publication program on educational learning environments continues today and has been supported by numerous grants from the Australian Research Council (similar to the National Science Foundation in the US).

My lifelong collaboration and friendship with Ken Tobin got into full swing when he worked as a colleague at Curtin University of Technology in the mid-1980s. In his chapter of this book, Ken describes some productive and award-winning studies that we undertook together both when he was a staff member at Curtin and during a later visit to Curtin while he was working in the USA. For example, this included research on exemplary teaching and on higher-level cognitive learning. Already, I have mentioned something of what was involved when Ken and I co-edited the *International Handbook of Science Education* during the years leading up to its publication 1998.

Although these several studies with Ken Tobin in Western Australia were conducted primarily within an interpretive framework, interestingly, each also had a minor component involving the use of a learning environment questionnaire. This enabled us to explore the potential of combining qualitative and quantitative methods in learning environment research (see the chapter entitled "Qualitative and Quantitative Landscapes of Learning Environments" in the *International Handbook of Science Education*).

A lot of the research and publication in which I have been involved has been with the group of 55 mentees who are my doctoral graduates. In cooperation with these doctoral students, I have been able to advance, extend and replicate learning environment research in many subject areas, at all levels of education and in many countries, including Australia, the US, Canada, Singapore, Korea, Indonesia, Brunei, Nigeria, and South Africa. The opportunity to travel to numerous other countries—to conduct research, visit schools, collaborate with colleagues, and try out research ideas in new contexts—has been a true highlight of my career and a source of much professional development and insight.

I also learned about the painstaking work involved in translating, back-translating and cross-validating questionnaires for use in other languages. But, through such work, traditions of learning environment research have been made accessible in the Chinese-speaking world, Indonesia, Korea, the Arab-speaking world, and the Spanish-speaking world through several doctoral students in Miami.

My research program on learning environments has been supported by around ten grants from the Australian Research Council (ARC). For example, these grants facilitated valued collaboration in the development, validation and use of some popular learning environment questionnaires including: the Constructivist Learning Environment Survey, CLES, with Peter Taylor (see *International Journal of Educational Research* in 1977); the Science Laboratory Environment Inventory (SLEI), with Geoff Giddings and Campbell McRobbie (see *Journal of Research in Science Teaching* in 1995); and the What Is Happening In this Class? (WIHIC), with Jill Aldridge and Campbell McRobbie (see *Journal of Educational Research* in 1999). Interestingly, the research with the SLEI and WIHIC pioneered the parallel development of "class" and "personal" forms of learning environment questionnaires and investigation of their properties.

I am very grateful to colleagues in Taiwan for the opportunity to collaborate with a team of Curtin researchers (especially Jill Aldridge) and a team of Taiwanese researchers (e.g., Jong-Hsiang Yang, Iris Huang, Huann-Shyang Lin, Hsiao-Lin Tuan and Hsiou-Chin She) in cross-national studies of learning environments in Taiwan and Australia. This involved classroom observations and interviews in each others' countries, the construction of narratives, many visits and rich conversations, and the translation, validation and use of the What Is Happening In this Class? in the Chinese language. This study opened up learning environment research in Taiwan and provided a model for international cooperation that has been adopted by the National Science Council in Taiwan. It provides a good example of cross-national research on, and the use of multimethods in, learning environment research (see the article by Aldridge, Fraser, and Huang in *Journal of Educational Research* in 1999).

I extended the cross-national research involving Taiwan and Australia to another cross-national study involving South Africa and Australia in cooperation with Jill Aldridge and Rudiger Laugksch. This study focused on the classroom and school environments associated with outcomes-based education and involved the development and validation of a classroom environment questionnaire in the Sepedi or North Soto language (see the article by Aldridge, Laugksch, Fraser and Seopa in *International Journal of Science Education* in 2005). We also researched the school-level environment in South Africa (see an article about to appear in *Learning Environments Research* in 2006).

Another ARC-funded study with Aldridge and others involved outcomes-focused and technology-rich learning environments in a new and innovative upper secondary school in Western Australia. This work is the focus of a book entitled *Outcomes-Focused Education* in preparation for publication by Sense Publishers.

Barry J. Fraser
Science and Mathematics Education Centre
Curtin University of Technology

KENNETH TOBIN

THE CHANGING FACES OF RESEARCH
IN SCIENCE EDUCATION

A Personal Journey

My dad was a baker's son who, because of the 1930s depression, did not finish high school. However, called to serve in the armed forces during the Second World War, he completed a high school equivalence certificate, enabling him to become a pilot. After the war he returned to baking and eventually became a successful businessman. Growing up I spent many hours doing various odd jobs at his elbows, listening attentively as he taught me to value education and hard work. His plan for me was simple—graduate from high school, go to university, teach, save for a few years, and return to university to become an engineer. I graduated from high school, obtained a teaching scholarship, applied to study for a science degree and was denied admission by a stern-faced administrator who had reviewed my scores on the leaving examination (an external examination taken at the end of grade 12; to signify high school graduation and potential for tertiary admission). He concluded, "You would not be successful in science. But you could manage history." I shook my head and informed the man that I would prefer to be a primary school teacher.

BECOMING A SCIENCE TEACHER

A little more than two years later, on a sweltering day in January, I entered the principal's office at Northampton Junior High. Northampton was a small rural township; a farming community located more than 300 miles to the north of Perth, the nearest city with a university. "You must be my math and science teacher. I am expecting you." Alan Jones, bare-chested and wearing short, tight football trunks, was delighted to have someone to teach two core areas of the curriculum. I received his greeting with mixed emotions. Western Australia has a central system in which a teacher can be assigned to teach in any public school in the entire state (a huge state with an area of about one million square miles). Although I had requested an assignment to teach science and mathematics in a rural high school, I was far out of field, having no university level education in science or mathematics. Furthermore, I was not yet twenty years old. I nodded to Alan in affirmation, signaling the start of a career in science education that would span more than forty years.

As the only science and mathematics teacher at Northampton Junior High I was virtually on my own as far as figuring out how to teach science. Being out of field I

K. Tobin, W.-M. Roth (Eds.), The Culture of Science Education, 47–58.

had to struggle to get the resources to support what to teach and how to teach it. I relied on the textbooks used by the students and ordered copies of the texts I had used in my final years of high school science. Also, my approach to teaching was shaped by recollections of my high school mathematics and science teachers. Constant invigilation from Alan ensured that I planned thoroughly. Every day without exception he arrived in my classroom with the cheerful request "What's on the menu Ken?" It was unthinkable that I would fail to produce my lessons for his inspection.

After two years in Northampton I was transferred to Applecross Senior High School in the suburbs of Perth. This appointment allowed me to begin part-time study toward the first of several degrees in physics. Because most of the other science teachers in my school were studying for advanced degrees or completing an initial science degree, there was time after school to meet as a group before commencing evening classes (four days a week). During such times we usually compared notes on what we were teaching, how we were teaching it, and ways to improve science education, especially for our students, but also throughout the state. Mainly due to the leadership of Ron Louden, the head of the science department, and enthusiastic colleagues my teaching of physics, chemistry, and general science advanced in leaps and bounds in my six years at the school.

After I completed a physics degree I took a break from teaching to travel to England, essentially to play tennis and possibly study for a doctorate in physics. While in England I taught the new Nuffield science curricula and A-level mathematics and chemistry. After two years I returned to Australia, assumed a curriculum development position for a year, and then taught for a year general science, biology, and chemistry at Applecross Senior High. Also, I commenced a research-oriented master's degree in physics, intent on obtaining a PhD. However, I was increasingly interested in researching teaching and learning and in 1974 I became a science educator, at what is now Edith Cowan University.

GETTING STARTED IN EDUCATION RESEARCH

John Lake, a Western Australian teacher educator, had just completed his doctorate at Rutgers University and was giving a research seminar on wait time and its relationship to verbal inquiry. I was enthralled—since I had decided to study the teaching and learning of physics in elementary grades. As he laid out his results my mind was darting ahead, considering how the use of an extended wait time could afford the learning of science. After the seminar I asked John what he thought about using a long wait time to increase science achievement. "It's too late. All the research on wait time that needs to be done has been done." His response felt like a brush-off and I became even more determined to research wait time in relation to science achievement.

My research on wait time was a quasi-experiment set in grades five through seven—in partial fulfillment for a master's degree in physics, a precursor to doctoral studies at the University of Georgia and follow-up research in Australia. The study was multi-faceted. I theorized that students' formal reasoning would mediate science achievement, located a pencil and paper measure of formal reasoning ability developed by a fellow Australian, and undertook a preliminary study to adapt and validate the instrument for use in Western Australian schools. Because the study was situated in Western Australian primary schools, where, at that time, science was not taught regularly or well, I adapted lessons from the *Elementary Science Study* to create a fifteen-week curriculum that began with *Ice Cubes* and included *Colored Solutions*, and *Sink or Float*. Appropriate achievement measures also were developed and validated. Finally, I addressed technical problems of measuring wait time—designing a band pass filter to pass audio frequencies while filtering out background noise.

I enrolled at Murdoch University for a doctorate in education—studying multivariate statistics with Barry McGaw, who had been an external examiner on my master's degree. There were no science education doctorates in Australia at that time, although science educators such as Barry Fraser were doing doctoral studies at Monash University on the east coast of Australia. After two years, with an eye on my career, Barry McGaw recommended that I pursue a doctorate in science education in the United States. I focused my applications on places where wait time research had been undertaken, for example the University of Florida, where Mary Budd Rowe was a professor. Mary suggested I consider a university with a larger science education faculty, pointing out that if she left for any reason I could be stuck without a supervisor. This turned out to be good advice since Mary left for a lengthy stint at the National Science Foundation soon after I arrived in the United States. Accordingly, because I was interested in David Butts' research on problem solving, I applied to the University of Georgia, recognized by many as the strongest program in science education at the time. I was accepted into the program as the first foreign student to be admitted.

Guiding my Doctoral Studies

"I want to do a PhD, not an EdD." My meeting with David was strained. Among my first words to him was a sentence that projected my concerns about whether or not an EdD would be accepted as a bona fide doctorate. There was no such thing in Australia and I hoped David would suggest I enroll for a doctorate in educational psychology—at the time the only PhD offered by the College of Education. David smiled and remarked, "If you want a PhD buddy, then go someplace else." With two young children and a spouse that was not an option and so I accepted the assurances from science education faculty that this EdD was the world's finest research degree in science education. Interestingly, and no real surprise to me, several years after I graduated the science education program obtained permission to offer a PhD in science education. In 2000 the University of Georgia's science education department presented me with the David P. Butts award for outstanding accomplishments of alumni.

I was fortunate to have Bill Capie and Russell Yeany on my doctoral committee because I had a very different background to most of my fellow doctoral students. Already I had worked for five years as a science educator in a university, earned a master's degree in physics, done education research, written a thesis, and completed the quantitative methods required for the doctorate. Barry McGaw had advised me to "avoid introductory courses in education statistics. Find the most advanced multivariate course and begin there." Doing as Barry advised required deviations from the blueprint, my persistence, and strong support from Bill and Russell to tailor the degree requirements to my goals. Fortunately these structures allowed me to obtain expertise in science education, publish my research in leading peer-reviewed journals, and develop a strong background in education measurement, statistics and research design. I should point out that a feature of Bill's style of mentoring was that he sat in several of the courses with me—modeling the necessity for faculty to be lifelong learners.

Becoming a Researcher

Finally Bill broke the silence, "so, what are you gonna be when you grow up?" Bill was taking me on a tour of school districts in southern Georgia and used the opportunity to broach a number of issues, including my program of study. I responded with confidence: I would teach science methods courses, develop science curricula for K–12 students, and conduct research on science teaching and learning. Bill paused for what seemed like a hundred miles before commenting: "No. You've got to choose one." After another long pause, I replied: "Then I'll be a researcher." From that moment onward my goals were set, the foremost being to become a researcher.

I learned a great deal from Bill. Following from my master's degree in which I explored formal reasoning in relation to science achievement, Bill and I created the *Test of Logical Thinking* (TOLT), an adaptation of a pencil and paper test developed by Anton Lawson. A feature of this ten-item test was its high reliability. We decided to use what we referred to as two-tier items—at the first tier a respondent selects a correct answer from given alternatives and, at the second tier, she identifies a reason to support the chosen answer. We identified the alternative reasons as those most commonly ventured for given answers during interviews of many students. For an item to be scored as correct on the TOLT it was necessary to provide both the right answer and its associated reason. The total score on the TOLT was a measure of formal reasoning ability, which we regarded as a continuous variable. Published in *Educational and Psychological Measurement*, the paper that describes the development and validation of the TOLT is widely cited in the *Citation Index for the Social Sciences* and the test is used worldwide and is available in numerous languages.

In the decade after graduate school a major influence on my professional growth was Russell Yeany, who was a source of wisdom during my doctoral program. Although I did not co-author work with him, he was a mentor in terms of how to be a science educator—a leader by example. For example, he encouraged me to build a squad of graduate student researchers at Georgia even though we did not have financial support. Also, he suggested I pursue a position at Florida State University, and supported my participation in the leadership of the National Association for Research in Science Teaching (NARST), as a member of the editorial board of the *Journal of Research in Science Teaching* (JRST), as a member of the Board, and then as President. Although I have not yet done it, he also advised me to get involved in serving research by taking a position with the *National Science Foundation*.

CHANGING FRAMEWORKS

I felt different after I finished my doctorate and I especially missed the scholarly community of the University of Georgia, where research was so highly valued. When I returned to Australia it seemed as if everybody expected me to occupy the same niche I had filled before earning my doctorate. Yet, I had changed and even though I was appointed as a senior lecturer at Curtin University, the writing was on the wall. Increasingly I felt the call to return to the USA, to address some of the pressing problems in science education and to join a larger community of science education researchers. But, before I left, I participated in two important studies in Western Australia.

What Happens in High School Science

In 1984 I traveled from Australia to the annual meetings of NARST and AERA with Rob Baker, who had studied ethnography with Harry Wolcott. During the long journey to New Orleans and back again we talked incessantly about different approaches to ethnography. As was my custom, I made two visits to international scholars—at the University of Texas and at Michigan State University. At Texas I gave a presentation concerning my research on wait time. As part of a quasi-experimental study I controlled the lessons taught by providing teachers with plans so that they would teach much the same thing during the study. Walter Doyle asked a question about what the teachers would have done if they had taught what they wanted to teach—that is, if they had taught their own curriculum. Although his question was easy enough to answer in terms of experimental design, I was interested in that issue too. I had tired of the reductionism of my research program and was swayed by Mary Budd Rowe's opinion that research needed to address the big issues in science education—macro effects such as high stakes testing that tended to drive the curriculum.

At Michigan State University I met Jim Gallagher, who had just retooled to become an ethnographer. He had sought Fred Erickson's assistance in learning ethnography and introduced me to Fred who gave me an advance copy of his soon-to-be-published chapter on interpretive research (Erickson, 1986). Grounded in hermeneutic phenomenology, interpretive research was an answer to many of the problems I had considered, especially in relation to being able to learn from what happened rather than focusing on a priori hypotheses using pre-developed and validated instruments. Interpretive research became a core part of my research methods from then until the present.

My research with Jim concerned the question "what happens in high school science classes?" This question derived from Walter's question and Fred's suggestion to always begin interpretive research with two general questions: what is happening and why is it happening? Substantively we studied classroom management, gender-related patterns of engagement, target student phenomena, and learning in laboratory activities. Together we learned how to do ethnography and, since I was at Jim's elbow, I learned a great deal about what it meant to be a participant observer and the emotional aspects of doing research with teachers. By studying what is happening from the perspectives of the participants we were often caught up in the highs and lows of being in classes and engaging intensively for extended periods of time. I have vivid images of Jim sharing his fieldnotes with participants; hence, a lesson to be learned is never to write anything you are not willing to share with others. Jim also cotaught in a peripheral way and, when we experienced zoo-like conditions in a classroom, we shed tears as we explored what was happening through the eyes of teachers.

Emerging from our research on what happens in high school science classes were target students—those who dominate interactions. This research provided insights into issues of equity and patterns of practice about which participants were not necessarily aware. Surprisingly, when asked about target students, the teachers and students could identify and accept them as the way things should be, often their presence being regarded as desirable. One of our papers on target students was the recipient of an award for the best paper to be published in JRST. Recently this paper was identified as one of the most influential articles published in JRST since its inception.

My collaboration with Jim solidified my participation in ethnography and classroom research. We created qualitative research groups at Michigan State University and the University of Georgia and began to meet as special interest groups at international meetings such as those of NARST and AERA. We also ran presessions at such meetings and in so doing many new researchers were introduced to qualitative methods. Finally, in an international sphere, we participated together and individually in places such as China, Japan, Taiwan, South Africa, and Central America to assist researchers to use interpretive methods in their research in science education.

INTERPRETIVE RESEARCH ON TWO CONTINENTS

At Curtin University I began to collaborate with Barry Fraser. Our exemplary practice study in science and mathematics education involved many researchers in Western Australia. The research combined ethnography with Barry's research on classroom learning environments. Through the study of exemplary participants we described what was happening in science classrooms taught by successful teachers, an important study because it examined *enacted curricula* in detail. The study was the first of many I would do with Barry and it was my initial foray into ethnographies that employed research groups. We published extensively from this study, including a book and numerous papers, two of which won best paper awards—with

Barry (NARST Best Paper award) and Patrick Garnett (Best Paper published in *Science Education*) respectively.

Curtin University created a visiting scholar scheme to enhance the international reputation of the university. Among the renowned visitors to attend were Jane Butler Kahle and Floyd Nordland. One of the collaborative projects in which I was involved with Barry also involved Jane who, like Russell Yeany, was a role model for me—someone from whom I have learned a great deal about being a science educator in a national and international sense. On this occasion the research team included Jane, Barry, Floyd, and Léonie Rennie—in a study that employed multiple theoretical frameworks, methods, and quantitative and qualitative data. Because of the complementary perspectives we employed, we adopted the metaphor of *windows* to describe our collaboration.

Researchers involved in the *windows* study met weekly and embraced multiple perspectives. From an ethical standpoint this study was a watershed in that, by the time we finished, we realized the pitfalls of doing research *on* teachers as distinct from research *with* teachers. Peter, one of the teachers in the study, was delighted to know that his teaching was to be featured in a book and, probably because of work pressure, he did not read the chapters we gave him prior to the completion of the study. When the study was finished and he read the chapters he was unprepared for the strong critique of his teaching—running the gamut from management style to sexist practices with female students. From this point forward we pledged to no longer do studies with teachers who were not part of the research group.

Back in the USA

In 1985 I returned to the University of Georgia on a senior Fulbright award. I created a research team consisting of graduate students and undertook several studies of the teaching and learning of science and mathematics in high school classes in rural Georgia. I became much more focused on radical constructivism and the theoretical work of Ernst von Glasersfeld, taught about interpretive research and used it in our studies of classrooms. Some of the participants in these studies have been among my ongoing colleagues, including Antonio Bettencourt, Mariona Espinet, Chao-Ti Hsiung, and Nancy Davis. Studies of learning to teach through peer coaching and alternative perspectives on teaching and learning set a context for me learning a great deal about ethnography and the frameworks used by teachers and researchers in their descriptions of teaching and learning. Also, my simultaneous involvement in studies set in the USA and Australia allowed me to see that my learning as a researcher was ongoing and informed my interpretations in each study in which I was involved at the time. That is, what I learned in one study framed what I learned in the other. Although I returned to Australia after a year-long stay at the University of Georgia, I was soon to assume a position at Florida State University.

Constructivism and Beyond

Even though I was situated at Florida State University for more than a decade, my research continued in both Australia and the USA, with two studies from each country being salient. A powerful study in which we explored: "Why Sarah cannot teach the way she wants to?" assisted me to understand better how teachers make sense of teaching and use metaphors and beliefs as schema to think about their teaching and plan changes in their classrooms. We also carefully studied enactment of changes and struggled to explore relationships between schema and practices. During this time I benefited from Ernst von Glasersfeld's writing and thinking on radical and then social constructivism as lenses for making sense of teaching and learning to teach. Inevitably my focus on constructivism led me into conflicts with scientists, philosophers and policy makers. Opportunists who used their opposition to constructivism for self-promotion rather than resolving problems and advancing science education surprised me. I declined to get sidetracked by responding to the critiques of those who wanted to attack constructivism largely based on what I regarded as intentional misconstruals of assumptions and intentions that could easily have been checked before rushing to press. Stakeholders in science and philosophy dug in their heels to the detriment of science education and advancing the agenda of learning from research. Interestingly, to this day some of the staunchest critics of constructivism have not moved discernibly in their scholarly writing, having hammered their stakes so deep as to immobilize themselves. Their steadfast positioning has had no impact whatsoever on me. I have focused on making sense of teaching and learning through the windows of a bricolage of theories and for me the arbiter of viability is whether or not what I learn is credible and useful in expanding our empirical and theoretical knowledge of teaching and learning. In 1993 I edited a widely cited book entitled the *Practice of Constructivism*. This book was based on a symposium presented at the annual meeting of the American Association for the Advancement of Science. The salience of this publication is that it connects constructivism to research, teaching and learning.

My research on metaphors and beliefs was a step toward working out how teachers conceptualized teaching and learning and used referents as objects for possible change, in the conceptual object (i.e., schema) and associated practices. The research was an important precursor to the use of sociocultural models for learning—in which cultural enactment is conceptualized as a dialectical relationship between schema and practices. When I mentioned to Mary Budd Rowe about what I was learning from my research on metaphor, she immediately replied to me that different metaphors were like master switches. As a teacher switched metaphors a different bank of lights came on. I regarded Mary as science education's great conceptualizer. Beginning with her work on wait time, Mary could see patterns that others missed and had a sense of the game that was of importance to practitioners and policy makers, not to mention researchers like me.

The culmination of more than a decade of my research on wait time was my 1987 publication in the *Review of Educational Research*, an article recognized by the American Educational Research Association (AERA) with its *Interpretive*

Scholarship award. Next year I received AERA's *Raymond B. Cattell Early Career Award* for my research on wait time and a new program of research featuring interpretive research. As a science educator I was pleased that the premier research organization in the world recognized the value of research in science education. Also, the recognition vindicated my resolution to pursue a research agenda on wait time despite John Lake's exhortation that all that needed to be done had been done.

I undertook a research appointment at Queensland University of Technology and for two intensive summers collaborated with Cam McRobbie on research in a suburban science class. The study was notable because of our extensive use of videotape, a focus on Chinese Australians who did not adhere to the model minority myth, and opportunities for me to move deeper into social theory, exploring linguistic imperialism and power and schema associated with the unquestioned use of cultural myths as referents for enacting science education.

My research had moved beyond equity issues associated with gender to include race and, in a study with Barry Fraser, David Treagust, and a teacher researcher called Henrietta Hoffman, we studied social class in relation to science teaching and learning. A feature of these studies was the use of learning environment questionnaires to obtain quantitative and qualitative data. This study was the first in which a teacher researcher was a fully-fledged member of our research team and it was a pity that the published version of the paper took so long to pass peer review that Henrietta opted not to participate as a co-author.

Steve Ritchie came to Florida State University on a sabbatical leave and so too did Peter Taylor. Both participated in extensive studies and coauthored key works with me. For example, for the first time in my research program, Steve and I used a middle school student as a researcher and began to see both the promise and the challenges of working with youth as researchers to reap the benefit of their unique ways of experiencing teaching and learning.

RESEARCHING URBAN SCIENCE EDUCATION

In my final two years at Florida State I was extremely frustrated. I had developed the uses of the Internet as a teaching and research tool and had moved my research focus to address science education in a large urban community in Miami, Florida. Also, in an extension of my collaborative research with Jane Butler Kahle, I undertook research in urban schools in Ohio. However, at Florida State University I found myself battling structures that were not conducive to research. I felt we were doing important research and development in Miami and yet the university was ambivalent and seemingly unappreciative. Furthermore, there seemed to be ongoing battles over the location of science education—in Education or Arts and Sciences—and an adherence to rules that did not fully acknowledge the autonomy of doctoral students in regards to when and how they fulfilled degree requirements. Administrators at Florida State University did not anticipate the globalization of graduate science education and their tendency to back off opened the door for Curtin University to take full advantage, offering more convenient ways for educators in Miami and other parts of the USA to undertake doctoral studies. Since leaving Curtin in 1987, I have continued to supervise Curtin doctoral students, mainly those situated in the USA. In part this is due to inflexibility in the degree requirements in my universities that often preclude teacher researchers from doing a doctorate while continuing to teach.

In 1997 I left Florida State and moved to the University of Pennsylvania (Penn), a major urban university and research center with a tradition of ethnographic research. Cristobal Carambo, a teacher researcher from Miami, joined me in Philadelphia thereby continuing our collaboration in urban science education.

Making Sense of Urban Science Education

With the goal of identifying new theoretical lenses through which to frame my research, I decided to study some doctoral level courses to obtain a deeper understanding of social theory that would equip me to better do research in urban classrooms, especially where the students were from racial and economic minorities. I was concerned with how teachers could cope with event-full urban classrooms. It did not seem possible to consciously process everything that was happening and then take actions that were appropriate, timely, and anticipatory. Also, how was it possible to assist teachers to take account of what students knew and could do, rather than focusing on overcoming deficits in their thinking? I knew that suitable models were available in the social sciences, but my accessing of these was eclectic. I took a course on African American psychology and then two courses in theoretical sociology. I learned so much that I decided to study for a second doctorate. However, my dean disapproved of me taking courses alongside doctoral students and prevented me from doing so in the School of Education.

At Penn I was fortunate to have wonderful postdoctoral associates, Cath Milne and Rowhea Elmesky, and some outstanding doctoral students, three being recipients of best dissertation awards (Gale Seiler, Rowhea Elmesky, and Stacy Olitsky). While I was at Penn, outstanding teacher researchers, such as Sonya Martin and Linda Loman, completed their doctorates through Curtin University while studying at the elbows of doctoral students, like Sarah Kate LaVan, enrolled at the University of Pennsylvania—all supervised by me and sharing the same research offices and seminars. This is an interesting example Curtin's initiatives in establishing norms that opened the doors for teacher researchers in the USA to get doctorates without having to spend years away from work and long periods overseas to satisfy residency requirements.

The highlight of my six years at Penn was my research program in urban high schools. The research was kick started by an auto/ethnography in which I taught in a challenging resegregated neighborhood high school in inner-city Philadelphia. I struggled to overcome my fear of the students, to learn about their culture and then to recognize the strengths of their practices and schema. Based on what I learned from the research I developed and used a plethora of sociocultural theories and learned a great deal about the necessity to adapt the cultures of teaching and learning in a process of successfully interacting across cultural boundaries associated with ethnicity, gender, social class and age.

My present approaches to research are an outgrowth of my immersion in cultural studies with Diana Crane at Penn and Joe Kincheloe at the Graduate Center of the City University of New York, and intensive ongoing collaborations with Michael Roth that had begun when he had come to Florida State University to study the teaching of physics for future elementary school teachers. Grounded in William Sewell's theories of culture, agency and structure we have explored science as a field of cultural enactment, studying how teaching and learning can be optimized and how enactment occurs from macro-, meso-, and micro-perspectives. We have also employed theories from the sociology of emotions.

Although Michael and I have never been employed at the same institution, we use the Internet (email and video conferencing) to collaborate daily. Our ways of collaborating reinforce the global aspects of collegiality, given the present technologies. Notable collaborations include the applications of sociocultural theory to science education and the co-founding of a new journal, *Cultural Studies of Science Education.* Our research on coteaching and cogenerative dialogues is beginning to have international impact and has spawned numerous articles, chapters in books, and three books. Also, our innovations are evident in others' published works and scholarly practices.

I use my research with Michael to raise a point of contention. What counts as research is disputed from time to time—evidenced in the presidential address at NARST during the 2006 meeting when the president compared a finding from Mary Budd Rowe's research on wait time to an outcome from what was represented as a recent qualitative study. I groaned when I heard his commentary because it reflects the tip of an iceberg that is probably weighed down by deep differences in epistemological and ontological commitments. Most often the stances toward what is and what is not acceptable research are not brought to light as explicitly as on this occasion. For example, the research I have undertaken with Michael is applicable to macro-issues such as those Mary Budd Rowe identified as priorities so many years ago—learning to teach, immigration, ethnic diversity, high stakes tests, and discipline problems in schools. Despite the fact that we enacted much of our collaborative research at Penn, there are few signs of its effects just three years after me leaving there. It appears as if few of my colleagues are sufficiently persuaded by what we learned about coteaching and cogenerative dialogue, for example, to continue its use in the teacher education programs. Dissemination has been to elsewhere, often as former doctoral students in their new universities enact what we learned from our research. Similarly, close colleagues like Kate Scantlebury have undertaken research and development while applying coteaching and cogenerative dialogues in their teacher certification programs and professional development in schools.

Research in the Big Apple

Since I moved to the Graduate Center in New York City, my research has involved teacher researchers and a continuation of studies on cogenerative dialogues, developing collaborative models of research informed by diverse perspectives. My goal is to study and overcome forms of inequity that have characterized science education throughout the world, but especially in cities like New York, where diversity is a defining characteristic. My approach to research is oppositional to prevailing policies in the United States, which are dominated by scientists' perspectives of what ought to comprise science education, and accountability models that assume teachers can be held accountable not only for their teaching effectiveness but also for the achievement of students. In contrast, my research is grounded in social and cultural theory, which assumes that instead of teachers establishing and maintaining control over students, it is imperative for them to build and sustain solidarity and for teachers and students to assume collective responsibility for practices and outcomes. A challenge for me is to reconceptualize science learning in terms of cultural production by exploring teaching and learning as praxis in terms of dialectical relationships between entities often considered as dichotomies.

Soon after arriving at the Graduate Center I was delighted to receive two career awards, one for my research in science teacher education, from the Association for Science Teacher Education and the other for my research on science teaching, a *Distinguished Teaching Scholar* (DTS) award from the *National Science Foundation*. The latter was noteworthy because I was the first recipient whose primary allegiance is education, as distinct from science. For as long as I have been involved in science education there has been a tendency for the research of science educators not to receive the weight of other forms of evidence and NSF's DTS honors previously had not been awarded to scholars like me. Perhaps my fortune was a sign of an overdue change to recognize the worth of research in science education.

THE ROAD AHEAD

My journeys in science education are ongoing. The road ahead involves research that catalyzes the potential of science to improve the social lives of urban youth throughout the global community. It is to be hoped that science educators can meet the challenges inherent in science for all, creating learning places in which science is reproduced and transformed as urban youth overcome a multiplicity of oppressions in a quest for autonomous and productive lives. If this is to occur, science educators need to confront forces that so seamlessly reproduce inequities associated with factors such as race, ethnicity, native language, and social class. Difficulties can be anticipated because well-intended efforts of the present and recent past have widened equity gaps and instead of being emancipatory, education has reproduced a layered society in which the beneficiaries of science education are the middle and upper classes. My vision is that this will change.

Kenneth Tobin
Presidential Professor
Graduate Center of the City University of New York

WOLFF-MICHAEL ROTH

IN SEARCH OF UNDERSTANDING

Or A Career as an Emergent Phenomenon

I wasn't supposed to be a science educator. But my early experiences would make me an advocate for learning through the eyes of the learners, science education and the needs of people in poverty, the role of choice and control over the objects of activity, means of production, and more generally, the learner's control over his or her life condition. Thus, my research objects would include open-inquiry, learning in the praxis of environmentalism, and science teaching and learning in urban schools serving mostly students from poverty, a condition that has framed my own life into my mid-20s. In particular the issue of learner control over the learner object would mediate my interest in cogenerative dialoguing, that is, the collective sense-making efforts involving all stakeholders, including students, to design the teaching and learning environment.

I WASN'T SUPPOSED TO BE A SCIENCE EDUCATOR

Where I grew up, nobody went beyond grade eight. Although I eventually became a science educator, there are experiences in my life that mediated the way in which I teach, research, and think about the world.

I grew up in poverty. We ate neither butter nor meat, for both were too expensive. To pretend I wore shirts, my mother attached buttons to a sweater so that she could button to it the shirt collars that she had cut from my dad's worn shirts.

As early as I remember and until fourth grade, I wanted to be a teacher. But to attend grammar school, I had to leave home and live in a boarding home. Nobody realized that a painless inner-ear infection made it impossible for me to hear. I did poorly. I still remember having trouble figuring out how much wallpaper it took given the dimensions of a room, windows, and doors. My teachers and classmates thought I was dumb, a village brute. I ended having to repeat fifth grade.

Grammar school did not come easy, and my parents helped me study for exams until ninth grade—about the time they had quit school themselves because of the war. My mother sometimes asked me, "How will you make it to college if you already have trouble understanding?"

In junior college, though majoring in mathematics and physics (I did not do as well as in other subjects), my favorite subjects were physical education, art, and literature. I preferred doing independent project work to anything else in all my subjects; I liked things "concrete," always attempting to express things in my own words. I also liked philosophy, but did not take courses in it, already being con-

K. Tobin, W.-M. Roth (Eds.), The Culture of Science Education, 59–71.

vinced that "if you want to lose interest in a subject, enroll in a course." This suspicion, which runs along the abyss between schooling and getting an education, has been framing my thinking and my relationship to schooling to this day.

Unsure about what I wanted to do, I ended up enrolling in a program for physics teachers with geography as secondary subject. However, my program advisor gave me two semesters before dropping out, because I had not selected mathematics as my minor area. I spited and said, "Then I am going to do a master's of science degree in physics."

I did not like taking courses and spent a lot of time outdoors—hiking, reading, meditating. Although I was not a top achiever, I found lectures boring. And when I did turn up, I was outspoken, pointing out any errors. But I loved the last year, while doing my independent research project.

BEING AND BECOMING A SCIENCE TEACHER

There are few jobs in Montreal in the late 70s and early 80s, and nobody wants to hire a physicist who has done his thesis on a topic in atomic physics. I begin to look for teaching jobs, but even certified teachers are laid off when they had less than twelve years of experience. But eventually I obtain a job in the far north. Already after my first year, I am hooked. I know I want to be a science teacher, work with kids, and get them interested in mathematics and science. My positive experience of becoming a science teacher by being a science teacher certainly mediates my subsequent research agenda on coteaching, essentially concerned with the phenomenon of learning to teach in and through the praxis of teaching, particularly by working at the elbow of another.

My First School

When I begin teaching in St. Paul's River, Quebec, I have no education courses or other formal training that prepare me for teaching. While I am getting myself ready to teach science in this three-room middle school (seventh through ninth grade), I am thinking about what I want to do. I know I do not want my students to experience what I have had to go through right until I was doing my graduate research. I want students to learn as I have done during the independent projects in secondary school and at the college level.

When I arrive at the school, I find that it has some basic equipment for a course on introductory physical science (IPS). The textbook is very thin, consisting mostly of instructions for experiments and investigations and of research questions to be conducted as extensions. I am to teach this course over a two-year period. I feel in heaven, as I can do exactly what I want to do and more so, what I want students to experience. I feel this way, even though I eventually settle for teaching in the basement, with ceilings between five and six feet high, without water—which students have to bring down in buckets downstairs—without fire extinguisher, and without chalkboard. For two years, we mostly experiment, extending what the textbook describes in every conceivable way; and there never is a lecture.

One year, I teach seventh-grade biology. I take the kids out once a week during the double period in the afternoon. My principal does not like it very much and asks me to write justifications, lesson plans, and many other documents. I am sure it is to make planning field trips a pain. But I want my students to go out and learn in nature about nature; and we do. My seventh graders, who have known little but taking notes and working with the textbook are learning to do random sampling, strip sampling, quadrat sampling. They study bare-rock succession, lake succession, and forest-fire succession. They study populations in transitional areas, from the river to the forest, the lake to the marsh, the mountain into the valley. Surprisingly to me at that time, students who have been at the bottom of the class in terms of their grade point averages do really well in this course. Even those students did well who exhibited first-grade reading levels on the Gates-McGintis scale. More than two decades later, I am able to show in *Authentic School Science* (Roth, 1995) and *Rethinking Scientific Literacy* (Roth & Barton, 2004) that open-inquiry learning environments, especially those situated in the community, allow such students to do well and develop identities as able citizens.

At that school, I have another experience that marks my approach to teaching and research for life. Earl, a student in my eighth-grade math class where five groups of students each learns at its own rate, is already repeating the grade. One day, Earl approaches me and says, "Sir, I don't feel like doing math." "What do you want to do?" I ask. "Read my novel," he responds. "Go ahead," I say, and he goes off reading his novel.

On the next day, Earl approaches me again saying, "I don't feel like doing math today." "What do you want to do instead?" I ask him. "Read my novel," he responds. As on the day before, I ask him to go ahead.

The pattern continued until Friday. At the end of the lesson, he approached me. "Sir, I feel I am far behind my group." I responded, "This is my sense, too." He promises, "Well, in three weeks I will have caught up with them."

It turns out that two weeks later, he is ahead of his group mates. On the last day of school, Earl approaches me and says, "You know what, I love mathematics. And you know why?" "Go ahead," I encourage and he continues, "Because after that one week when I was reading my novel, I knew that you wouldn't make me do math when I didn't feel like it." Earl has become the central organizing figure for what I understand to be the need of a learner: control over the conditions when to learn what.

My Second School

After that year, I move to another town where, after taking some evening and summer courses, I taught ninth-grade general science and tenth- through twelfth-grade computer science. My experiences repeated although many of the students would fill the ranks of the unemployed—the going rate was 75 percent among the 18- to 25-year olds, and over 40 percent of the students never finished high school including most of the students in my homeroom, the lowest ranked of the academically streamed homerooms in this school.

Again, I teach science through laboratories. I am struggling though with the fact that students do not automatically get the science out of the activities, and I find that having them work in groups allays some of the problems. The students appear to enjoy science, and many of them come back to school after dinner to work on this or that project.

For the computer science students, I have a policy that they can come and go as they please under the condition that they stick to the weekly or bi-weekly learning contracts we sign. It turns out that a number of students, previously and concurrently caught and punished for dealing drugs, really do well in my course, which gives them the freedom to work and learn at their pace and leisure. Like Earl, these students like to work knowing that they are not forced to do something when they do not feel like it. Here too, many students return in the evenings—in some instances, I drive them home, some into the neighboring villages, at 10 P.M. and sometimes as late as midnight.

Eventually Lloyd Ryan, one of the assistant superintendents, comes to see me. Working at night in the neighboring school board office, he has noticed my students spending a lot of time in the evening at school. He asks me what I am doing and how I am teaching that makes students *want* to come to school. Lloyd eventually suggests that I should have more impact, and that this requires me to have at least a master's degree in education. I tell him that I do not like going to school. He convinces me, however, not only to take the graduate record examination—which he arranges to be held in our town so that I do not have to drive the 400 kilometers to the next official testing site—but also to enroll in a PhD program at the University of Southern Mississippi, which at the time offers a summer program for teachers, and has a residency requirement of only two terms, one of which can be fulfilled during the summer. Equally important to me is the fact that science education is part of the School of Science and Technology.

THE RESEARCH BUG

While doing my doctoral work and conducting other studies, I catch the research bug. I no longer intend to return to my old job but to pursue a university career.

During my first summer in graduate school, I take, among others, a foundations course with Marlene Milkent. Each week we read a book and then discuss it, the first one being *The Structure of Scientific Revolutions* (Kuhn, 1970). Many of my classmates do not like "the heavy reading schedule," but I love both readings and discussions. Marlene begins to mock my eagerness to contribute and continues to stir my interest by suggesting to me other readings. At the end of the summer, I say to her, "I am going to do my PhD and I am going to do it with you." Marlene accepted me. During the year back home, I decide to become a fulltime student.

I also catch the research bug. Although Marlene is not interested in conducting research herself, she sparks my interest in it. Following on from our daily interactions, we even decide to conduct a study together: A pre-post control group design to test the effect of a computer program, which, using a considerable number of stems, creates word problems with randomly generated values every time the program is run. Encouraged and supported by Marlene, I design a study of proportional reasoning—a classical topic that North American science educators such as Anton Lawson, Robert Karplus, and others took over from Jean Piaget—but from a neo-Piagetian perspective. Neo-Piagetian perspectives, as those promoted by Toronto-based Robbie Case and Juan Pascual-Leone, use information-processing ideas—e.g., limitations on mental processing and mental storage—to explain the stages in Piaget's developmental model. In science education, Anton Lawson is an inspiration for me, and my first several articles are guided by his writing.

Graduate school also meant poverty—I spent months in a tent or in a car, eating once a day, because I did not have the money to do otherwise. I lived with the down and out, and saw the havoc a Western society can wreak among its most vulnerable members. This experience, too, framed what I would be researching and thinking later.

DISAPPOINTMENT IN ACADEMIA

My first appointment at Indiana University turns out to be a big disappointment, and I leave after two semesters. Many aspects of my lifeworld conspire making it impossible for me to stay in academia, and I eventually decide to return to high school to do what I know best: teach science.

During that year, I have come to know the work of Kenneth Tobin, whom I have seen at the 1988 NARST conference at the *Lake of the Ozarks* resort. Ken is reporting on the qualitative research he and others (e.g., James Gallagher) have been doing and about constructivism. I am struck that with all my reading—my supervisor Marlene Milkent jokingly referred to me as the only person she knew who was keeping *Science Education* and the *Journal of Research in Science Teaching* as his or her bedside literature—I have not really noticed Ken's work. At Indiana University, I meet weekly with Tom Duffy, an instructional designer, and Don Cunningham, a semiotician, with whom I not only am able to talk about the bad experiences I am having but also about constructivism and semiotics.

I begin to realize that I have done research in an area that is doomed. I feel that just as I have read in *The Structure of Scientific Revolutions*, my neo-Piagetian paradigm is on its way out. Given the structure of academe—publish or perish—there is no way that I have time to retool *and* publish sufficiently to make the tenure barrier. Therefore I leave academe to take a position as department head of science and physics teacher at Appleby College. This move turns out to be beneficial, because it allows me to retool, enact a new research agenda, and establish an extensive research record grounded in the everyday praxis of teaching and learning.

Throughout my years at Appleby, I am in contact with Ken Tobin, who is an active supporter, invites me to contribute to an edited volume, to serve as a reviewer for the journal he edited, and even to apply to openings at Florida State University. To a large extent, his mentorship mediates my eventual move back to the university.

RETOOLING AND A NEW RESEARCH AGENDA

During the summer following my first year at Appleby College, I am invited to teach a summer course *Physics for Elementary Teachers* at the University of Victoria. While teaching, I happen to read a number of books that turned my world upside down, and which had a strong mediating influence on everything I will be doing as a researcher subsequently more than any literature in science education. The books include *Cognition in Practice* (Lave, 1988), *Laboratory Life* (Latour & Woolgar, 1979), and *The Manufacture of Knowledge* (Knorr-Cetina, 1981). These readings lead to further readings, which ultimately became salient in one or the other study I was going to conduct. Returning to the school, I decide to use a camera and videotape students engaged in a variety of the activity structures I have been using as a teacher—open-ended inquiry, concept mapping, group discussions—to find out what happened when I was not with students and how to improve the way in which tasks were presented.

Based on these readings, I begin to use the metaphors of *cognitive apprenticeship* and *authentic science*, two themes that had become central to much of the research I am doing between 1990 and about 1998. Together with other teachers, I arrange classroom environments in which teachers interact with students using *cognitive apprenticeship* as a referent. My goal is to make it possible for students to do science in the way sociologists and anthropologists are reporting it, that is, the way I have experienced science while doing my graduate degree in physics. These readings strike a cord with me, because they allow me to come home to my own experiences. My research on students in open inquiry environments culminates in my *Authentic School Science*, where I bring together a series of studies I have conducted, among others, with my colleague and future graduate student Michael Bowen. My writing at the time uses Ken Tobin's writing as a model, for example, organizing findings in terms of *assertions* that are then substantiated with evidence from the data sources.

My experiences of teaching ecology and field research in my first two schools have led to an interest in this subject, which really lies outside my root disciplines, physics and mathematics. A study of learning in an eighth-grade open-inquiry biology classroom that Michael Bowen teaches becomes crucial to my research in the sense that it opens up new avenues. Based on an extensive record of videotaped inquiry and discussions, and interviews and tests I assemble during my spare periods, we are able to show how problem posing changes as learners become familiar with their research object. Furthermore, we are able to show how processes at the classroom level mediate within-group learning, and how group processes mediate individual learning. The articles are accepted in *Journal of the Learning Sciences* and *Cognition and Instruction*, which serve audiences other than science educators. This research therefore constitutes a turning point in a double sense: I am beginning to research knowing and learning in biology and ecology and I am moving outside of science education proper.

Another important study concerns concept mapping and the social construction of knowledge. I have used concept maps since 1985, but I am not convinced that they help all students. Feeling that students encounter difficulties constructing them, I begin to videotape students. With the article "The social construction of scientific concepts or The concept map as conscription device and tool for social thinking in high school science" (Roth & Roychoudhury, 1992) I have made a breakthrough not only for myself in terms of thinking about knowing and learning but also in terms of introducing a new way of theorizing learning in school science.

My readings in the sociology of science eventually lead me to *Discourse and Social Psychology* (Potter & Wetherell, 1987) and *Discursive Psychology* (Edwards & Potter, 1992). The two books change the way in which I think about language more so than *Talking Science* (Lemke, 1990), which is used as a referent by other science educators. I am taking a pragmatist turn and begin to recognize in language not a neutral tool for externalizing thoughts but a means to constitute sociomaterial reality. I write a series of articles, which are grounded in a teaching experiment and interview study on the nature of science, but end in a reconceptualization of language, what counts as knowledge, the nature of science, and identity.

My reading interests increasingly move outside of the science education discipline. While I continue to keep up with what happens in my root discipline, the works that mediate my thinking and research are from other fields. Among others, my push for rigorous approaches in interpretive inquiry, courses in which I teach in my current position, eventually also lead to my being appointed to coeditor of a major online journal in the field, *FQS: Forum Qualitative Sozialforschung / Forum Qualitative Social Research*, which receives much of its funding from the Deutsche Forschungs Gemeinschaft, which is the German equivalent to the U.S. National Science Foundation or the combined Canadian Social Sciences and Humanities Research Council and Natural Sciences and Engineering Council.

FURTHER TRANSGRESSIONS

In 1992, I realize that I have to leave Appleby College. There are an increasing number of conflicts with the administration arising from the fact that I act on behalf of students when they are to be punished without proper justification; and I have picked up for teachers who have received their pink slips without the proper procedure having been followed. I take a position as statistician at Simon Fraser University.

My search for understanding how people learn science continues to take me across disciplinary boundaries. Unsatisfied with the descriptions of science learning from a third person perspective, which I feel to be inappropriate as they do not take into account just what is salient to a specific learner at the moment, I begin to venture into phenomenology. In part, phenomenology, which I have come to know about nearly twenty years earlier through *Sein und Zeit* (Heidegger, 1977), becomes a salient topic through the ethnomethodological studies of science, with which I become familiar through the work of Harold Garfinkel, Lucy Suchman, Michael Lynch, and Eric Livingston. Through their work, I come to understand the tremendous competence that we enact on a daily basis and through which we make the world what it is. Through this work, I also become familiar with conversation analysis and the work of the applied linguist Charles Goodwin. His interest in gesture, reported in the context of studies of perception or the organization of pointing, leads me to take a closer look at the role of gestures in science learning. My own research continues to have a centrifugal aspect, which takes me ever farther away from traditional science education concerns. The more I learn, the more I realize that science education research and practice are based on poorly theorized commonsense constructs.

My interests in phenomenology are related to the underdog theme of my life and to my fundamental aversion to determinism and deterministic theories of human life. Karl Marx noted in his eleventh thesis on Feuerbach—which to me is the single-most important aphorism guiding my life—that philosophers intend to understand the world, when the real point is to change it. But to know what others can do to change their lots, I need to understand how the world looked through their eyes, what their actual rather than theoretical action possibilities were. I come to realize that I cannot truly help students unless I know what is salient to them and what they are including in their reasons for acting. By the second half of the 1990s, I am convinced that much of educational research has failed exactly on this point: it is too much concerned with making students run the curricular maze to get to some endpoint like rats in the labyrinth rather than focusing on what the real (learning) needs of the students are.

In 1997, I begin my work as the Lansdowne Professor of Applied Cognitive Science in the Faculty of Education at the University of Victoria. Because of the endowment, the professorship comes with a reduced teaching load. This frees me up to follow the increasingly centrifugal nature of my interests and activities. With my graduate student Michael Bowen, I conduct ethnographies and interview and think-aloud studies among scientists, which lead us to publish in very different disciplines, including applied linguistics, linguistics, history and philosophy of science, sociology, and social issues.

In the course of this research, I become dissatisfied with the theories I have used thus far: They either reduce knowing to the individual or to the social; and they dichotomize knowing and material being in a material world. During 1998, while working on *At the Elbow of Another* (Roth & Tobin, 2002) where I attempt to integrate our work on praxis into a suitable theoretical frame, cultural-historical activity theory becomes salient. In a short period of time, I write a number of very different studies on coteaching from an activity theoretic perspective. Initially, I am drawing on a triangular representation that activity theorists tend to use to highlight the mediated nature of human activities. Ken is not too keen on using such representations for our book. He also tends to write his theoretical accounts more simply than I, always being tempted to bring to bear the entire extent of my readings. That summer turns out to be crucial in terms of finding my own style of writing, mediated by the collaborative writing with Ken, and it becomes the starting point of a long process of understanding activity theory. This interest—which eventually leads to my current editorship of *Mind, Culture and Activity*—has continued to date. But perhaps it has become increasingly theoretical, as I want to work on the aporias of the theory and its use, and to push its development.

Working with Ken also takes me into the science classrooms of urban schools. Here, working with students most of whom live below the poverty threshold, I come to be confronted with my own childhood and living with the barest minimum. The research I conduct is mediated and my interest fueled by these early experiences.

Over this period, I develop to such an extent that the cultural-historical and sociocultural community has become a new home for me—at least for now. My most recent interests have been in developing dialectical theories of culture, knowing, learning, identity, and so forth. For the past several years, therefore, the nineteenth century philosophers Georg Hegel and Karl Marx, and the (dialectical) philosophers Jacques Derrida, Emmanuel Levinas, Jean-Luc Nancy, Didier Franck, and Paul Ricœur constitute the dominant influence on my work.

SCHOOLING AND SCIENTIFIC LITERACY

Over the years I have come to realize that the nature of schooling in general and the conceptualizations of scientific literacy in particular not only reproduce inequities but also further aggravate them. My own research agenda of the early 1990s—concerned as it was with cognitive apprenticeship—has further contributed to the conservative agenda of science education to reproduce a particular vision and culture of science rather than to challenge and change it. But as my work has extended into areas outside the classroom, linking *apprenticeship* to everyday activities including environmentalism, allows me to develop a position that many consider as radical. My repeated experiences of poverty and school failure, a realization that the needs of many students are not met, an increasing awareness for the political and deterministic nature of schooling, and an increasing understanding and sense of revulsion against the manipulation of the natural and social worlds by big transnational businesses contributed to the trajectory that my ideas about school science and scientific literacy evolved. My experiences, both of poverty and relative success also have made me suspicious of the claims that in a democratic, capitalist society, "everyone has a chance to make it."

In the early 1990s, I am gung-ho with respect to bringing more students into science. But I realize that students learning needs are not met if science educators persist in making every student think about the world in the same way. I become aware of the manipulation of society by multinationals such as Monsanto, who proliferate the use of genetically modified organisms and high doses of pesticides and herbicides, and the production of non-viable ("suicidal") plants that bear infertile grain. As a gardener who satisfies his entire vegetable needs year-round, I am suspicious about the agenda of making more students think like Monsanto scientists. I become more interested in providing learning experiences for students that are modeled on everyday activity, where the scientific perspective is but one of those required to solve the problems humanity faces.

This emerging sense is first articulated in "Deinstitutionalizing school science" (authored with Michelle McGinn). Michelle and I argue that there are other activities in human society that perhaps are better referents for designing school science activities. In my work, it leads to implementing several design experiments where seventh-grade students are learning science while participating in environmentalism that was already established in their hometown. My doctoral student Stuart Lee and I arrange for environmentalists, politicians, biologists, and others to come to school, as we make it possible for parents and other visitors to participate in supervising and scaffolding students while they pursue studies of their own design around the general topic of the health of a local watershed. Concurrent with the design experiment, Stuart and I also conduct several ethnographic studies one focusing on an environmentalist group, another on the controversy over access to drinking water, another one concerned with watershed stewardship, and yet another one among a network of environmental groups concerned with eelgrass, an indicator plant for the health of coastal areas.

These studies allow me to rethink science and scientific literacy along multiple dimensions. Following environmentalists or those involved in the water controversy, I realize that scientific literacy needs to be rethought in terms of a collective practice. The water controversy in particular allows me to understand that environmental issues in particular are so complex and require so much diverse knowledge that we need not a minimum knowledge about atoms and molecules on the part of every stakeholder, but rather the competence to interact with others whatever their area and level of expertise. Such competence allows scientific literacy to emerge in and as collective praxis: every required aspect of science (facts, theory, method) can be brought to bear just as the knowledge of any other domain including law (environmental and social justice), medicine, Traditional Ecological Knowledge, politics, and so forth.

Today, I am convinced that our conceptions of scientific literacy, even in their rethought form, are not advanced enough to capture the essence of the continually emerging nature of literacies, the spawning of new literacies within existing literacies. As Gilles Deleuze and Félix Guattari suggested, existence is a chaos that moves at infinite speed of birth and disappearance; only dynamic conceptions of culture can keep up with this speed. Schools—in their focus on the reproduction of knowledge, and as instruments for the reproduction of middle class culture, individualism, capitalism, and exploitation—are the antithesis of institutions allowing a democratic society of the future to emerge and evolve. They are designed to produce difference, as Michel Foucault noted, to fill the ranks of those in (structural, equilibrium) unemployment and low-wage jobs. Looking ahead, I see my efforts as a science educator increasingly moving toward the abandonment of the traditional way of organizing schools and school science, and to move toward a model of liberal arts education for life in and toward a truly democratic society, where students are in the position to co-determine what it is they need to know for expanding their control over their life condition and the requisite room to maneuver they desire.

FROM APPRENTICESHIP TO THE LABORATORY AS EMERGENT CULTURE

The aforementioned books on situated cognition and science studies provide me with directions as to the development of ideas about apprenticeship in general and about apprenticeship as graduate experience and ethnographic field method in particular. Thus, I not only find the metaphor appealing for designing school science and for analyzing what happens therein, but also for thinking about and organizing my work with graduate students. In the early 1990s, I view graduate students as individuals who aspire to be part of a community of practice, research on knowing and learning, and my role as a mentor and facilitator assisting students to move through a trajectory from *legitimate peripheral* to *core participation* (Lave & Wenger, 1991) conceive it. For my research among scientists and later on among environmentalists and in a variety of workplaces, it becomes a metaphor for doing ethnography through participating in the ongoing activities.

This approach works initially, framing my relationships with Michelle McGinn (research methods, Brock University) and Michael Bowen (science education, University of New Brunswick). It leads to productive trouble when Stuart Lee, now a policy analyst with Environment Canada, completes his PhD under my supervision. While working on a jointly authored piece in 1999, he finds my editing of his writing oppressive and inhibiting true scholarship. It turns out that we have repeated opportunities for addressing the attendant issue head on in the course of writing two articles on autobiography in science education and on apprenticeship as auto/ethnographic method. It is at that time that I recognize the limiting nature of the apprenticeship metaphor, because it tends to emphasize the reproduction of culture rather than its productive renewal and change. I begin to learn how to edit the writings of my students (and collaborators) in a way that no longer truncates their agency by changing texts in ways that are consistent with their styles rather than imposing my "natural" style. Earl has returned, as I realize that my graduate students need to have control over their conditions as much as he.

Initially I have been working with only one or two graduate students at a time; but multiple research grants have allowed me to increase their number and the formation of a research laboratory. I think of this laboratory as a culture, which develops in time, and which is produced and reproduced through the practices that are independent of individual members. New individuals join and learn to reproduce the culture as much as allowing them to produce new cultural forms.

The graduate students do not only write with me, but also participate in each other's writing projects where appropriate. Our weekly meetings and joint analysis sessions lend themselves to fostering collaborations into areas and disciplines other than the root discipline—Yew Jin Lee (science education, National Institute of Education, Singapore), not only has conducted studies of knowing and learning in fish hatcheries but also has participated in studies of learning in seventh-grade classrooms and elementary school science. This is possible in part because in my research group, individuals do not measure or compete against one another to establish hierarchies; they do not rank order one another according to institutional categories such as prior degrees or length of studies. Thus, a doctoral student (e.g.,

Lilian Pozzer-Ardenghi) may teach data analysis to a new postdoctoral fellow or introduce him to the particulars of a study site. Similarly, a new master's degree student may be responsible for organizing peers, doctoral fellows, postdoctoral fellows and professors for laboratory meetings.

My research group now is interdisciplinary and international. It includes studies on ethical issues as professionals such as dentists move from dental school to private practice, how people learn on the job as electricians or hatchery workers, how scientists learn during laboratory work, how environmentalist groups and networks realize their goals, how people become ship officers, how kindergarten children come to know about science, and so forth. Today the best way of characterizing my work is: studies of knowing and learning across the lifespan, in formal and informal settings. It has included students and postdoctoral research fellows from Brazil, Canada, Denmark, France, Germany, Korea, Netherlands, Singapore, Taiwan, and USA.

CODA

When I began thinking about what to write in this chapter and in particular what to choose as a title, "Against all Odds" comes to my mind first. But then, upon reflecting, I realize that such a title would be unfitting, for it would pretend that our society provides opportunities to all students, including those growing up in poverty. But this is not the case. As we see in the pop music or sports cultures, the number of individuals who "make it to the top"—including the jobs, regular income, and glory—is quite small. My schoolmates from the early years left school after eighth grade and became farmers, tradespersons, and housewives. I have been the only one from a class of twenty who has had the opportunity to go on to college and university. I discard the title and motivation as inappropriate.

My career has taken many twists and turns and my interests have taken my research into many different directions. My ultimate search has been that for understanding human nature, not as explained in deterministic models of psychology or sociology, but as it is lived and experienced in everyday praxis. Ultimately, even if my biography may look disparate to an outsider, it is my lived experience that holds it together. This lived experience has left traces in my body, thereby mediating the action possibilities I may have tomorrow. Today the best way of characterizing my career, and, because of my life as a singular plural, the culture of science education, is through the term *contingency*: from the potential that my life actualizes it draws a potential that it appropriates. Thus, in paraphrasing Deleuze and Guattari (1991), I would stay that the actual is not what I am but rather what I become, what I am in the process of becoming, that is to say, the Other, my becoming other; the present on the contrary is what I am and, thereby, what already I am ceasing to be. Who I am and consequently, the aspect of science education culture I enact, has a contingent nature, as contingent as the languages we speak, our notions of selfhood, and our community, to paraphrase several claims that the pragmatist philosopher Richard Rorty has made.

Today I understand that an individual career is a concrete realization of culture, and culture is produced by individual careers. Each encounter with another person therefore is both crucial and not crucial in the continuous emergence of one's career—but if there is a single most important person to my own realization of a science educator's life, it would be Ken Tobin, with whom I have come to share a substantial part of my scholarly life. For my intellectual development, Jean Lave and Jacques Derrida may have been the most important scholars; Martin Heidegger's *Being and Time* certainly has been the single-most important book.

Wolff-Michael Roth
Lansdowne Professor, Applied Cognitive Science
University of Victoria, Canada

STRUCTURING SCIENCE EDUCATION

The five auto/biographies in this part A of our book on the cultural history of science education include individuals who, collectively, have been on the forefront of the field for nearly five decades. Each auto/biography not only tells us what *this* person has done but also the realization of the concrete possibilities science education provided to its collective membership at a particular point in its history. In this epilogue to part A, we pull out five main ideas of the science education history and culture by discussing the five autobiographical contributions in terms of (a) the professional networks that they have created and sustained, (b) the vanishing of shaping forces in science education, (c) scholarly leadership positions, (d) contradictions that shape the field, and (e) the commodification of science education. Because both editors are part of this first group, we decided to use the following strategy for writing the section involving one or the other. When the person featured is Ken Tobin, it is the voice of Michael Roth reporting to the readers; the reverse is the case when the person featured is Roth.

PROFESSIONAL NETWORKS

The sociological notion of *invisible college* has pointed to the importance of social networks to the professional lives of scientists and to the organization of professional cultures. The term refers to the informal network that links scholars from around the world in addition and even in the absence of formal networks that link their institutions. The *invisible college* constitutes in fact a major means of communication in all sciences; and postdoctoral fellows are often exchanged between members of the college until these fellows become members not only of some faculty but also of the college itself.

Although the five scholars have not talked about invisible colleges, the role of linkages between key figures in the field of science education certainly runs through the five narratives. For example, James J. Gallagher's biography shows his preference to build, sustain, and nurture social networks involving institutions and individuals. He values his work with doctoral students and postdoctoral associates and collaborated with them successfully over many years. Intellectually, Gallagher's valuing of conceptual understanding might link to his graduate student connections to Jerome Bruner and his efforts to relate science to technology and society possibly derive from his post doctoral experiences with Paul de Hart Hurd. As a graduate student Gallagher worked with Chick Ahlgren, who was so influential in formulating *Science for All Americans* and the associated *Project 2061,* establishing a basis for Gallagher's ongoing collaborations with these institutions. Jane Butler Kahle also established important connections with high profile gender equity researchers at the University of Wisconsin and through these social bonds she accessed key personnel—such as, Peter Hewson (science education), Norm Webb (evaluation), and Andy Porter (policy, teacher education). These researchers were associated with a highly successful center that was home for numerous national projects and Kahle's alliance with them provided an infrastructure for collaborative projects supported by external funding, and productive outcomes.

In the early 1980s, Barry Fraser realized the importance of creating social bonds with key international figures in science and mathematics education. As the director of the *Science and Mathematics Education Centre* (SMEC) at Curtin University, he created a visiting scholar scheme that brought prominent scholars to Australia to collaborate with faculty at the center. Through this scheme researchers from many countries spent significant time at SMEC, interacted with students and faculty, and created social bonds that served the creation of large international research projects situated in Australia, Asia, Europe, and North America. These connections included many of the researchers whose biographies appear in this book, including those included in the first section and who collaborated with Fraser and his colleagues at SMEC and the affiliated National Science Education Centre (especially for women). Fraser's approach was to support substantive visits from scholars who were aligned with one or more of his faculty. Accordingly, the links were likely to be substantive and strengthened by reciprocal visits, co-authorship, and ongoing collaboration. Furthermore, the network grew when graduate students from Curtin, like Kate Scantlebury and Cath Milne, came to the United States to work with Kahle and Tobin respectively, took permanent jobs in the United States, and created their own research groups.

Using the term *traversals* (Lee & Roth, 2003), sociocultural scientists have come to theorize the movements of certain individuals across institutional boundaries that lead to the emergence of networks linking people and institutions. Ken Tobin is one of those individuals in science education through whose movements and relationships stable networks came about. His social network among researchers is grounded in relationships with the faculty with whom he interacted at the University of Georgia and colleagues he met during his active involvement in conferences during graduate school and his early career—especially the annual meetings of the American Educational Research Association, the National Association for Research in Science Teaching (NARST), and the Association for Science Teacher Education.

Michael: Through Tobin's publications in, and avid reading of, key journals in educational research and research in science education people interested in his work recognized him and he knew who to look out for at conferences. Indeed, it was at one of these meetings that he met for the first time each of the scholars in this section of the book and most of the authors in the remainder of the book. The network around him was strengthened by the visiting scholar scheme created at Curtin University, a graduate student research group he formed during a sabbatical leave at the University of Georgia, and ongoing collaboration with some of his former doctoral students. Unlike Gallagher who sustained productive relationships with peers from his days as a doctoral student, none of the doctoral students from Tobin's peer group became prominent researchers and he lost touch with all of them. Also, his doctoral committee, though highly prominent when he was in graduate school, faded in terms of their shaping of the field. However, they got him started on what turned out to be the right path while Tobin was in graduate school and it is likely that they shaped the field mainly through their work with doctoral students. Even so, the relative lack of impact of their research is puzzling. A possible solution to the puzzle may lie in the theory and methods underpinning their research. A common feature of these key figures from the University of Georgia during the final quarter of the 20th century are their positivist roots and associated foci on creating and validating models in terms of quantitative relationships among variables. Science education, however, had moved on to permit rich interpretive studies of the events in science classrooms; many science educators now are less interested in statistical descriptions of knowing and learning but are more concerned with understanding the nature of the specific needs of particular (types of) students. The relative decline of the University of Georgia as an influential science education center may be associated with this shift in the field. Below we provide a thumbnail sketch of each of the five scholars on Tobin's doctoral committee as an illustration of these issues, the vanishing of once prominent individuals and schools, issues that we continue to pursue throughout this book.

THE VANISHING OF SHAPING FORCES IN SCIENCE EDUCATION

Sometimes there are individuals, groups, or whole departments of science education that have shaping influence on the field although their work comes to be little cited and they themselves vanish into the background. Sometimes, universities decide to cut tenure line positions or professorships, such as when the number of physics education chairs at the University of Bremen, Germany—the institution where Manuela Welzel (see part E) graduated—was decreased from three to one, essentially wiping out what was a thriving center of physics education that had included Hans Niedderer, Stefan von Aufschnaiter, and Hannelore Schwedes. Generally, however, this does not mean that the schools or individuals no longer exist; they often continue to exist but with a lesser role as a force that shapes the field as a whole. For example, Anton Lawson has shaped science education in the 1970s and 1980s, but his work on formal reasoning has all but disappeared from science education; some followers, like Michael Roth who wrote his dissertation on a topic that paralleled Lawson's work but in the area of mathematical physics, shifted to create or follow new and emerging trends, whereas those continuing to do such work increasingly found it impossible to publish because of the non-salience of this work to current science education concerns. That is, because of a shift in culture, the acceptability of forms of research also changes. This also has been the case, for example, with several key science educators who worked in the department of science education at the University of Georgia. William Capie, Tobin's major professor, created teacher assessment systems, validating measures against criteria including student achievement. Although he served as major professor for several other science educators, Capie quickly moved out of science education in the mid-1980s and retired at an early age.

James Okey, president of the National Association for Research in Science Teaching (NARST) in 1979 and a prominent science educator at the time, was well known for his research on process skills and pioneering uses of technology to improve learning. After obtaining several large grants from the U.S. National Science Foundation (NSF) and at the peak of his career, Okey retired to assume a career as a chef.

Russell Yeany, editor of the *Journal of Research in Science Teaching* and NARST president, became a role model for Tobin. However, Yeany was lured into administration and his impact in science education diminished as his administrative duties increased. During Tobin's sabbatical at the University of Georgia in 1984 Yeany became Head of the Department of Science Education—leading what was then a large and prominent science education doctoral program. Yeany encouraged Tobin to form a research squad using doctoral students as colleagues—pointing out the legitimacy for doctoral students as participants in research activities outside of their paid assistantships. The participants in the studies we undertook at the University of Georgia allowed the group to form social bonds that transcended several countries including Spain (Mariona Espinet), Taiwan (Chao-Ti Hsiung), Portugal (Antonio Bettencourt), and the United States (Elizabeth Swanson). A key lesson from this collaboration is that productive research groups can be created and sustained in the absence of federal funding. Yeany was prominent in science education and his premature departure, first into administration and then into retirement was a loss to the field. Although Yeany's research in science teacher education was pioneering at its time, it is infrequently cited in the literature today.

Michael Padilla arrived at the University of Georgia as an assistant professor at about the same time Tobin arrived as a doctoral student. Unlike his colleagues Padilla had an interest in Jean Piaget's theories in relation to science education. He was mentored by Yeany and Okey and progressed through the ranks together with Joe Riley. All four of these science educators got involved in administration, though Padilla and Riley continue to be active in science education to the present day (though retired from the University of Georgia). Padilla obtained several large NSF grants associated with the professional development and initial certification of teachers and his professional activities were oriented toward the National Science Teachers Association—an association of which he is currently president. Riley retired and continued his international activities in science education—presently in Singapore.

Why has the direct impact of these researchers—being as they were from what has been at one point a leading science education program in the United States—been diffuse? Perhaps theoretical and methical changes in the field of science education contributed to a relative lack of impact. The issues these individuals pursued with their research no longer are in vogue; and with the disappearance of the individuals, their research agendas and paradigms also disappear just as it is described in *The Structure of Scientific Revolutions* (Kuhn, 1970). It is also possible that these scholars changed goals from a focus on their own research to the research of their doctoral students. In much the way Fraser highlighted the leveraging effect of doctoral education on a field, it is possible that scholars in science education at the University of Georgia focused their efforts on the production of the next generation of science educators—including, but not exclusively, researchers. Finally, the shift of productive scholars into administration can be a lure that gradually erodes impact in a field such as science education. At an early stage of his career, when Tobin was struggling as a Head of Department at Florida State University, David Berliner, a prominent teacher educator was invited as a keynote speaker in a dean's seminar series. During a conversation Berliner remarked about Tobin's involvement in administration: "Didn't anyone tell you the best job in America is as a tenured full professor?" Tobin took this to mean that the best job is being full professor doing research and teaching graduate students rather than administration. It is a lesson Tobin was to learn well and from that moment on he has been sensitized to the ways in which administrative duties can erode the time available for productive scholarship.

We do not mean to imply that the only way to support scholarship in science education is by doing research. Fraser clearly shows this through his experience with doctoral education. Kahle assumed prominent leadership roles at the National Science Foundation and the Biological Sciences Curriculum Study and shaped science education through her administrative roles as well as her research. Also, she was a powerful leader in professional organizations such as NARST, where her constant concern was equity—especially gender equity; and she was a role model for many science educators, especially (but not only) females.

There are variations among the five scholars in terms of the extent to which they have been involved in administration. Fraser has been a centre administrator continuously for 25 years. Gallagher was an administrator in the early 1980s, Kahle was dean of education for a short time, an administrator of her own research center at the University of Miami, and program director for funding science education projects at the National Science Foundation. In contrast, except for short periods as a head of department and Director of Teacher Education, Tobin has avoided administrative positions; and Roth, with an endowed chair, has been in a research position for a decade, ever since he went to the University of Victoria, but did not hold any major administrative position other than co-chairing the university research ethics board.

SCHOLARLY LEADERSHIP POSITIONS

An important aspect of the impact a scholar has on the field comes through participation in leadership positions of the organization and funding agencies or through editorial positions in leading journals and book series. The five scholars have assumed leadership roles as NARST presidents (Kahle, Tobin, Fraser), in funding agencies (Fraser, Kahle), as journal editors (Fraser, Gallagher, Roth, Tobin), as book series editors (Tobin, Roth), and as handbook editors (Fraser, Tobin). Also, in an international sense the five science educators have been active in research undertaken in several countries. Instantiating *traversals* across national boundaries, institutions, and research groups, such traveling and connecting individuals may well shape the trajectory of the entire field much in the same way that even minor influences change the spatio-temporal trajectory of any chaotic system and even change the location of its *strange attractors* that constitute and shape the field.

The role editors take on is crucial as it can shape how forms of writing (genres) evolve, what kind of research comes to be accepted (e.g., statistical versus interpretive), and the topics typically covered in a journal or book series. Some editors take a "bean counting" approach adding up the recommendations of the reviewers (e.g., accept with major revisions and reject = reject) without actually taking on the responsibility of deciding whether the reviewers may have erred or are completely on top of the task. Some editors force authors to implement all changes, whereas others leave it to the authors to make an informed decision about the usefulness of reviewer comments with respect to their potential of improving the study. The field of science education has seen variations in this respect. The former editor of the *International Journal of Science Education* (prior to 1987 called the *European Journal of Science Education*), Richard Kempa, accepted articles when reviewers deemed them suitable without asking authors to modify their manuscript. His motto has been, "Let the author author his or her manuscript." One of the recent journals, *Cultural Studies of Science Education*—for which we (Tobin, Roth) serve as coeditors-in-chief—has a similar policy, which we expect to open up opportunities for exploring new forms of genre and scholarship rather than regressing to a point that all articles more or less look and read the same.

In any field, some topics, theories, or methods are tied to certain individuals. In science education, the questions surrounding gender equity are tied to the name Jane Kahle, who trained a number of science educators who became scholars in their own right in the field of gender equity and feminist approaches—Kathryn Scantlebury being but one example. Kahle's greatest contributions have been through her continuous research, especially in studies that have focused on improving the quality of science education at a state level and in large urban areas in the USA. For about 30 years she has systematically examined gender equity in science and in so doing has focused attention on potential hegemonies that have reproduced disadvantage for females. Perhaps Kahle's greatest legacy is her constant attention to gender equity in science and the steady improvement in numerous indicators of equity. The gender equity movement was international in scope and has paralleled many realms in which the roles of women changed to be more comparable to those of men. It also allowed an increasing number of female science educators to emerge and to make a mark on what for the longest period has been a male-dominated field. We return to this issue in part D, "The Reduction of Gender Barriers."

In the way research on gender equity is tied to the name of Kahle, the name of Barry Fraser is tied to learning environments research. This form of research began in science education and then became a discipline within the social sciences, with its own journal and committed scholars. Throughout his career Fraser built on his uses of questionnaires to obtain quantitative measures of attitudes and perceptions of the learning environment—of students, teachers and researchers. Over time there has been a plethora of instruments that now comprise a database of scales that can be used in research and evaluation, available for grades K–12 and college. Notably, as the fields in which they were used changed, Fraser and his colleagues have used new theoretical underpinnings and have fine tuned instruments to obtain reliable measures for individuals and groups of individuals.

CONTRADICTIONS THAT SHAPE THE FIELD

In cultural-historical approaches to sociology and social theory, inner contradictions play an important role because they become the driving forces for change. Here we are not thinking about contradictions between merely external factors but rather contradictions that operate within the current unit of analysis, which, in the present instance, is the field of science education. Inner contradictions, to be operative, need to be available at the level of consciousness to lead to change. Having opposing trends within a field in itself is not sufficient for quick change to come about; rather, open discussion of the differences and incompatibilities sharpens the collective understanding of the issues at hand. The great debates in science education during the 1980s and 1990s focusing, for example, on the difference between qualitative and quantitative research methods or the debate opposing those with constructivist agenda to others pursuing a realist agenda, are examples of moments when the inner contradictions of a field are raised to the level of consciousness.

Every field goes through major changes at certain points in its history. The transition from purely statistical approaches to doing science education research to interpretative approaches constituted one of these major changes—others being, for example, the change from behaviorism to Piagetian approaches, or from Piagetian approaches to (radical, social) constructivism. In retrospect, such transitions are characterized by sometimes-fierce open debates and less open processes during peer review in which competing paradigms strive to be accepted or to dominate the field. One of these major shifts, the change from purely statistical research to interpretive approaches generally and ethnography more specifically, also involved the five individuals featured in this part.

James Gallagher had a hand in bringing ethnography into science education and he briefly touches on the violence he experienced at the hands of senior science educators when he presented qualitative work at an annual meeting of NARST. Fortunately we see less of this sort of hostility at public meetings these days, but the epistemological and ontological wars that underpin the battles between interpretive and statistical approaches are pervasive and also are addressed throughout this book. Gallagher and Tobin collaborated to create qualitative research programs at Michigan State University, Curtin University, the University of Georgia, and later at Florida State University. In 1987, Tobin and Gallagher published two articles in the *Journal of Research in Science Teaching* and the *Journal of Curriculum Studies* that became exemplary and transformative for others in the field. Their

79

collaborative studies created significant interest at annual meetings of NARST and the editors of key journals were receptive to ethnographic research. Even so, many involved in the peer review process were unfamiliar with ethnography and applied inappropriate criteria in recommending to the editor of journals. Fortunately, scholars like Kahle and Fraser, who typically have employed designs that incorporated quantitative data, strove to include ethnography in their studies. Despite his roots in statistical approaches, Roth's teacher-researcher studies in the early 1990s, all of which were interpretive, readily were accepted by the major journals in the science education field.

Gradually the tide began to change, though never entirely—at the 1992 NARST conference in Boston, Massachusetts, an ad hoc meeting brought together more than one hundred researchers discussing what appeared to be a bias against interpretive approaches. Gallagher and Roth together with Bill Kyle and Sandra Abell co-authored an editorial published in the *Journal of Research in Science Teaching* in which they argued for a mature discipline of science education that *integrates* research paradigms rather than suppresses one over another (Kyle, Roth, Abell, & Gallagher, 1992). This editorial was an attempt to sublate the exclusionary opposition between quantitative and qualitative approaches, which constituted an inner contradiction in the field, and to advocate an issues-based approach to the problems in science education to be solved drawing on research methods best suited to deal with the issue. Nevertheless and probably because of the close connections between science and science education and the positivist leanings of many scientists, there have been strong forces pushing the uses of quantitative data and refusing to acknowledge the legitimacy of educational research unless it employed a "scientific" design—often equated with quantitative data and quasi-experimental designs. These pressures were most evident in competitions for federal funding in the United States, making it difficult for scholars to employ sociocultural theories and innovative research methods not associated with the statistical manipulation of numerical data.

In any field of the social sciences, some individuals shift paradigms, leading to the change of the field, whereas others continue doing what they always have done. Thomas Kuhn therefore came to the conclusion that paradigms do not shift because people shift what they have been doing and thinking but because those who do not change will eventually die out. Thus, some science educators began doing statistical research only to become major advocates and practitioners of interpretive approaches.

Michael: For example, Tobin began his research with quasi experiments of the teaching and learning of elementary and middle grades science. From the outset he collected recordings of what happened in classrooms and analyzed audiotapes measuring a range of teacher and student variables. He also developed observational protocols to measure "transactions in science" and used sophisticated statistical models to gain insights into how to improve the quality of teaching and learning. Over time he became frustrated with the reductionist aspects of defining and measuring variables and the inability of this sort of research to handle the unfolding exigencies of social life. Ethnography was an appealing alternative and largely through his collaboration with Gallagher and with assistance from Fred Erickson, he evolved a way to do interpretive research in science classrooms. Now, for more than 20 years he has continued to develop qualitative genres of research and in the past 15 years or so he has collaborated with me to create models for doing research that are appropriate for micro-, meso-, and macroscopic levels of social life as they apply to science education. Tobin's substantive foci have been situated in urban classrooms for the past decade and his theoretical frames from cultural sociology have increasingly embraced the centrality of emotions to teaching and learning.

Debates at the annual NARST meetings in the late 1970s and early 1980s were characterized by critical exchanges between scholars within the quantitative paradigm. Senior scholars from the USA were highly critical of some international colleagues for running multiple t-tests and large numbers of correlations—reporting only those that were significant and not adjusting the type I error rate to take account of the large number of tests. One could feel the emotional intensity in the critiques and there was an understandable reluctance to get involved in NARST on the part of junior scholars and some international scholars for whom English was not a first language. There also was a pattern of senior scholars interrupting speakers to make a point or ask a challenging question. Before this pattern declined it seemed to get even worse as the paradigm wars of the early 1980s erupted. Leading science educators using a Piagetian framework heatedly clashed with those adopting an Ausubelian framework that emphasized the role of prior knowledge in learning. Probably as a consequence of the unpleasantness of some of these interactions, several prominent Australian science educators stopped participating in the annual meetings of NARST and a strong group of conceptual change researchers was established as a Special Interest Group within the American Educational Research Association. These tensions also were reflected in the published literature and set a stage for the disputes between science educators using numerical data and those using qualitative data, and subsequently those incorporating constructivism into their work and those who embraced realist ontology. However, some individuals embraced both approaches from early on. For example, even though most of Kahle's research was quantitatively oriented she did interpretive research and valued the complementarity of qualitative and quantitative windows into science education. Similarly, Fraser has consistently used and advocated a mix of interpretive and quantitative data in his research on learning environments.

As any other field, science education offers opportunities and possibilities for quite unusual careers to emerge and develop. Thus, after receiving their doctorial degrees, science educators generally pursue university degrees and become researchers or return to teaching K–12 science and a few may teach science at the junior college level. But science education as a field provides possibilities for different career trajectories, such as the ones Roth realized when he returned to the classroom rather than immediately getting on a tenure track.

Ken: It turned out to be a remarkably astute move when Roth went against the current of conventional wisdom when, after a disappointing postdoctoral experience he returned to the classroom where he retooled as a teacher researcher and developed his competencies as an interpretive researcher. What a bold move so soon after completing a doctorate—learning new social theory and research methods undertaking research that not only was different from what he had done in his doctoral degree but also radically different than anything else done in science education. He became a trailblazer as he embarked on one of the most productive bursts of scholarship in our field. It is one of the possibilities others in the field, too, have made, though the workload at the school level often weighted them down so that they could not or did not contribute to the literature—other successful teacher researchers include Jim Minstrel, well known for his classroom work on teaching physics and Elizabeth Finkel, who co-authored with Margaret A. Eisenhart the influential *Women in Science*.

From the outset Roth was an unusual scholar within the science education community. I remember him as unusually attentive to what happened in a NARST presentation, eager to pursue deeper understandings and not so concerned with making a public impression. I recall being involved with him as an emerging scholar as he reviewed papers, contested ideas using email, and raised questions about the status quo.

His rate of publication was unprecedented in science education and, like scholars such as Michael Scriven, he published in journals outside of the specialization in which he initially made his name—in his case science education. Initially, peers generally and editors specifically greeted Michael's papers and publications with considerable enthusiasm. He explored the uses of theoretical frameworks from outside of the dominant paradigm and showed science educators how to use sociocultural perspectives to make sense of science education. Although it is hard to attribute who first introduced particular scholarly works to our field, it is fair to say that Roth regularly used the works of French, German, Russian, and Scandinavian psychologists and philosophers in his published works. Not surprisingly, some of science education's gatekeepers resisted the changes associated with Roth's perturbations to the field. Complaints arose from powerful others—including assertions like: "he is publishing too much," "publishing too much of the same work in different journals," "uses jargon to dress up old ideas," and "does not pay attention to what others have done." Accordingly, it took several years for Roth's high levels of productivity to be accepted within the science education community.

In the cultural history of a field, there sometimes are constellations that appear to require innovations; such innovations, however, never are entirely new but already enabled by the current state of the art of a discipline. Thus, for example, when Tobin, Kahle, and Fraser persuaded Falmer Press to publish their book length research in the mid 1980s it was evident that there was no structure for publishing book-length research in science education. That is, there was a contradiction in the sense that some scholars and scholar groups pursued research suitable for being reported in monograph lengths but publishing houses did not have science education as a category. Tobin negotiated a series with the American Association for the Advancement of Science (AAAS). When AAAS Press was closed down, Kluwer Academic Publishers picked up the series, which continues today as the Science and Technology Education Library—a series that publishes book length manuscripts in science education—published by Springer Verlag, which recently has bought out Kluwer. Nowadays there are numerous options for publishing books on research in science education—for example, we both are editors for book series (*SensePublishers*) that publish research in science education, opening doors for researchers to write longer manuscripts.

A salient contradiction in science education as in many other fields comes from the contrast of number and length of manuscripts submitted to the science education journals collectively and the available space. Yet the productivity of scholars is measured in terms of their output so that there exists great competition for a limited number of journal pages. Thus, many journals have adopted a maximum word limit—in contrast, other journals such as *Cognition and Instruction* have published 100-page articles. Interestingly, in regard to length of manuscripts, some of the gatekeepers of science education have mounted a campaign to force shorter manuscripts. One such attack occurred at the 2006 annual meeting of NARST and efforts have been taken to continue the campaign at the subsequent annual meeting— led by former editors of the *Journal of Research in Science Teaching,* some of whom are arguing for forms of research that seem more aligned with positivism and realist ontology. Yet this latest perturbation comes at a time when there have been substantial changes in the ways in which scholars access journals. Paper journals are becoming old technology and almost all of the key journals in science education have electronic versions and the length of manuscripts is becoming less of an issue. As peer reviewers and graduate programs focus on the need for providing solid theoretical frames to support research and for all research to be nuanced by what has been done previously the research is becoming more complex and manuscripts are longer. In addition, Tobin and Roth have created *Cultural Studies of Science Education,* a journal published by Springer Verlag. This new journal publishes fewer articles per issue and has an increased number of pages per volume— allowing for lengthier manuscripts and associated scholarly debates about those articles.

THE COMMODIFICATION OF SCIENCE EDUCATION

As any other scholarly field, science education has not eschewed the constraints and opportunities associated with capitalist markets, a major shaping force of society and culture. These markets, at the present day entirely profit oriented, themselves are, as Karl Marx showed, a product of cultural-historical developments, both drive and are driven by other developments in society. Thus, book and journal publishers take on projects based on market shares and potential for net profits.

Universities offer programs attracting doctoral students and future scholars from other countries to complete their degree in return for tuition fees; and U.S. universities provide nine-month contracts to their professorial employees, thereby enticing them to get research or teacher enhancement grants as a way of supplementing their income during the summer months. In other countries, the field develops differently, as, for example, teacher enhancement is funded directly through school systems and professors hold 12-month appointments, changing the way in which scholars may allocate their time, commitments, and interests.

The scholarly arena in science education has not escaped the increasing pressure of the capitalist markets concerned with squeezing profits wherever possible, certainly at expenses that we have not yet fully fathomed. However, cost-cutting measures have forced university libraries to limit and even decrease their subscriptions even in the face of an ever-increasing number of journals. In science education, too, new journals have joined the ranks of the more traditional ones, *Science Education, Journal of Research in Science Teaching, International Journal of Science Education,* and *School Science and Mathematics.* Thus, *Research in Science Education,* the *Journal of Science Education and Technology, Science & Education,* the *International Journal of Science and Mathematics Education,* and *Cultural Studies of Science Education* have been created within the last two decades. However, some journals feel the financial pressures and live at the brink of extinction, such as the *Canadian Journal of Science, Mathematics, and Technology Education,* which was initially funded through a grant from Imperial Oil, but which has not found a sufficiently large subscriber base to guarantee its survival. Association with a large publisher appears to be crucial for any journal to be sustained.

Much of the money available to support science education has been associated with curriculum development and teacher preparation and enhancement. Research has to be done in conjunction with such activities and in most cases is a small proportion of a budget. At the National Science Foundation in the USA, for example, the relative proportion of the budget available to support research is small. Accordingly, many science educators are encouraged by institutional reward structures to seek external funding that supports scholarly activities that are not conducive to doing research. It is not unusual to find that most of the research from such projects is associated with doctoral dissertations that are not published in peer reviewed journals.

Universities pressure science education professors to get grants, which mediates the amount of time they have available for conducting and writing research. This is particularly the case in the US, where science educators aim at receiving grants for teacher enhancement, with small often-negligible research components, rather than for conducting research on, for example, science learning processes (which in the US often goes to psychology departments). For example, Jim Gallagher regrets that he did not publish as much as he might have and did not publish a paper from his dissertation. We regard this as a contradiction since he has such an illustrious career and has been productive in doing research and disseminating in other ways. Here we raise questions about the structures that afford one of science education's most talented scholars not disseminating research through publication to the international communities in which he has been so prominent. A corollary question points to the relative weight given to publications—apart from their role in promotion and tenure. One possibility to consider is that for many universities and individual professors (especially in the US) the end game is external funding. Was there an imperative to obtain external funding and thereby create an infrastructure to support doctoral students and postdoctoral fellows? Our experience has been that prior to earning tenure publishing is close to the only game in town in research universities; however, the situation changes after earning tenure and promotion. From then on the pressure is to obtain external funding and in science education most funding is available for purposes such as systemic reform, professional development, and curriculum development. In the next section we examine some of the economic issues associated with setting institutional priorities and hence changing the structures in which scholarship occurs. Here we explore the issue of publishing from a dissertation—which we regard as a necessity, not an option.

Ken: One of the possible ways of changing the failure to publish from a dissertation has shown to be fruitful in the case of Roth. He changed the requirements for his students to obtain an MA or a PhD degree: the former have to submit, under his guidance, three publishable papers to international journals and conference proposals for international conferences, whereas the latter have to submit six articles and conference proposals. They also are co-authors on papers and articles first-authored by Roth or another student or postdoctoral fellows. Once finished with their studies, they therefore not only have a tremendous and highly competitive publication record but also sufficient experience to make it on their own in academia. Another example of a leading scholar placing high priority on graduate students publishing their research is Curtin University's David Treagust who has virtually set aside his own writing agenda to focus on coauthoring manuscripts with his graduate students based on their thesis research. Together these cohere and build on those that have come before to comprise an impressive longitudinal program of research on conceptual change in science education that has now extended for two decades.

Michael: My understanding is that you, too, had written a dissertation that consisted of nine articles, eight of which you did publish. In my context, professors in education denote this approach as *the science model*. However one denotes it, to me its advantage lies in advancing research programs rather than having a field in which each new PhD candidate researches something of his or her interest that may not be connected at all with the collective interests and (theoretical, empirical) needs of the field as a whole.

The commodification of science education as product to be marketed and sold abroad mediated what individuals in the field have come to do and led to the emergence of centers where many converged to seek degrees. For example, Peter Fensham's appointment as the first chair in science education in Australia is an illustration of the trends in the United States occurring in other countries a few years later. Fensham, a chemist from the University of Melbourne, like so many of his

United States counterparts, took a job in science education and initiated what became a leading center for science education. As Fraser describes, he was to coordinate his full-time doctoral studies in science education at Monash University with curriculum evaluation activities in the *Australian Science Education Project* (ASEP). However, at that time doctoral programs in science education were not accessible in all parts of Australia. Many Australians, including Gregor Ramsay, who was the director of ASEP, returned to Australia having completed doctoral degrees in institutions in the USA, where large curriculum projects were underway. Accordingly, Australian educators, such as Tobin, knew about doctoral programs in the US through returning Australian scholars and U.S. scholars who had taken positions in science education in a rapidly expanding science education field.

By offering their programs internationally, universities attracted foreign nationals, who frequently pay much higher than the regular tuition fees, and thereby add revenue to their coffers. In this way, science education programs and degrees have become commodities that could be sold far beyond the boundaries of the city, state, or country. In the 1970s and early 1980s, the trend was for Australians to travel to the US to complete doctoral degrees and then to return to Australia to assume or resume positions in science education in Australian colleges and universities. (We return in greater detail to the career trajectories of individuals from other countries to the US and back home in part C, "US-Trained Science Educators Abroad.") Tobin was an example of a scholar who fit that trend. Roth also did his doctoral studies in the United States, traveling from Canada to Southern Mississippi, and then undertook a postdoctoral appointment in science education at Indiana prior to returning to Canada where he first becomes a science department head at a private college-preparatory school and then professor. The University of Southern Mississippi was one of those institutions able to draw a large contingent of master's and doctoral students from around the world, particularly in its *Summer Program of Graduate Studies in Education*, which, with over 300 graduate students per year, was able to sustain a considerable campus life in the normally low-key summer months. Most of its science education graduates—who received their degrees from the School of Science and Technology—are now preparing science teachers or teaching undergraduate science courses at colleges and universities all over the US.

In 1987 the Western Australian Institute of Technology in Australia became Curtin University and began to offer doctoral programs in science education. Even prior to 1987 Curtin had nurtured a strong international program based on its research master's degree. However, with the advent of the doctoral degree the internationalization of Curtin's science education program made it the largest graduate program in science education in the world and produced scholars who assumed faculty positions throughout the world. Also, large numbers of scholars throughout the world became involved in Curtin's research agenda: learning environments (Fraser), conceptual change (David Treagust), gender equity (Leslie Parker), and informal science (Léonie Rennie). In addition, Curtin faculty and graduates were recipients of a large number of best paper, best dissertation, early career, and distinguished career awards. A trend for Australian scholars in science education was reversed with large numbers of doctoral students in Miami, Florida, and New York.

Also, collaborative studies with leading scholars in Taiwan brought attention to the research agenda and international presence of Curtin science educators. Today, the innovative program at Curtin every year is training hundreds of science educators from around the world in return for tuition fees that allow the Centre of Science and Mathematics Education to persist and exert its influence.

As we see in the next section of the book, Curtin was not the only university that recruited overseas students. Australia's Monash University, England's Leeds, and Germany's Kiel also were prominent centers for science education that attracted international collaboration long before Curtin was a key player in science education. In France, the science education group around Andrée Tiberghien at the *Centre nationale de recherche scientifique* in Lyon has been able to attract many doctoral students especially from southern Europe.

Part B

SCIENTISTS BECOME SCIENCE EDUCATORS

BECOMING SCIENCE EDUCATORS

This section includes the autobiographies of four leading scholars who, like those in Part A have contributed significantly to science education through their research and the support they have given to others. We collected these autobiographies into one category to feature an important way in which science education is organized around the world and the particular kinds of career trajectories that are possible: Individuals with training in one of the sciences become science educators. A common aspect in these instances is that Svein Sjøberg, Peter Hewson, and Penny Gilmer—like Jim Gallagher and Michael Roth featured in Part A—are scientists who made a transition into science education. Reinders Duit was not a physicist by training but has worked in a system where professorships in science education are given to individuals who have obtained a PhD degree in one of the sciences. Having completed a second doctorate in science education, Penny Gilmer continues to work as a biochemist and has her primary appointment in a department of chemistry and biochemistry at Florida State University.

Cultural-historical approaches to human nature have made salient that culture and identity are path dependent, that is, the nature of culture or identity of a person at one moment in time cannot be understood independently from what they have been before. That is, science education as a field, offers the possibility to be a science educator. The identities members of the field develop are marked by the participation of scientists-become-science educators, who thereby reproduce the possibilities for scientists to become science educators. This path dependence is a possible explanation for the rather more conservative approach to educational thinking that one can find in science education when compared to the kinds of theorizing done in other educational subfields, such as curriculum theory. Another area in which path dependence and conservatism are observable can be found in the genres of writing educational research and the theories that members use. Thus, whereas curriculum theorists may use poetry, voice over, or dialogues, science educators are predominantly using straight narratives. Postmodernist and chaos theories have made their mark on theorizing curriculum, but these approaches have been absent in science education research. If science educators have chosen a different route, then they find themselves often confronted with a peer review process that rejects their work.

Another result of the close relation with science and the path dependence of the culture of science education can be seen in the foci of research and in the way manuscripts are treated in the peer review process. Thus, there is an explicit focus on science content, science pedagogy, and science learning. If research conducted in science classrooms focus on communication or transactional processes without also showing *how much* students learned science, then journals in the field are disinclined to publish the work—an experience we both have made time and again, for example, with respect to our work on coteaching and cogenerative dialoguing or with respect to studies based on cultural theory or the sociology of emotions. Although understanding communicative processes is fundamental to understanding why, what, and how students learn when confronted with science curriculum materials, there is a resistance in the field to publish research that shows what makes learning science possible in the first place. The close association science education has with science and the migration of scientists into the field is one possible explanation for this resistance.

There are differences, however, across continents with respect to the frequency of scientists in science education, which can be understood in terms of the respective cultural-historically developed institutional structures. In Canada and the US, for example, science education normally is taught in colleges and faculties of education. In the 1980s when Michael Roth pursued his doctoral work, the University of Southern Mississippi (USM) was one of only eight universities in North America to have a science education department in a school of science and technology. (In fact, at USM there were two types of science education degrees, one conferred through the School of Science and Technology, the other through the Faculty of Education.) In other countries, however, science educators are trained within specific science departments, most at regular universities but some at what in Germany are called *Pädagische Hochschule,* universities specializing in teacher training. Thus, Manuela Welzel (this volume) is a professor of physics education in one such university, whereas she received her PhD in a science faculty. The history of the Science and Mathematics Education Centre (SMEC), at Curtin University, began in a physics department. Later, the school wide importance of science education was acknowledged when SMEC was administratively relocated in a larger unit, the School of Applied Sciences. In the mid 1980s, when Barry Fraser became Director, SMEC was reconstituted as a university wide center, cosponsored by the Divisions of Humanities and Engineering and Science. Having responsibility for graduate degrees in science and mathematics education, scientists retained strong feelings about their stakeholder status, while acknowledging the salience of the humanities.

The relation to science and the entry of scientists into the field shapes science education and even leads to an ambiguous situation. Scientists, even if they have had no training in cognition, development, or pedagogy will find it easier to garner large grants and foundation stipends than their colleagues that have received their training through faculties of education. For example, at the time of this writing, the University of British Columbia has attracted the physics Nobel Prize winner Carl Wieman and committed $12 million over a five-year period to improve the quality of education for science students. Similarly, in the United States, organizations like the National Science Foundation have made it clear that scientists from Colleges of Science must be collaborators on projects if they are to be competitive. Accordingly, well known scientists have been an integral part of successful grants. An example is Jane Kahle teaming up with the physics Nobel laureate Kenneth Wilson to secure a large grant for the systemic reform of science education in Ohio. Funds to support research on learning science also appear to flow more easily into U.S. psychology and cognitive science rather than into mainstream science education departments.

The close relation with science and scientists not only has its upsides but also comes with considerable downsides as well. For example, there is a trend to educate students to standard (canonical) science and therefore to the reproduction of the science fields rather than to reflection and critique. The reconstructed forms of scientific journal publications are used as a model for science instruction rather than, for example, the naturalistic studies that philosophers (remember the debate concerning color between Johann Wolfgang Goethe and Isaac Newton) and interested folk have conducted for centuries—including science as practiced in environmental activist groups, science as apparent in community-based environmental stewardship programs, or ecological knowledge such as that held by aboriginal groups around the world (see Elizabeth McKinley, this volume). Scientists and scientifically trained science educators find it more difficult to accept the efforts of scholars such as McKinley, Glen Aikenhead, or Pauline Chinn, all of whom raise awareness for the important roles traditional ecological knowledge might assume in science education (cf. Sjøberg's [this volume] concern with respect to indigenous knowledge).

The result of its interdisciplinarity may also be the source of the belittling attitudes members of other disciplines have toward it. Thus, scientists feel they know more science, (educational) psychologists know more about learning and development, pedagogues know more about general classroom management, and so forth (cf. the way in which Peter Hewson and Svein Sjøberg [this volume] got their first jobs in education without having school experience or an educational background). This attitude was recently communicated to one of us by an educational psychologist who, believing that one cannot master two forms of practices, intimated that interdisciplinary scholars likely are more shallow than real (cognitive, natural) scientists. Science education will be a mature field once it can stand on its own, recognized for its strengths (and its weaknesses), once it is recognized as being able to stand on its own, soliciting and engaging with others in interdisciplinary efforts designed to bring requisite resources to the problems at hand.

There are other commonalities in these four auto/biographies that repeat the themes we articulate in Part A. For example, Walter Westphal, a physicist turned physics educator few science educators today will know, links the narratives that Svein Sjøberg, Reinders Duit, and Peter Hewson provide. Walter was at the University of British Columbia, where Peter had something like a conversion experience; Walter also led for many years the physics education group in the Institute for Science Education at the University of Kiel of which Duit was an integral part, and where Sjøberg visited early in his career before becoming a science educator himself.

SVEIN SJØBERG

SCIENCE EDUCATION

An Interdisciplinary Field

I was a rather typical Norwegian boy in the 1950s and early 1960s. Norway was busy rebuilding after WWII, and our heroes were the scientists and engineers—in addition to the national heroes in skiing and skating, of course. At a young age, I decided to become a physicist as well as a world champion in speed skating. Now, very much later, I realize that things did not develop as planned. They seldom do.

I became a relatively decent speed skater and spent five years of my life literally going in circles around a 400-meter ice track. My career in science lasted somewhat longer. I trained as a physicist but soon changed my trajectory. Some decades later I find myself as a professor in science education, a field that was nonexistent in my part of the world when I was young. Since I, more or less, was the first in this field in the Nordic countries, it might be that my story of converting from "pure" science to science education has some elements that might be of interest also for others in the field.

FAREWELL TO PHYSICS?

In 1970, I received my university degree in experimental nuclear physics and got a scholarship to work towards a PhD in this field. My future was supposed to be the measurement of half-lives of nuclear energy levels in Cobalt 57 studied by alpha-gamma reactions in a cyclotron. But I had started to get a growing unease about the project, and this was the source for a reorientation of plans.

As student, I had been engaged in a rather critical and radical group that worked to reform the emphasis in the contents of the university studies in science. We thought the focus was too narrow, merely a concentration on the well-established concepts, laws, and theories. We lacked an emphasis on the philosophical, ethical, human, and cultural sides of science in our studies. A few of our professors had such interests, and they had nurtured our reading about such issues. For my own part, I read the philosophical as well political writings of Bertrand Russell, the debates between Niels Bohr and Albert Einstein, and also became involved in debates (and actions) concerning nuclear power as well as nuclear armament. I became aware of organizations like *Pugwash* and the *Union of Concerned Scientists*.

As a result, I felt more and more uneasy with the prospects of pursuing pure science in a world where science was used and misused and not always to the betterment of people's lives. At the same time, my fascination with science and my love for science was still the same. I did not, however, see how to combine this

K. Tobin, W.-M. Roth (Eds.), The Culture of Science Education, 95–106.

love-and-hate relationship with science. On the one hand, I still saw the history of science as a brave story about radical thinkers' attempts to push back ignorance and superstition. To me, the great scientists were rebels who fought against authority and tradition, be it secular or (more often) religious. On the other hand, my own work with gamma detectors in the lab seemed so far away from such ideals! I became, however, more and more concerned about teaching science, mainly about the philosophical, ethical and social aspects that I had missed in my own school and university studies.

Then something happened: The physics department at the University of Oslo decided to set up a new unit, a laboratory for school science. The background for this was that high school teachers missed a link back to the place where they had received their training. They asked for advice and in-service courses and they raised questions on matters concerning new developments in physics, about experiments, equipment, safety in the lab, etc. The physics department offered me the job to build this new unit. After considering the offer for a day or so, I accepted. This ended my career in physics, and I started wandering into unknown territory. We had no proper name for the new field, and as far as I knew at that time, this was no academic field, but more a matter of practical action.

EMERGING NEED FOR THEORY AND REFLECTION

Our Laboratory School for Physics Education was established in 1971. Our activities were many and varied. We gave advice to teachers and schools on experiments and equipment, we ran in-service courses on physics content and safety in labs, and we developed a system where our department of physics could receive school classes to be introduced to current research activities etc. Very soon, I was invited by the Ministry of Education to become a member of official curriculum committees for various levels of the school system. It was apparent that few university scientists had shown such interests, and the few people who did, were warmly welcomed. There was no need to "lobby" or fight your way into such positions. I became involved in making official, national curricula for many levels of our school system, including teacher training.

For me, this rapid rise to a kind of power raised some problems. My knowledge of physics was acceptable, but my knowledge of schools, teaching and learning was more than weak. I had, throughout my studies, worked as a part-time teacher, and my father was a science teacher who later became a school director and administrator, also on a national level. But I strongly felt the need for a stronger theoretical educational background in addition to my science background. I turned to the departments of education at my university. But I soon came to realize that there were few if anybody who had any interests related to the actual teaching contents of schools, let alone to science and mathematics. Their own background, research, courses and teaching were on general aspects of education: general psychology of teaching, learning and motivation, or the schools as an organization, children with special needs. This strengthened my impression that science education was not an academic field, but simply an area of practical activity, where the actors could pick and choose from general theories of psychology—if they felt the need for that sort of theorizing.

INSPIRATION FROM ABROAD

But I soon discovered that there was a world out there, in other countries, with people working more focused and professionally in the area of science education. Some events became eye-openers for me: I happened to attend an annual meeting of the *Association for Science Education* (ASE) in Leeds. To my surprise, this association of science teachers had a long tradition, thousands of members, professional journals, and strong links with academic groups doing research and development in addition to their activities like teacher training. This first trip to Leeds turned out to be the beginning of a long-lasting connection. The experience was followed by study trips to other countries and institutions. I went to a newly opened institute for science education (IPN) associated with the University of Kiel (see Duit, this volume). I became aware of a long German tradition in the field of science education and of "Fachdidaktik" (didactic of subject matter). I benefited from my understanding of German, and was able to read their literature and take part in their discussions. This early experience at IPN has also been long lasting, and I am now a member of the Scientific Committee of IPN. I gradually got to understand that science education was indeed an academic field that was professionalized and had a rather long international tradition, that there were academic journals, research institutes, associations, and an international invisible college of active researchers. I did some reading on my own, but soon decided that I really needed a more thorough and systematic introduction into this (for me) new field.

A NEW START? RENEWED ACADEMIC STUDIES

A grant from my own university and a scholarship from the British Council gave me the possibility to study in Leeds, and I took an MA in education. This study introduced me to important areas like sociology of education, curriculum theory, educational psychology, and to research methods in the social sciences. I worked in a very interesting environment, the *Centre for Studies in Science and Mathematics Education*. My supervisor and main contact in Leeds was David Layton, who has ever since been an inspiration for my work. I remember in particular his historical account for the shaping of science curricula (Layton, 1973). Inspired by this, I wrote a small thesis on the development of the physics curriculum in Norway as part of the MA. When I was in Leeds, David developed the idea for a new journal: *Studies in Science Education*. Since then, the articles in this journal have been on the reading list for most courses in science education worldwide. It was, of course, a great honor when I, much later, was invited to be a member of the advisory board of this journal.

The year in Leeds was a new start. I met many people who later became mentors as well as good friends. Rosalind Driver was returning from the US where she had completed her PhD under Jack Easley. Somewhat later, they wrote a seminal article on "Pupils and paradigms" (Driver & Easley, 1978). I also met Edgar Jenkins, who later became a mentor and good friend and research partner. Edgar followed David Layton as the editor of *Studies*, and also as editor of another influential publication, the UNESCO series *Innovations in Science and Technology Education*. Although my own focus in the studies in Leeds was on science education, I had great and inspiring lecturers who really changed my thinking about a series of issues; these included Michael Young, Basil Bernstein, Douglas Barnes, and Lawrence Stenhouse.

GRADUAL DEVELOPMENT IN NORWAY

My MA in education from Leeds (1975) and the courses, conferences, and seminars in the following years gave inspiration and perspectives for the development of academic studies in science education at my own university. In the years to follow, we also invited many of my new international friends and contacts to help us advertise and develop the emerging field of science education. David Layton came several times, Rosalind Driver even more often.

Within the Faculty of Science we gradually developed studies to a master's degree in the hitherto nonexistent field of science education. The themes for the dissertations in the period from an early start in the mid 1970s are symptomatic for the development of the underlying thinking as well as my own development and growth into the field. The first titles of the master's theses were strongly related to the subject matter in science: The analysis of textbooks, critical scrutiny textbooks for possible misunderstandings in their presentations of, for example, scientific concepts, laws and theories. Somewhat later Michael Shayer and Philip Adey influenced us—through their work on the conceptual demands of teaching science, which took a strongly Piagetian direction.

Twice I received an HM King Olav's award for academic dissertations for work in this field: In 1978 for my work on the use of educational psychology in the teaching and learning of science; in 1983 (with Svein Lie and others), for work on gender and science education (e.g., Sjøberg & Imsen, 1987).

MY PIAGETIAN STAGE

An important influence in science education became the theories of Jean Piaget. In 1982, I finished my PhD thesis in which I explored the relationship between the theories of Jean Piaget and physics as a discipline and a school subject. The choice of Piaget seemed obvious for a person with my background: Piaget had a science background, used a language with strong scientific connotations, employed mathematical logic to formulate his ideas about stages, and used classical experiments from school science to elicit children's thinking (e.g., Inhelder & Piaget, 1958). Piaget admired physics, considering it to be the prime example of a mature discipline. All this made Piaget an obvious choice for me! Through my work with the original texts of Piaget and translations and popularizations, I came to appreciate the philosophical and epistemological program that underpinned his work.

Eventually my own interest turned away from the stage theory of Piaget and reoriented towards its constructivist aspects. There is good reason to revisit "the early Piaget" from the early 1930s, where he wrote in detail about children's ideas. Many of these ideas were rediscovered when mainstream science education started to become interested in children's "misconceptions" and "alternative paradigms." A key actor in the development of constructivist theory was Rosalind Driver, long before it became mainstream and the dominating paradigm.

GENDER AND SCIENCE EDUCATION

In the Nordic countries (Sweden, Denmark, Norway, Iceland and Finland), the feminist movements (in plural!) have been strong for many decades—associated with politics and influential in shaping social issues. The feminists were particularly active in the student organizations where I "grew up" intellectually. According to all international statistics, the Nordic countries lead the world on most indicators of gender equity. The United Nations' Development Programme (UNDP) publishes the influential Human Development Report every year (e.g., UNDP, 2005). This report contains many indicators for all countries in the world, and it has also developed a gender empowerment measure. This is a combined measure for many aspects of gender equity similar to level of education, participation and salaries in the labor market, and positions held in politics and the economy. The five Nordic countries top this list. But there is a paradox: Despite the overall high level of gender equity, the choices of school subjects, studies and careers are more gendered in the Nordic countries than in most other countries.

One of the first issues we tried to address was to understand (in order to remedy) this situation. We soon came to realize that the reason for girls opting away from the "hard" sciences (mathematics, physics, and technology) had to do with the image and the more or less implicit values and (lack of) perspectives of school science. It seemed that girls opted out of science based on the same sort of objections that I had against science curricula and science teaching, that is, the lack of historical, social, cultural, philosophical and ethical perspectives. Hence, working for a better science curriculum and more gender equity became two sides of the same coin in my "project." Already around 1980 we started a project on girls and Physics, and with my colleague Svein Lie, we took an active part in the establishment of the association for Gender and Science and Technology. The first conference was held in 1981 followed by a second in 1983. The gender perspective has been strong in the various initiatives and projects that I have worked on throughout my career.

LARGE-SCALE COMPARATIVE STUDIES: SISS, TIMSS, AND PISA

In the period 1982–86, I was the national research coordinator for *The Second International Science Study* (SISS). SISS actually was the first of such large-scale studies that Norway took part in. The political and ideological struggles around participation and interpretation of such projects is interesting but beyond the scope of this chapter. While involved with SISS I also learned to respect statistical techniques and the complicated logistics used in such studies. Although I became rather skeptical towards some of the educational interpretations and political misuses of such studies, I became convinced that Norway could benefit from participating. I lobbied and argued for Norwegian participation in the follow-up studies, including the *Third International Mathematics and Science Study* (TIMMS) and the *Programme for International Student Assessment* (PISA).

Norway now takes an active part in these studies; my colleagues at Oslo University are responsible for the running of the projects in Norway. They (Svein Lie and others) are also key persons in item development. At times, there are internal difficulties in our science education group, because I am taking part in public debates about the detrimental educational consequences of these studies. My main critical points are political and ideological in nature: Results from these studies are often trivialized to become one-dimensional league tables, they create simplistic images of the overall quality of the school system, and they constitute a pressure to harmonize and universalize science curricula and testing. Such pressures run contrary to the need to contextualize curricula and to build on the culture, needs and interests of the learners.

FORMAL ESTABLISHMENT AND CONFLICT

By 1985, our work had started to be well established and we had master's degree students not only in physics education, but also in chemistry and biology—where they established school laboratories similar to the one in physics. Our faculty later decided to form a *Centre for Science Education* at Oslo University. I got the job to lead this new development. At that time, Doris Jorde (see this volume) moved to Norway. She had a PhD in science education from a U.S. institution and came at the right moment. In the years to follow, we have been working closely together in building science education as an academic field. In fact, we complement each other quite well, both in contents and research methods. She is an expert on classroom interactions, studied with qualitative methods. My interests are more on the social, cultural, ethical, and political aspects of science education, and as a researcher, I am better qualified in quantitative methods. Our studies in science education were now accepted to be a regular part of the study program in the faculty of education, and we had a broad portfolio of activities. Then an administrative decision changed the plans. The university decided to reorganize several of its activities and formed a new department for teacher education. In this process, they moved our center to this unit, which later became a department with a new faculty of education. We strongly opposed this decision, but in vain. Being in a faculty of education felt strange, because our students and the degrees that we awarded still belonged to the Faculty of Science and Mathematics.

This experience can exemplify the problematic nature of the academic field of science education. It is in part natural science, in part social science. And it can often be seen as part of teacher training. This challenge of having multiple identities and loyalties is something we have to live with in our field, and I am aware that different organizational solutions exist in different countries, even with different setups in each country.

For me, this rearrangement was difficult. I had been urged by my faculty to build a new unit and a new activity that the faculty needed and wanted. When the job was done, we were moved to another faculty, away from our science home and background. The price for interdisciplinary work seemed to be a loss of identity. For some time, we fought against this decision, but I became so frustrated with fighting bureaucracy that I finally grasped an offer to leave the university for another job. That proved to be very rewarding.

INTERNATIONAL INVOLVEMENT THROUGH UNESCO AND NORAD

I have worked most of my academic life from a base at a University, but the exception is the period 1990–93, when I worked as educational advisor for the national commission of UNESCO and for the *Norwegian Agency for Development Cooperation* (NORAD). My task was to analyze and give advice on educational initiatives—for example, by UNESCO—and on policy issues and projects relating to Norwegian aid to the education sector. These years were extremely interesting, and gave new perspectives as well as many new contacts all over the world. Among other things, I learned to see the economical and financial aspects of educational systems, and their role in national and human development. In 1993, I returned to Oslo University, now as professor of science education—actually the first of its kind in Scandinavia. I returned with many new ideas and many of my later commitments and projects were spin-offs from this period, in particular activities oriented towards developing countries. I used some of the international meetings to build projects that later developed into research cooperation.

TWO PROJECTS: SAS AND ROSE

One example of a project that grew out of my years with UNESCO and NORAD is the *Science and Scientists* project (SAS) in the period 1994–2000. Together with Jayshree Mehta from India and Jane Mulemwa from Uganda, we developed this project, a survey among young learners (age 13) on various aspects relating to their views on science, scientists and school science. *Science and Scientists* was meant and designed to be a small, exploratory study, but many scholars from different countries wanted to join the study, and it grew beyond our expectations (and resources). More than twenty countries—amounting to 20,000 students—joined the study. For many of our international partners, this was the first involvement in international research cooperation, and many of our colleagues wrote national reports and articles (e.g., Sjøberg, 2002). There was a strong gender agenda underlying SAS, and most of the researchers were female, also from developing countries.

The SAS study may be seen as a pilot study for the much larger and better-planned study, *Relevance of Science Education* (ROSE). This project started in 2001 and is likely to continue for the years to come. It has funding from various sources, mainly from the Research Council of Norway. *Relevance of Science Edu-*

cation is an international comparative study that taps into the diversity of interests, experiences, priorities, hopes and attitudes that children in different countries bring to school (or have developed at school). The underlying hope is to stimulate an informed discussion on how one may make science education more relevant and meaningful for learners in ways that respect gender differences and cultural diversity. We also hope to shed light on how we can stimulate the students' interest in choosing science and technology-related studies and careers. *Relevance of Science Education* has, through deliberations that involved science educators from all continents, developed an instrument that tries to map out attitudinal or affective perspectives as seen by fifteen-year old learners. There are about 40,000 students from 35 countries taking part in ROSE. About ten PhD students from different countries will base their thesis on ROSE data (e.g., Schreiner & Sjøberg, 2004).

IOSTE AS INTELLECTUAL HOME

There are many international organizations and associations in the field of science education. I am member of many of these, and I have been rather active in some of them over the years, but consider the International Organization for Science and Technology Education (IOSTE) as my intellectual home. IOSTE developed during the cold war out of the need for international understanding and dialogue, and became a meeting place for Science and Technology (S&T) educators who shared some values and beliefs and a vision that S&T education should face the real challenges of our time, including sustainable development and the empowerment of people to take active part in all aspects of life in society. Science and technology as vehicles for betterment of life and for development of critical citizens in living democracies are key common concerns. Key persons in IOSTE have been Peter Fensham, Glen Aikenhead, Jayshree Mehta, and Jim Gaskell. The biannual international symposia of IOSTE are important events in my life, and they are more truly international than many other conferences in S&T education. Participants come from all over the world, including developing countries, and the conferences are also hosted in places that indicate this profile. I have been a board member of IOSTE for many periods, and I also served as IOSTE chair in the period of 2002–04. Professional associations like NARST and ESERA are oriented towards academic research and do not have an explicit value commitment except the promotion of high-quality research. IOSTE has clear values and the symposia leave room not only for the presentation of research but also for discussions on matters of political, social, and cultural importance relating to S&T in society and in education. There is certainly a place for both kinds of organizations, but I enjoy and am inspired more by IOSTE.

WIDENING THE RESEARCH PERSPECTIVE: RECENT PHD STUDENTS

For me, science education is a wide and interdisciplinary field of research, development and action. It resists being captured by a strict and normative definition, either from the inside of the field or (even worse) from people standing at the outside and wanting to impose restrictions on what counts as science education.

Let me illustrate the great variety of the field with examples from my own research students and their PhDs. In the period 1996–2001 I had a grant from the Research Council of Norway on science education for citizenship. The project resulted in three PhDs and I also wrote a book with the project title, and this is now the standard textbook for science teacher education in the three Scandinavian countries, where the book is slightly adapted to national needs in cooperation with colleagues in Sweden and Denmark. The three PhDs on the project are Erik Knain, Stein Dankert Kolstø, and Marianne Ødegård. Knain wrote his PhD (in Norwegian) on "The silent voice of science education" (e.g., Knain 1999), Kolstø (2001) wrote a PhD entitled *Science Education for Citizenship: Thoughtful Decision-Making about Science-Related Social Issues*. This thesis is in English and is based on a series of articles, that is, the common format in science, but still not so frequent in science education in most countries. Ødegård (2001) presented her PhD thesis, *The Drama of Science Education*, which is about how public understanding of biotechnology and drama as a learning activity may enhance a critical and inclusive education.

The great diversity of approaches and perspectives are even more evident in the most recent PhDs written by students of mine. Astrid Sinnes (2005) wrote a thesis entitled *Approaches to Gender Equity in Science Education: Two Initiatives in Sub-Saharan Africa Seen Through a Lens Derived From Feminist Critique of Science*. This topic was partly chosen because I had been involved in several initiatives in Africa to promote gender equity in science education, that is, to improve material conditions, to remove barriers, to increase participation and achievement etc. Camilla Schreiner (2006) wrote her thesis based on the ROSE project: *Exploring a ROSE-Garden. Norwegian Youth's Orientations towards Science—Seen as Signs of Late Modernities*. In the thesis, Camilla uses theories based on contemporary sociology and youth research, in particular theories of late modernity. As can be seen, the diversity of research problems, theoretical positions and research methods is great. If there is a common denominator, they do address problems felt to be important, be it socially, politically or educationally.

MENTORS AND INSPIRATION

I have already mentioned the inspiration I got in my Leeds period from getting to know people like David Layton, Rosalind Driver, and Edgar Jenkins. Somewhat later I met Peter Fensham for the first time. I liked and admired the way his ideas about science education always were underpinned by a set of human values. I know that I was just one of many young people who have benefited from his generosity and personal warmth. We first met at a U.N. conference on science and technology and the future human needs in India. After working together there on issues of social responsibility of science, he invited me to spend a sabbatical at Monash University. I stayed with my family in a small flat in his house. In Peter's writings, but even more so in conversation, I found ideas that I could easily embrace, be it on the significance of science for all, on the concern for the environment and many other aspects with a social and ethical underpinning. The cooperation with Peter developed into a personal friendship, and we have spent time together also in the mountains in Norway.

Of special significance for my own development was John Ziman. Although his main interest was not in science education proper, he made important contributions to the thinking in this field. I got to know John through various committees and in professional meetings. After an impressive career in physics, John spent the last decades of his life exploring issues on the interplay between science and society. Another important mentor has been Glen Aikenhead. I have learned a lot through our friendship and cooperation. I am not sure that I share his views on the epistemological status of indigenous science, but I do share his value commitments to working with marginalized groups. I have also had the pleasure to work closely with Joan Solomon, in particular on STS issues and as an international committee for the great Portuguese initiative *Ciencia Viva*. This national initiative is probably the most ambitious attempt in Europe to promote the culture of science in a country that was very little developed after the fall of the old, semi-fascist regime in 1975. The key person for doing this is Jose Mariano Gago. He was a professor of physics at CERN and has been Minister of Education and Science in Portugal for several legislative periods. He has pushed the same agenda in the European Union. On his initiative, the European Union (EU) made a thorough report on the situation for S&T in education in Europe. Both John Ziman and I were members of the team leading to a report with the telling title "Europe Needs More Scientists!" It has turned out to be very influential in shaping EU policies in this area.

SCIENCE AND SOCIETY ACTION PLAN IN THE EUROPEAN UNION

Although science in schools has been the major concern of most of my career, I have always had a somewhat wider interest relating to many aspects of science and technology in society. In recent years, I have had many opportunities to reorient myself in that direction. The European Union is putting a lot of effort into joint research and development. The current *Frame Programme 6* includes several initiatives under the umbrella of an action plan for science and society. When this was at a planning stage, I was invited to present a background paper for the European ministers of education and research. I am pleased to see that many of the ideas that I put forward are also visible in this plan. Later, when the action plan was put into operation, I was invited to be a member of the expert advisory committee for this initiative. The science-and-society initiative covers aspects like ethics, gender equity, public understanding and dialogue, science communication, and science education. In addition to being a program of its own, these aspects are also meant to cut across all the different thematic areas of research in the EU. The work in this group has been extremely rewarding, and has again given me new perspectives on issues that I consider to be of high importance.

It has also implied that I have been engaged in several important projects and initiatives of great political and educational significance. One of these is the Euro-

Eurobarometer study, a large survey of the entire EU population. In 2005, two of these studies addressed peoples' attitudes, interests, values and knowledge related to science and society.

Many of the questions raised in the large-scale Eurobarometer surveys are identical to the questions that we posed in ROSE. The influence is two-ways. When we developed ROSE, we borrowed items from previous Eurobarometer studies, and later, I could argue for the inclusion of ROSE items in the 2005 Eurobarometer. We are now involved in very interdisciplinary research cooperation to utilize the rich data that we now have on values, interests and perceptions related to S&T among students as well as the adult population.

CONCLUSIONS AND REFLECTIONS

The writing of this brief autobiography has prompted me to reflect in ways that one seldom gets time to do. One has to take stock and try to find patterns in one's professional and personal life story. Some concrete outcomes of my career so far are easy to summarize. I gradually grew into a new field that did not exist in my part of the world. Now it is rather well established in Oslo as well as in other places in Norway. A similar development has taken place in, for instance, Denmark and Sweden, and I have been involved in many of these processes. In Oslo, we have now graduated more than 100 individuals who have a master's degree in science education and about fifteen with a PhD. I have been involved in many of these activities, and also pushing for these issues in media and public debates. It is of côurse rewarding that my book in science education is widely used as a basic textbook in all three Scandinavian countries.

I have received several recognitions of the value of my contributions. It was very moving that Nordic colleagues involved in science education research produced a joint book to honor me on my sixtieth birthday (Jorde & Bungum, 2003). This book is a kind of state-of-the-art of the field of science education in Norway, and contains research from 25 of my colleagues, mainly former students of mine. In 2005, I also received an award from the International Union of Pure and Applied Physics for my contribution to physics and science education. As is evident from this chapter, I also have a long series of international commitments and engagements, in fact only a few of these have been mentioned.

There has never been a career plan behind this development. I have simply followed my instincts and interests as they have shifted. This is a kind of luxury one can afford when one enters a new field that has no clearly defined traditions, research paradigms, authorities and no defined theoretical canon. This was the case for the field of science education in my country some decades ago.

The process of my professional development may indeed seem very haphazard. But when I try to see this in perspective, I realize that there is some continuity behind it. In fact, I have in the last years returned very much to the perspectives that colored my years as a student: the interest for science in a wider social context, including social, ethical, cultural and indeed political dimensions.

The field of science education internationally is to a large degree dominated by "retrained scientists" like me. Our activities often reflect our own original training, culture, world-views, etc. The science education community is often apolitical and not very oriented towards culture, ethics, and social concerns. This orientation is often also reflected in the choice of research problems as well research methods. The dominating research has been (and perhaps still is) an orientation towards teaching and learning in a relatively narrow sense, the understanding of scientific concepts, often based on research incorporating a standard design of pre-test, treatment, post-tests and statistical testing of learning gains. In the last decade or so there seems to be more variety and openness in the research agenda. Science educators are discovering the social and political aspects of science (pure and applied) as well as science education. People with a political and social science background often get a feeling of déjà vu. We seem to witness a rediscovery of the debates of the 1970s (at least in Europe) based on ideas from Paulo Freire, the Frankfurt School and critical theory, and the cultural and gendered nature of knowledge, including science concerns about equity and discrimination, education as reproduction of power structures and class. I welcome such a shift of focus. Science education research consists of more than classroom studies of teaching and learning. A problem is that the newly converted have a tendency to take new ideas to their extremes. The recent discoveries of cultural, feminist, and other critiques of science can easily be exaggerated. In the rejection of naïve positivism, many jump to the other extreme, embracing all sorts of relativism and subjectivism. Some also embrace indigenous science in ways that seem to be based more on wishful thinking than on epistemological analysis. Many consider such positions as politically radical stances, which may be counterproductive and reactionary.

Being part of this worldwide community of educators, researchers, and activists has been (and still is, of course) exciting and rewarding. Science is not just a school subject and an academic discipline. It is shaping our material world as well as our worldviews and ideas. Science and technology have Janus-like faces: they can be evil and good; they can both save and take lives; and they can liberate minds and improve material conditions, but can also be used to cause destruction, oppression and domination. As science educators, we have to face this wide array of challenges and find our own purpose and meaning; and we have to set our own agenda for research based on the concerns and priorities that we value. Of course, we do not always have the possibilities to freely make priorities in the way I suggest here, we are all constrained also by material resources and more or less political priorities outside our own range of influence. But the room for independent action is often larger that we think.

Svein Sjøberg
Professor of Science Education
University of Oslo

A MODEL OF EDUCATIONAL RECONSTRUCTION AS ORIENTATION OF SCIENCE EDUCATION RESEARCH

A Personal Note on the Development of Science Education Research that Aims at Improving Practice

In the 1960s I studied physics and mathematics at the University of Kiel (Germany) to become a Grammar school teacher. I was lucky enough to attend the first seminars provided by the physics education staff of the IPN (Institute for Science Education) that opened in late 1966. The German Volkswagen Foundation funded the institute in an effort to improve science instruction in German schools. Clearly, the founding of the institute was also a response to the "Sputnik Shock" that led to projects for improving science education in many countries, with a particular emphasis in the USA. The need to improve scientific literacy was generally seen as essential in order to compete in the cold war. In Germany, however, improving scientific literacy was also embedded within a strong movement towards general literacy fuelled by the idea that education is a key human right.

The early members of the IPN physics education department staff, Stefan von Aufschnaiter and Hans Niedderer were of the opinion that the best way to provide a solid base for improving physics instruction would be to start as early as possible. A curriculum development project was launched in 1968 that should cover grades five to ten, the lower secondary level in German schools. I was invited to carry out my master study within the first piloting year of that project. The topic was the simple electric circuit as taught in fifth grade. Empirical research on students' pre-instructional conceptions were included drawing not only on written tests but also on what now is called a performance test. Students were asked to build up the illumination of a doll's house. That was my start into science education research, into nearly forty years of research on improving student understanding of key ideas of physics.

I would like to point to an important issue from the outset. It was a matter of chance that I was in the right place at the right time to join the right people in their work. To work and grow in an institute that was starting a new research field was challenging and rewarding. The hierarchies were rather flat then: all members of the team were eager to learn from the others. Later in my career I again was lucky enough to meet and join colleagues who significantly contributed to the development of my thinking about science education research. To name just a few, Peter Fensham visited the institute in the early 1970s repeatedly. I am grateful for the discussions with him. In the early 1980s David Treagust visited the IPN and we

K. Tobin, W.-M. Roth (Eds.), The Culture of Science Education, 107–120.

discussed matters of the role of students' pre-instructional conceptions and conceptual change. We became good friends and have closely cooperated since then, mostly on conceptual change issues. In 1990 Shawn Glynn stayed a year with us at the IPN. That was the start of another fruitful cooperation, especially concerning the role of analogies in teaching and learning science. In the middle of 1995 we had the privilege to host Michael Roth for a couple of months. We used the time to carry out a joint study on learning processes. He introduced Michael Komorek, Jens Wilbers, and me to the art of carrying out video-based studies. Briefly summarized, the development of my views on and my practice of science education research has been significantly influenced by close cooperation with many colleagues. I owe them very much. Much of my professional identity is determined by the projects that I have conducted with a considerable number of colleagues.

In the following I illustrate the development of my conception of science education research. It is solidly grounded in the German *Didaktik* tradition and has been further developed by adopting various theoretical frameworks, mostly of constructivist origin. My actual way of thinking may be outlined by the key ideas of a model of educational reconstruction that developed over the years and was explicitly elaborated in the 1990s—in close cooperation with Ulrich Kattmann, Harald Gropengießer, and Michael Komorek.

A BRIEF SKETCH OF THE GERMAN TRADITIONS OF BILDUNG AND DIDAKTIK

Hermeneutics

It is essential to point out first, that traditional German pedagogy was strongly embedded in hermeneutical epistemological views as established by Wilhelm Dilthey (1833–1911). It appears that this tradition is the reason that behaviorist ideas had a much smaller impact on the educational system in Germany as compared to the predominance of this view in the USA.

Bildung and Didaktik

The German terms *Bildung* and *Didaktik* are difficult to translate into English. A literal translation of *Bildung* is "formation." In fact *Bildung* is viewed as a process. *Bildung* stands for the formation of the learner as a whole person, that is, for the development of the personality of the learner. The meaning of *Didaktik* is based on the notion of *Bildung*. It concerns the analytical process of transposing (or transforming) human knowledge (the cultural heritage) like domain specific knowledge into knowledge for schooling that contributes to the above formation (*Bildung*) of young people.

Two conceptions of *Didaktik* deeply influenced my way of thinking about science education. The first conception is Wolfgang Klafki's *Didaktische Analyse* ("Educational Analysis") published in 1969 (Fensham, 2004). His ideas rest upon the principle of primacy of the aims and intentions of instruction. They frame the educational analysis, at the heart of which there are the five questions (Table 1).

Table 1. Key questions of Klafki's (1969) Educational Analysis (Didaktische Analyse)

1.	What is the more general idea that is represented by the content of interest? What basic phenomena or basic principles, what general laws, criteria, methods, techniques or attitudes may be addressed in an exemplary way by dealing with the content?
2.	What is the significance of the referring content or the experiences, knowledge, abilities, and skills to be achieved by dealing with the content in students' actual intellectual life? What level of significance should content have from a pedagogical point of view?
3.	What is the significance of the content for students' future life?
4.	What is the structure of the content if viewed from the pedagogical perspectives outlined in questions 1 to 3?
5.	What are particular cases, phenomena, situations, experiments that allow making the structure of the referring content interesting, worth questioning, accessible, and understandable for the students?

Another significant figure of thought within the German *Didaktik* tradition is the fundamental interplay of all variables determining instruction proposed by Paul Heimann, Gerhard Otto, and Wolfgang Schulz in 1969 (Figure 1).

In my view the most important issues of the German Didaktik tradition as outlined in Table 1 and Figure 1 are the following. In planning instruction (by the teacher or curriculum developers) the science content to be learned and students' cognitive and affective variables linked to learning the content have to be given equal attention. The science content is not viewed as "given" but has to undergo certain reconstructive processes. The science content structure (e.g., the force concept) has to be transformed into a content structure *for* instruction. The two structures are fundamentally different. In the first step, the "elementary ideas" with regard to the aims of instruction have to be "detected" by seriously taking into account student perspectives (e.g., their pre-instructional conceptions). It is obvious that key ideas of the later constructivist perspectives of teaching and learning science were already part of the German *Didaktik* tradition.

Traditions of Student-oriented Science Instruction

Traditionally in Germany after primary level (grades 1 to 4) a streamed school system is prevailing with three levels of schools. Until the end of the 1960s traditions of student-oriented science instruction, however, played a significant role nearly exclusively at the primary level and in the lower streams of the German

Intentions (aims & objectives)	Topic of instruction (content)	Methods of instruction	Media used in instruction
Why	What	How	By What
Students' intellectual and attitudinal preconditions (e.g., pre-instructional conceptions, state of general thinking processes, interests, attitudes)		Students' socio-cultural preconditions (e.g., norms of society, influence of society and life on the student)	

Figure 1. On the fundamental interplay of instructional variables

109

system. The German Gymnasium (grammar school) preserved, for instance, the tradition of being quite strictly oriented toward the sciences. Ideas of the student-oriented German "Reformpädagogik" of the 1920s were adopted primarily at primary level and to a certain extent also in the lower levels of the school system. This also holds for a certain tradition of empirical research concerning students' pre-instructional conceptions. There are early studies in the 1930s and later in the 1950s and 1960s. But these studies nearly exclusively concern primary level students or students of the lower level of the school system.

In the early 1970s in Germany many professorial chairs for science education were established at universities as a major response to the above movement to improve science education. It is interesting to note that two clearly separated groups of science educators developed, those being primarily oriented toward a particular science discipline (e.g., biology, chemistry, or physics) and those attempting to bring science issues and student perspectives explicitly into balance. Edgar Jenkins observed a similar split of the science education research community also in various other countries.

Martin Wagenschein

In 1962 Martin Wagenschein, a physics teacher and teacher educator, published "Die pädagogische Dimension der Physik" (The pedagogical dimension of physics), which has been rather influential in Germany, Austria, and the German speaking part of Switzerland. Today there still are Martin Wagenschein Associations preserving his ideas. His thoughts also played an essential part in developing my ideas of good physics instruction. His aim is to make physics concepts and principles as well as physics processes and views about physics understandable for students. His approach to teaching physics, as developed in the above book as well as in his other publications, is very student oriented and consistent with many of the ideas later developed within constructivist approaches. Teaching science should be in the form of Socratic dialogue providing the learner a voice and taking him or her serious discourse partner. It should also be oriented at the genesis of student thinking about the issues addressed. Several examples of how to deal with students' preconceptions are discussed.

OVERVIEW OF MAJOR RESEARCH AND DEVELOPMENT EMPHASES

In this section I briefly summarize the development of my views of efficient science instruction and my conception of research that allows improving practice.

IPN Physics Curriculum

As already mentioned my work in science education started with the evaluation of a pilot instructional unit on the simple electric circuit for fifth-grade students. This work was theoretically grounded in the above-mentioned German tradition of *Bildung* and *Didaktik*. But we also adopted various other theoretical frameworks, among them Jean Piaget's work, especially his stage views of cognitive development, David Ausubel's seminal work on the role of students' pre-instructional conceptions, and Thomas Kuhn's views of conceptual change in the history of science. The development of the pilot instructional units was the beginning of a curriculum covering physics instruction in grades 5 to 10. Of course, the already existing science curricula in the USA and the projects in the UK supported by the Nuffield Foundation were taken into account. Also more behaviorist approaches of curriculum development and evaluation were adopted. However, they were embedded in the German tradition of *Bildung* and *Didaktik* and hence issues of student learning always played a major role in the curriculum design. Content for instruction was carefully constructed, by taking into account aims and student perspectives.

Further, international conferences organized by the IPN in the early 1970s enabled us to become familiar with ideas of leading science education researchers around the world. Walter Westphal, who became head of the physics department in 1972, provided experiences with projects on improving physics teaching and learning he carried out in Vancouver with Galen Erickson and Peter Hewson. Karl Frey, a curriculum specialist, the managing director of the institute for many years, facilitated access to the international state of curriculum theories. In other words, there was a research climate in the IPN that allowed for the blending of German traditions with more recent views in different cultural backgrounds. My PhD Thesis was part of our attempts to closely link development and research. The long-term retention of key concepts of heat achieved in one of our units for sixth-grade was investigated. It turned out that sets of concepts closely linked after instruction were best remembered by the students. I was also engaged in a study on the effects of the new IPN Physics Curriculum for grades 5 and 6. Major effects identified concerned significant influences on curricula for early physics instruction in several federal states in Germany as well as on research on teaching and learning physics in various places in Germany. The units of the IPN Physics Curriculum in grades 5 to 8 were very much oriented toward making students familiar with basic physics phenomena and to facilitate basic understanding of key concepts that afford interpretation of the phenomena. For the subsequent grades 9 and 10 another emphasis was chosen. The aim was to make students familiar with the role of science to understand the impact of science and technology on society. In other words, there was a kind of STS orientation. Especially, this part of the curriculum was hotly debated. There was a rather strong resistance, especially, of the stakeholders of the Gymnasium tradition. There was, for instance, a unit on the pros and cons of nuclear power stations developed toward the end of the 1970s by Helmut Mikelskis and Roland Lauterbach, where the nuclear industry tried to intervene.

The Development of Content Structure Diagrams

A brief note concerns theoretical considerations on the concept of *science content structure*. This is again a term that may be misunderstood in English. It does not merely denote science subject matter of a certain topic but includes the structure, that is, the relations to all the other concepts a given concept might have. The basic idea is that concepts may not be understood one by one independent of the relations to other concepts but that it has to be taken into account that concepts have to be seen as theory laden. This points to a major problem of learning science. To understand a single concept always means to understand also the theory in which this concept in embedded. As a result the content structure for instruction has to be carefully designed in such a way that students may be introduced into the theory and the particular concepts. Drawing on Robert Gagné's "learning hierarchies" introduced in the 1960s I developed a method to plan the content structure for instruction by designing logical flow diagrams. The arrows between the boxes denote that the content issue of the previous box is a prerequisite for learning the content issue of the subsequent box. The arrows, hence, denote logical dependencies *and* issues of learning as well. Interestingly, the same kind of diagrams are used in the *AAAS Atlas of Scientific Literacy* (2001) in order to facilitate coherence and consistency of science content of larger domains across the curriculum (K–12). We now use variants of such concept structure diagrams to investigate the content structure of video-documented lessons.

Research on Students' Conceptions and Conceptual Change

Investigating students' pre-instructional conceptions and developing theoretical perspectives that allow conceptualizing teaching and learning science has been my major concern for about 20 years. In the end of the 1970s a group of German researchers interested in investigating the role of students' pre-instructional conceptions in teaching and learning science formed. Walter Jung, the most influential physics educator in Germany at the time, was the key member of the group. Other members were Helga Pfundt, Ernst Kircher, Christoph v. Rhöneck, and me. We developed a program for dealing with students' pre-instructional conceptions. International workshops facilitated close cooperation with leading researchers from various countries. Helga Pfundt started a bibliography on students' alternative frameworks that only after she died in 1984 appeared for the first time in print. The bibliography has been continuously updated since then.

The development of research on teaching and learning taking into account students pre-instructional (alternative) conceptions may be briefly summarized in the following:

- Investigation of various student conceptions on the level of science concepts and principles (some people ironically called it "misconception hunting") started in the 1970s, was very strong in the 1980s, and is still flourishing.
- Investigations of student conceptions of science processes and views of the nature of science started in the later 1980s and are still a major research field.
- Investigation of student views of teaching and learning (meta-cognitive views) also played a certain role in the late 1980s and 1990s.
- Investigation of various teacher conceptions (of science concepts, processes, the nature of science, and meta-cognitive views) started in the middle of the 1980s and has become

now one of the major research fields of science education. Especially studies on teacher professional development have played a significant role since the late 1990s.

- Studies an improving instruction by "conceptual change" approaches have been developed since the early 1980s. Still research on conceptual change is a major research field—in science education and in psychology as well.
- Theoretical orientations were Piagetian (especially the idea of intimate interaction of assimilation and accommodation), Ausubelian, and cognitivist in the 1970s. In the 1980s and 1990s various constructivist views developed. Initially radical or moderate (individualistic) constructivist views of teaching and learning constituted the major orientation; later, during the 1990s, variants of social constructivist or social cultural views were integrated. It appears that multi-perspective epistemological views are now predominating—as no single epistemological view may address the many features that the phenomenon of learning includes.
- Finally, research methods initially were predominantly individual interviews in the beginning. In the 1990s technology made it possible to carry out (video-based) learning process studies. More recently, practice studies (also often video-based) investigating the practice of "normal" science lessons have become more frequent.

My own work fits in the above steps of development. In the early 1980s I carried out a study on teaching and learning the energy concept comprising critical analysis of the paths to the energy concept in the literature and empirical studies on students' conceptions of the key features of the energy concept: conservation, degradation, transformation, and transport. Christel Jenelten-Alkofer contributed to this work with a PhD study on the emergence of energy ideas during cognitive development based on the work of Piaget. Later in the 1980 Sofia Kesidou more fully investigated students' conceptions of heat and energy degradation in her PhD study. The role of analogies in teaching and learning science played a major role for some ten years starting in 1988 with an empirical study on teachers' use of analogies in close cooperation with David Treagust. Subsequent theoretical work also included cooperation with Shawn Glynn.

In the early 1990s a program on "educational reconstruction" of non-linear systems started that lasted nearly fifteen years. Major concern of this program was to analyze the "educational significance" of this new topic for science instruction, that is, the question whether this topic is worth teaching and also included studies on whether this is possible. Initially, the strange behavior of a chaotic pendulum was the focus, later various other non-linear systems were also analyzed. After his PhD comprising a learning process study on understanding chaotic systems Michael Komorek directed this project. Further major studies include a close cooperation with Michael Roth in carrying out a video-based learning process study (Duit, Roth, Komorek, & Wilbers, 1998) and a PhD study by Jens Wilbers on the role of analogies in understanding chaotic systems (Wilbers & Duit, 2005), and a PhD study by Dimitris Stavrou on understanding the interplay of chance and deterministic laws in non-linear systems. Throughout these years and since then also reviews of research and the further development of theoretical positions played a major role in my work (e.g., Duit & Treagust, 2003).

Video-based Studies of the Practice of Introductory Physics Instruction

Unsatisfying results of German students in the international monitoring study *Third International Mathematics and Science Study* (TIMSS) at the end of the 1990s changed the climate for science education research in Germany substantially. It became evident that—in international comparison—German students did less well than expected. The level of scientific literacy achieved appeared to be seriously deficient. As a response various initiatives to improve science and mathematics instruction were launched. A large-scale quality development project to improve the efficiency of science and mathematics instruction was started in late 1998. Some 180 schools in most federal states participated. The IPN organized this five-year project. Many science and mathematics educators played a significant role in these attempts.

Further, the German Science Foundation (DFG) funded a large-scale research program on possibilities to improve instructional quality. Like the above quality-development program Manfred Prenzel, the managing director of the IPN, directed this research initiative. This six-year program started in early 2000. Within this program we have carried out a project on investigating the practice of German introductory physics instruction (Prenzel et al., 2002). There are three two-year phases. In the first phase, the piloting phase, thirteen teachers from two federal states participated, in the second phase 50 teachers from four federal states. These teachers stemmed from schools selected randomly. There is also a parallel study in Switzerland directed by Peter Labudde to facilitate comparison of physics instruction in two countries with different cultures of physics instruction.

A video-based study to compare mathematics instruction in the USA, Japan, and Germany was carried out as part of the TIMSS. We used a similar, however, more elaborated design. Video-documented physics lessons played a major role with additional data from student and teacher questionnaires and teacher interviews. The study was carried out by members of the education sciences and the physics department of the IPN: PhD studies of physics education issues were carried out by Maike Tesch (on the role of experiments), Ari Widodo (characteristics of constructivist oriented design in the lessons) and Christoph Müller (teachers' views on teaching and learning physics). Maja Brückmann is still working on her attempts to reconstruct the content structure of physics lessons in Germany and Switzerland.

There are various results of these studies that may not be presented here. Most important for me were the experiences while interviewing the teachers. It turned out that most of them in the first phase and nearly all of them in the second phase were not (well) informed about research findings on the role of students' pre-instructional conceptions. They further hold rather loose views about teaching and learning. Their thinking about physics instruction was very content dominated. These findings shocked me quite substantially as I had tried to inform teachers by various means for a long time about the essential role student conceptions play. In a way, these experiences gave me another push to redirect research interests towards attempts to change teachers' views about good instruction and their instructional behavior.

The *Physics in Context* project co-directed by my colleagues Silke Mikelskis-Seifert and Manfred Euler and myself provides the frame for investigating the development of teachers' views about good physics instruction. Major means to make teachers familiar with the recent state of research on teaching and learning physics are brief summaries of research findings, so called "piko-letters." However, preliminary findings show that a couple of teachers do not even use these short summaries written in easy access style. As many other studies have also shown, it is very difficult to change physics teachers' rather content oriented thinking about physics instruction. It appears that we still fail to "persuade" or "convince" them that the view we are proposing is more efficient than their views developed in long years of practice.

In an additional project—which is part of the third phase of the above physics video-study—my colleague Manfred Lehrke, our doctoral student Claudia Kastens, and I attempt to develop teachers' views by analyzing video-documented lessons. Over a year about ten teachers analyze their own video-documented lessons and lessons of their colleagues. However, it turns out that it is rather difficult to get teachers involved. It seems that many teachers are afraid to allow colleagues to look into their classrooms.

THE MODEL OF EDUCATIONAL RECONSTRUCTION

Much of my work concerned the development of a model of educational reconstruction. This model includes the major facets of my conception of science education research that provides help in improving practice. The model briefly sketched in the following was developed in the 1990s in close cooperation with Ulrich Kattmann and Harald Gropengießer and my IPN colleague Michael Komorek (Duit, Gropengießer, & Kattmann, 2005). Initially, the model has been developed as a theoretical framework for studies as to whether it is worthwhile and possible to teach particular areas of science (such as in the above project on educational reconstruction of key ideas of non-linear systems). It draws on the need to bring science content related issues and educational issues into balance when teaching and learning sequences are designed. It provides a framework for instructional planning by curriculum developers but also by teachers in school. As I show more fully below, the model also provides a frame to indicate the major domains of science education research.

Key Features of the Model

The model is based on the above outlined German tradition of *Bildung* and *Didaktik*. Key features of the above also sketched work on clearly distinguishing the science content structure and the content structure *for* instruction are also included. However, the model is also embedded in more recent constructivist epistemological frameworks. There are two facets of this epistemological orientation. First, student learning is viewed as students constructing their own knowledge on the grounds of the already existing knowledge. Second, also science knowledge is seen

115

as a human construction. We presume that there is no true content structure of a particular content area. What is commonly called the science content structure is the result of a consensus in a particular science community. Every presentation of this consensus in the leading textbooks is an idiosyncratic reconstruction of the authors informed by the specific aims they explicitly or implicitly hold. Accordingly, the science content structure for instruction is not simply given by the science content structure. It has to be constructed by the curriculum designer or the teachers on the grounds of the aims affiliated with teaching the particular content. In other words, the science content structure has to be reconstructed from educational perspectives. That is the very essence of the term "educational reconstruction."

There are three intimately linked components of the model of educational reconstruction (Figure 2). First, the *analysis of content structure* includes two closely linked processes: *clarification of subject matter* and the *analysis of educational significance*. Clarification of subject matter draws on content analyses of leading textbooks and key publications on the topic under inspection but also may take into account its historical development. Interestingly, also taking students' pre-instructional conceptions into account that have often proven not to be in accordance with the science concepts to be learned contributes to more properly understanding the science content in the process of subject matter clarification. Experiences show that the surprising and seemingly strange conceptions students own may provide a new view of science content and hence allow another, deeper, understanding. Traditionally, science content primarily denotes science concepts and principles. However, recent views of scientific literacy claim that also science processes, views of the nature of science and views of the relevance of science in daily life and society should be given substantial attention in science instruction. All these additional issues also need to be included in the process of educational reconstruction.

Second, *research on teaching and learning* comprises empirical studies on various features of the particular learning setting. Research on students' perspectives including their pre-instructional conceptions and affective variables like interests, self-concepts and attitudes play a particular role in the process of educational reconstruction. But many more studies on teaching and learning processes and the particular role of instructional methods, experiments and other instructional tools are also available. Furthermore, research on teachers' views and conceptions of the science content and students' learning are an essential part.

Third, *development and evaluation of instruction* concerns the design of instructional materials, learning activities, and teaching and learning sequences. The design of learning supporting environments is at the heart of this component. The design is, first of all, structured by the specific needs and learning capabilities of the students to achieve the goals set. Various empirical methods are employed to evaluate the materials and activities designed, such as interviews with students and teachers, for example, on their views of the value of the designed items, questionnaires on the development of students' cognitive and affective variables, and also analyses of video-documented instructional practice. Development of instructional

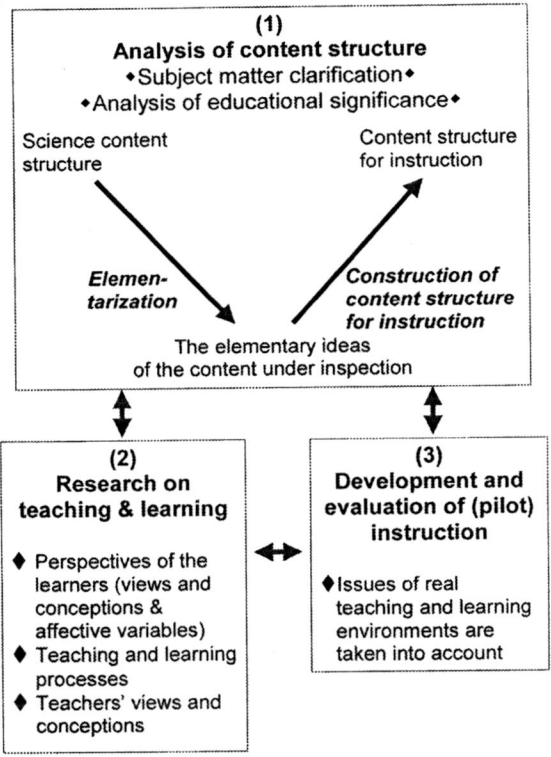

Figure 2. The Model of Educational Reconstruction

materials and activities as well as research on various issues of teaching and learning science are intimately linked.

Educational Reconstruction: A Frame for Science Education Research

The model as presented in Figure 2 covers major domains of science education research and development. The systemic nature of the model, that is, the intimate interplay of the components, points to the necessity that all the features displayed have to be taken into account in every research study. If studies on the effect of certain multi-media devices are carried out, for instance, considerations on the analysis of the content structure have to be included in some way. If student conceptual changes are investigated and new conceptual change approaches are developed and evaluated this work will only become part of normal practice if also normal teachers' readiness to use these approaches is investigated. The model also points to the rather complex and challenging features of science education research. Science education research is interdisciplinary in nature. For successfully carrying out research many competencies from various domains beyond science are needed: Philosophy and history of science views provide the background for a critical analysis of content structure; pedagogy and psychology (as well as other social sciences like linguistics) provide frameworks for empirical and normative studies. The model is, however, too limited to cover all domains of science education research. It is focused at research on the actual teaching and learning processes. These processes are embedded in a sequence of wider contexts, the contexts of the particular school and the society in general. Hence, research on curricular issues and science education policies has to be seen as an important additional domain.

CODA—VISIONS AND THE PRACTICE OF SCIENCE INSTRUCTION

Looking back on nearly 40 years of being a science education researcher it is most interesting for me that there appears to be a certain continuous development of ideas over the years. In the first years of my work many ideas were already there that later became more explicit and more elaborated. In a way in the early 1970s I was already a constructivist without knowing what constructivism is—simply because key ideas of constructivism were embedded within the German tradition outlined and in the work of other people we also studied, most significantly in the work of Jean Piaget, David Ausubel, and Thomas Kuhn. Interestingly, becoming a true constructivist in the 1980s und 1990s opened the eyes for a new interpretation of these "old" sources. Also ideas that we later made explicit in elaborating the model of educational reconstruction continuously developed. Reading existing papers on the necessity of distinguishing science content structure and content structure *for* instruction from the perspective of the model it is hard to believe that it took so long to summarize major views of science teaching and learning in the model. In retrospect it is also interesting that the model once stated gave my research interest and emphases a different direction. So far my research was focused on student learning—studies based on the model made it clear that a more holistic view is necessary, namely also to include views and preferences of the normal teachers out there into the research projects.

Final remarks concern visions of science instruction in school, universities and in informal settings on the one hand and visions of science education research on the other. In both cases there may be serious clashes between visions and the reality—of "normal" instructional practice and research practice. Still there is a large gap between research findings and their application in actual practice. In the 1990s intensive discussions on the "awful reputation of educational research" (Kaestle, 1993) resulted in the insight that research facilitating improvement of practice needs to be of an applied kind where research and development are close linked – like in approaches of *design research* (Cobb et al., 2003). It appears that the above model of educational reconstruction falls in this category. Nevertheless, research and development practice show how difficult it is to implement well-established research findings into practice. To actually facilitate improving practice, research should not be restricted to what works in some sort of laboratory settings (even if they are near to actual practice) but also should include studies on the major problems and deficits of normal instructional practice. However, the visions of good and efficient science education research may not meet research practice as multiple competencies in science and a substantially large set of other disciplines are necessary. Science education research that allows improving practice is complex—often too complex for a single researcher?

POST SCRIPTUM

As a response to the first draft of the present chapter Michael Roth commented that I write in a somewhat distanced voice, that the person of the author is very much in the background of the text. I agree. It appears, however, that this way of writing reveals much of my way of thinking about my work in science education. As a researcher I try to keep control of what I am doing by rational analysis as far as possible. Of course, that does not mean that affective issues are not involved. Research on and with humans, that is, teachers and students, is only possible if deep empathy of the particular person is included. In both cases, students and teachers, I want to get to know major features of their thinking—on the background of certain theoretical perspectives. I do not intend to reveal in which way their thinking and acting are deficient. I think that engagement in the students' alternative framework movement in the 1970s and 1980s was so attractive for me for this reason. The major concern of that movement was to dismantle students' ways of thinking, and not blaming them for their errors and mis-conceptions. Of course, research also needs to include a deep empathy for the particular science, in my case primarily physics. I am convinced that this domain provides essential figures of thought that are essential for the future citizen personally and for successful participation in society.

In the beginning of the 1990s Piet Lijnse wrote a note on research on teaching and learning science at the IPN. He characterized my way of thinking as being near the Anglo-Saxon pragmatic tradition as compared to the members of another group he called "lost in deep German thinking." I think Piet was partly right. There is a certain dose of pragmatism in my thinking about science education research. However, as outlined more fully above, my thinking is deeply rooted in the German tradition. There is a blend of the German and the Anglo-Saxon traditions that appears to be characteristic for me.

The German traditions of *Bildung* and *Didaktik* put the main emphasis on the person of the learner, on her or his personal development. I share this emphasis. For a considerable number of science educators the emphasis appears to be on science. I never professionally worked as a physicist. My master's thesis was not on physics but already on physics education. I think, that early focus on the education side led to my position of giving equal emphasis on the learner and on science (with a certain primacy of the learner) in my thinking about science education research and practice. I had to pay a price for this emphasis. A couple of times I failed in applying for chairs in physics education in Germany. It appears to me that a major reason was that the physicists who dominated the selection process preferred candidates being more on the science side of the spectrum. Interestingly, the series of disappointments affiliated with these failures in the end turned out to be a great advantage for me. I could continue my work as a researcher at the IPN with few teaching demands—being in my view still the best place to carry out science education research in Germany.

There have been, of course, a couple of further disappointments over the years. The most influential for me being rejection of articles submitted to journals, proposals of papers submitted to conferences and research proposals submitted to funding organizations. Further, there have been several cases of battles with educational scientists concerning aims and methods of empirical science education research. It has become clear several times that science education research sits between many chairs, that it is attacked on two fronts, by the scientists and by educational researchers. I think my somewhat rationally controlled and distanced view of the own work helped to deal with these kinds of disappointments. Also here I tried to understand the arguments and in many cases—however by far not in all cases—it turned out that the attacks were well grounded and that it was possible to improve the own point of view. In a nutshell, I think that my distanced way of thinking about science education research allows keeping emotions affiliated with the daily work under control and facilitates to turn disappointments into advantages.

Reinders Duit
Leibniz Institut für die Pädagogik der Naturwissenschaften
Kiel, Germany

PETER W. HEWSON

CONTINUITY AND CHANGE

From Physicist to Science Educator

I recognized that my career path would be that of a science educator in 1977 when I applied for my first sabbatical leave. I had started out my professional life as a physicist, but in reflecting on what I wanted to accomplish in a sabbatical, I realized that I was more interested in how students learned physics than in exploring the frontiers of physics itself. How did it happen that I moved away from the career path of a physicist? What is the legacy from my physics foundations that I have carried with me on this journey of transformation? What are the influences that drew my attention to and raised my interest in science education? What nourished my professional development in, and sustained the evolution of my interests in, and my activities related to science education? What contributed to the significantly different direction my career has taken?

PHYSICS FOUNDATIONS

Physics fascinated me for two reasons. On the one hand, it explicitly deals with the natural world, by focusing on simple systems that yield to detailed, exact analysis. On the other hand, it uses the language of mathematics extensively. I was good at mathematics and I found a great deal of satisfaction in deriving clean, exact solutions of problems. Geometric proof was challenging but rewarding, and calculus was a revelation to me. I knew, of course, that the real world was messy, but making approximations, discarding small quantities, dealing with large numbers of particles were less appealing to me. I developed a view of the world that stressed the importance of deep, unique, exact, confined, causal accounts of natural phenomena. An important part of this physics vision is the notion of an isolated system, in which all the important interactions happen within its confines rather than across its boundaries.

I started my university education at the University of Cape Town in my native South Africa, where I studied theoretical physics and applied mathematics. I then went to Oxford University (England) to study theoretical physics for three years, and completed my doctoral thesis entitled "On the interactions of $\neq$-mesons with nuclei." This was followed by two-and-one-half years as a postdoctoral fellow in theoretical physics at the University of British Columbia (Vancouver, Canada), the site of the Tri-Universities Meson Facility (TRIUMF) an obvious fit with my doctoral studies. I returned to South Africa in July 1971 to take up an appointment in

K. Tobin, W.-M. Roth (Eds.), The Culture of Science Education, 121–131.

the Department of Physics at the University of the Witwatersrand (colloquially known as *Wits*) in Johannesburg.

It is perhaps no surprise that my doctoral study focused on the implications of a model of a confined system—the interaction between a $\neq$-meson and the nucleus of an atom. While the exploration of the model's implications required some fairly sophisticated mathematics, the model itself was quite simple. In retrospect, I was engaged in Kuhnian normal science. Courses on scientific methods or the philosophy of science were not part of the core curriculum, and I simply immersed myself in doing physics. As a result, my notions about the nature of physics developed in quite an intuitive fashion. Yet they are important, because they played out implicitly in much of the early work that I did in science education, and continue to have appeal to this day, albeit now as part of a much broader array of epistemological ideas.

SCIENCE EDUCATION AWARENESS AND INITIAL INTEREST

I had no inkling that my future career would be in science education when I arrived in Vancouver in October 1968, to take up a postdoctoral fellowship in theoretical physics at the University of British Columbia (UBC). My mentor, Erich Vogt, and I soon agreed upon a new project and I settled into my new surroundings with all the excitement and trepidation that accompanies such a golden opportunity. I was on a straight path to becoming a physicist at a research institution or university. But several events during my first year at UBC took me in a different direction. I married Mariana Thomas in July 1969, Walter Westphal, and Walter Boldt set up a collaborative project between physics and science education called the Physics Education Evaluation Project (PEEP), and I took on the responsibility of teaching a tutorial section of a freshman physics class.

Mariana grew up and was educated in South Africa. She completed her diploma as a science teacher at Oxford, where we met. She taught high school science in London, she was a research assistant in a zoology laboratory in Oxford, and for the eighteen months prior to our marriage, she taught science education at the University of Botswana, Lesotho, and Swaziland in southern Africa. After her arrival in Vancouver, she was admitted to a master's program in science education at UBC, with Walter Boldt as her advisor. As a result, references to Jean Piaget, Thomas Kuhn, and others began to find their way into our dinner conversations, along with positrons and interactive mainframe computing.

Walter Boldt completed his doctorate at the University of Illinois, under the supervision of Jack Easley. Walter Westphal was a visiting professor from Germany in the UBC physics department with research interests in hydrodynamics. Their meeting at a party led to many productive conversations, as a result of which, they set up PEEP in which Westphal taught a non-calculus based introductory physics course, and Boldt supervised a team of science education graduate students who looked at the educational aspects of all components of this course: the large lectures to some 500 students, the associated laboratories and tutorials, and the assessment system. In structuring the project, Boldt drew on Flanders' interaction analysis, Bloom's taxonomy of educational objectives, along with the work of his Illinois mentors, including B. O. Smith's classroom and curriculum frameworks and Robert Stake's evaluation model. I was initially involved with the project in two ways. My tutorial section was one of several included in the interaction analysis study, and Mariana interviewed several of my tutorial students to study their cognitive development from a Piagetian perspective. Initially I approached all of this in a dilettantish fashion, as light relief from the serious tasks of physics, but in time, through social interactions with PEEP project members, and my attending project meetings, my interest in a broad array of educational ideas began to evolve.

I had little idea that my new awareness of, and interest in, science education would be anything other than a sideline. I was expecting to return to a position in physics research at the South African Council for Scientific and Industrial Research (CSIR) in Pretoria. A CSIR bursary that carried a service commitment had funded my undergraduate degree. As the end of our time at UBC approached, I contacted the CSIR about fulfilling my commitment. What I heard back was a surprise—since there wasn't a position for me, I would be released from my commitment, provided I worked in South Africa for the next six years. I sent out resumes to South African universities, and received the briefest of notes from the head of physics at Wits, to the effect that there was "quite a possibility of a position." I accepted a three-year temporary position.

LEARNING TO TEACH PHYSICS

On arriving at Wits in July 1971, I was immediately immersed in, and consumed by, the tasks of teaching physics to a class of some 200 first-year medical students. There was a well-established system in place: I joined a team of lecturers who, over the years, had created an extensive, working curriculum of lectures, tutorials, and laboratories. My colleagues, who were teaching parallel sections of the course, were willing to share their teaching materials with me. Our conversations were always about the mechanics of these courses: improving tutorials, introducing new experiments and equipment into laboratories, writing better examination questions, and so forth. It did not cross my mind to ask questions about the system itself—it was what I had experienced as an undergraduate, and I had no grounds for dissatisfaction. I saw my primary task to be writing lecture notes and giving lectures. While I could have taught from my colleagues' notes, I found in those first few years that I needed to write my own. After all, if I did not understand the logic of what I was doing, how could I do the job that was expected of me? The corollary was that if I presented the material in an ordered, systematic, reasoned fashion, my students would understand it as I did. Even though it took me the best part of a year to feel comfortable giving large lectures, I was easily socialized into learning to teach an introductory, undergraduate physics course. Most of my energy went into the preparation of lecture notes, and I grew to experience an important educational reality: teaching a topic is an excellent way to learn it.

IMMERSION IN TEACHER EDUCATION

About the time I arrived in Johannesburg, a series of events was playing out that had a profound influence on my career. The South African Parliament passed legislation requiring all white secondary teachers to be educated at the university level. While many other South African universities had already established teacher education programs, this was not the case at Wits. There was a good relationship with the nearby Johannesburg College of Education (JCE): prospective secondary teachers completed their degrees at Wits, and went to JCE for their teaching credentials. The new legislation required that all this had to change, with the establishment of a Faculty of Education on the Wits campus. The first dean was a geographer, and with the support of other subject departments, he decided that methods instructors would be located in subject departments across campus. The alternative choice would have been in the department of education, whose interests at the time were largely focused on the academic study of education—philosophy, psychology, sociology, and history of education—with little emphasis on the practice of teacher education. This led, within the first year of my appointment, to a lecturer's position being advertised in the physics department with responsibility for teaching methodology to prospective physics teachers. From my perspective as a temporary lecturer, it was an obvious application to make. The selection committee offered me the position, presumably because they thought that a sound content foundation in physics more than compensated for my lack of qualification and experience as a high school teacher. Recognizing my lack of school experience, the committee gave me nine weeks to spend in local high schools, and trusted that I would be able to learn the rest on the job.

There were different contributors to my professional development in science teacher education: I was a member of a community involved in setting up a new teacher education program, I networked with local teachers, and I obtained my high school teaching diploma. Most importantly, at home, Mariana was a constant source of ideas, support, and reassurance.

The community that designed and implemented the new Wits teacher education program was a large group of new appointees, not only in the education department with interests in teacher education, but also in subject departments around the campus as methods instructors. We were drawn from high schools, colleges of education, and university departments. There were professional development advantages for me in being a member of this community, rooted in the challenges and excitement of being involved in a new program. We were all committed to making a success of the new enterprise, and met regularly to design, reflect on, and improve the program. There was a great deal of pooling of ideas, and working collaboratively, and paying attention to the relationship between the goals of the program and its specific courses and activities. I learned a lot from this community, not least of which was the importance of respecting the very different perspectives and experiences of its members. It took longer for me to become introspective of my own expertise as a physicist, and recognize both my contributions to the discourse of the community, and my shortcomings in understanding education.

In retrospect, the program that emerged had highly innovative features. These were a direct product of a diverse community committed to an integrated experience for its students. For example, we contextualized educational theories, policies, and practices by creating Witsbridge, a fictionalized high school, within which we developed a series of case studies that exhibited the contradictions, dilemmas, and contrasting priorities of practicing schools.

I networked with local science teachers. I had worked closely with several in my initial nine weeks. I met many others as I went around the schools observing students doing their teaching practice. I attended regular teachers' meetings at the Johannesburg College of Education. To be sure, there were grumbles that the university had seen fit to come down so heavily on the side of content expertise at the expense of school experience, but those lessened over time.

After several years I enrolled as a student in the Wits teacher education program, not only because of the legitimacy that holding a teaching diploma would provide, but also because of the opportunity it would provide me for the systematic study of education. It was a fascinating year in a number of ways. Intellectually I was introduced to the philosophy of knowledge and education through the writings of Hamlyn and Peters: exploring the meaning of learning and teaching was a revelation to me. It was not only the ideas; it also introduced me to a type of argument that I had not previously experienced as a physicist. It was also a year of confusing identities. In some courses such as physics, I taught prospective physics teachers. In others such as the philosophy of education, I was a fellow student with them. In others such as the methods course in science I not only taught it, but also had to get credit for the course. I soon found that learning to teach the methodology of teaching physics was a different matter from learning to teach physics. This was so for several reasons, some rooted in the Wits context of the time, while others were more general. Since the university was setting up a new program, there was no existing curriculum for me to draw on, a factor compounded by my lack of background knowledge and experience. I also became aware from the literature that curriculum in this area was much more contested than in introductory physics.

BECOMING A SCIENCE EDUCATOR AND CONCEPTUAL CHANGE

When the time came to plan my first sabbatical, I realized that my interest in becoming a physics educator had superseded my sense of being a physicist. The question I was interested in pursuing was how it was possible for students to come to understand a structured, logical, and hierarchical domain such as physics. On the one hand, my experience was that my carefully constructed physics lectures that were so filled with meaning for me, did not necessarily make sense to the medical students I was teaching. On the other hand, general prescriptions such as "Teach from the known to the unknown, from the simple to the complex" provided little guidance to me. After my early introduction to Jean Piaget, I was reading the work of David Ausubel and I found his focus on meaningful learning to be an appealing idea. The pragmatic question was where to spend my sabbatical. I wrote to several people but without success. Finally I decided to write an open letter to the National

Association for Research in Science Teaching (NARST), outlining my interests. I had one reply, and it was life changing: Joe Novak invited me to spend my sabbatical at Cornell University.

Joe Novak facilitated the opening of many doors for both Mariana and me when we arrived in Ithaca in January 1978. He invited me to sit in on his graduate classes. He had contacts with the physics department that led to my teaching a section of a freshman course. He became an informal supervisor for Mariana who two years previously had started a doctoral program in science education at Wits. He suggested I talk to Ken Strike and George Posner. Ken and George used the history and philosophy of science to formulate the conditions under which disciplinary conceptual change happened, and wondered what parallels there might be with student learning in science. This suggested that we look at how students dealt with a prototypical conceptual change in science. We settled on the transition from Newtonian mechanics to Einstein's special relativity, because we assumed that the conclusions of special relativity, that is, moving clocks run slower, would be obviously counterintuitive for students, and that accepting these ideas would require students to undergo conceptual change. There was a unit on special relativity in the freshman course I was teaching, and I interviewed a number of students in my tutorial section about their understanding of, and reactions, to relativistic principles and results. These interviews provided a range of instances that illuminated how these students struggled to find meaning as they sought to understand and interpret relativity's counterintuitive outcomes. Epistemological commitments and metaphysical beliefs were embedded in their answers, and demonstrated that the analogy between disciplinary change and student learning was both plausible and fruitful. The results of these interviews thus influenced the final formulation of the conceptual change model of learning as a rational activity that happened within a conceptual ecology of quite different kinds of knowledge.

It was a heady time to be in an environment that was awash with ideas about cognitive learning, information processing, buggy algebra, constructivism, and attention to subject matter knowledge. Mariana and I resonated with the focus on cognition and qualitative methodologies that seemed a welcome alternative to behaviorism and quantitative research styles. We attended AERA (American Educational Research Association) and NARST meetings for the first time. While initially overwhelming, the opportunity to listen to educational leaders, and to meet others with like-minded interests, not only from North America but also from Europe and Australia, was particularly motivating. The slow development of a consciousness that we were members of an international community was empowering.

The primary article that came from my collaboration with George Posner, Ken Strike, and Bill Gertzog was "Accommodation of a Scientific Conception: Towards a Theory of Conceptual Change" (Posner, Strike, Hewson, & Gertzog, 1982). It has been, and remains to this day, a seminal article in the field of science education and beyond. It is by far the most extensively cited of all my publications. Interestingly, it took a while to catch on. It was only published three years after we had completed it, and its citation rate during the first decade after its publication was only half of what it has been over the last decade. It has been cited in more than thirty journals, in fields ranging from science education, through educational psychology and the sciences, to medicine, nursing, and the humanities. Its focus on change as an individual, rational process has, quite appropriately, been criticized as only part of the story. Yet the advances in our understanding of the motivational and attitudinal, and the social and cultural aspects of learning have served to complement, rather than supersede, the central ideas of the conceptual change model. These have certainly been generative of other lines of inquiry, with implications for teaching, teacher education, and multicultural education, to name a few.

It was not surprising that, after a sabbatical that stretched out more than 18 months, we did not relish the thought of returning to a South Africa that seemed isolated, both intellectually and politically. Back at Wits, I came up against the difficulty of putting into practice ideas that seemed exciting and self-evident to me, but that others found both controversial and unrealistic. Yet there were opportunities to grow in South Africa. I used conceptual change ideas to write computer programs that diagnosed and addressed alternative conceptions of motion. When prospective science teachers in my methods classes rejected notions of focusing on students' ideas, the importance of considering how teachers learn slowly began to crystallize. Mariana completed her dissertation on alternative conceptions and conceptual change-influenced teaching approaches with African students. She joined a multi-disciplinary group of social science researchers at the National Institute of Personnel Research, where she was able to pursue topics such as culturally influenced conceptions of heat held. Together we started an informal seminar that focused on students' ideas in science. And we co-authored several articles, one arising from her doctoral study, another on conceptual conflict, and a third exploring the implications of conceptual change for teaching and teacher education.

The international community was also a critical part of this phase of our professional lives. Mariana and I had opportunities to travel overseas to attend conferences and present papers. The most noteworthy year was 1983. At the AERA meeting, an international group of researchers in science education proposed the establishment of the Cognitive Structure and Conceptual Change SIG. I attended a three-week summer workshop at La Londe in southern France where I was able to meet and talk with physics educators from Europe and around the world. Most significantly, I was offered a faculty position in science education in the department of Curriculum and Instruction at the University of Wisconsin-Madison.

COMING TO THE UNITED STATES: TEACHING FOR CONCEPTUAL CHANGE
AND TEACHER EDUCATION

Coming to Madison in the depths of a Wisconsin winter in January 1985 was a time of change for all of us. While it was the third time we had come to North America, we were now landed immigrants rather than visitors. While I retained the teacher education component of my job, moving out of a physics department meant giving up undergraduate physics teaching and taking on the teaching of graduate courses in science education. It was particularly hard on Mariana who left a good research position within a close, supportive community for an uncertain, soft money world in an unfamiliar institution. We did continue with the collaborative work we were doing in South Africa in a project that conceptualized teachers' conceptions of teaching science. The process of introducing students' ideas, learning as conceptual change, and teaching for conceptual change into initial science teacher education coursework, drew our attention to the significant differences between current teaching practices and those advocated in teaching for conceptual change. This in turn pointed to the importance of considering what it is that teachers need to know (i.e., their conceptions of teaching science) and be able to do in order to facilitate students' conceptual change learning. This led to a series of co-authored articles, the most notable of which was "An Appropriate Conception of Teaching Science: A View from Studies of Science Learning," recognized as the Outstanding Paper of 1988 published in *Science Education* (Hewson & Hewson, 1988). This line of inquiry in teachers' conceptions generated further studies for both of us, albeit in different directions. Mariana found a place in a project investigating faculty development in medical education, and I embarked on studies of high school science teachers and science teacher education as a result of collaborations with my faculty colleagues.

Recognizing the similarities between students learning science and teachers learning to teach science suggested the need to identify the prior knowledge that teachers bring with them to the classroom. This led to an understanding of the role that teachers' prior knowledge plays in their understanding of what it means to teach, and to a recognition of the difficulties involved in constructing and if necessary reconstructing their conceptions of teaching science within the context of their classroom practice. These ideas were the foundation of two collaborative studies I conducted with faculty colleagues. The first of these studies (with Bob Hollon) looked in depth at the practice of experienced science teachers to understand the relations between their thinking and their instruction. The second (with Ken Zeichner and Bob Tabachnick) investigated the use of action research, a significant form of reflective practice, as a means of facilitating prospective science teachers learning to teach for conceptual change in the context of elementary and secondary science teacher education programs. This second study culminated in a set of articles, published in a single issue of *Science Education*, the first being entitled "Educating prospective teachers of biology: Introduction and research methods" (Hewson et al., 1999). The paper set documents the influence of course- and field-work components of the programs on different students as they progressed from methods courses to student teaching. It also includes invited commentaries on the study's significance in providing a far-reaching description of the complex interactions of various components of a teacher education program in learning how to teach.

The advantage of my moving to a department with an established graduate program in science education was the opportunity it provided me to work with graduate students. In particular, this facilitated further development of the conceptual change model and its implications for classroom teaching. Richard Thorley explored ways in which the concept of status, encapsulating the conditions of conceptual change, could play a larger, more explicit role in classrooms. Sister Gertrude

Hennessey integrated status language into her elementary classroom and demonstrated the powerful metacognitive role that this played in student learning. Mike Beeth immersed himself in Sister Gertrude's classroom to understand the nature of the learning environment she created to support students' conceptual explorations. John Lemberger demonstrated that in a classroom designed from a different perspective—that of modeling—status determinations played an essential role in students' successful problem solving. Hyun-ju Park showed that the components of students' conceptual ecologies that they brought to their understanding of natural phenomena were broad and extensive. One example of this line of work is "Status as the hallmark of conceptual learning" (Hewson & Lemberger, 2000), in which we argued that deep conceptual learning, whether or not it involves major restructuring of students' fundamental beliefs, is characterized by consideration of the conceptual status of focus concepts and the ideas and beliefs in which they are grounded.

We recognized from the outset that the conceptual change model of learning, in common with all theories of learning, has implications for teaching, and pursued these in a parallel agenda. Studies rooted in conceptual change learning informed the conceptualization of what it means to teach for conceptual change. Drawing from a wide array of studies carried out in many different countries, we formulated guidelines that characterize teaching for conceptual change. These were published in "Teaching for conceptual change" (Hewson, Beeth, & Thorley, 1998), and focus on teachers providing explicit opportunities for students to express their ideas, to be metacognitive, to examine the status of their own and others' ideas, and to provide the justifications that underlie their thinking. While clearly a subset of all that is involved, these continue to be generative for graduate students exploring not only course design and analysis, but also the preparation of teachers to teach in culturally diverse schools.

INTERNATIONAL COLLABORATION

In 1994 Mariana accepted a position at the Cleveland Clinic Foundation in Ohio as Director of Faculty Development. For me, the Ohio connection opened another unexpected door when Jane Butler Kahle invited me to the Miami University of Ohio to work for a week with a group of South African researchers. This was an activity resulting from conversations between the NSF (National Science Foundation) in the United States and the NRF (National Research Foundation) in South Africa. This led to my directing a project to develop graduate and post graduate research capacity through collaborative links between South Africa and the United States. I found the project to be a natural fit with my expertise and a reconnection with my origins. South Africa's science education research community is smaller, of more recent origin, and commands fewer resources than that of the United States. This does not, however, mean a one-way flow of ideas and expertise: both partners benefited. While a primary purpose of the collaboration was to support recently established South African doctoral programs in science (and mathematics) education and their doctoral students, the major outcome of the collaboration was the conceptualization and establishment of an annual research school in South Africa in 2003, with structures and purposes that could inform more established programs. These activities were outlined in "Building education research capacity: Collaboration between the United States and South Africa" (Hewson, Curtis, Schneckloth, & Damonse, 2005). In retrospect, this project held some surprises for me in terms of the complexity of negotiating projects that respected the needs, resources, and contexts of different communities, the enjoyment of working with enthusiastic, capable people on difficult problems, and the unexpected structures, such as the research school, that emerged.

REFLECTIONS

Growing up and working as a white South African in the *apartheid* era was a complex, often contradictory, experience. Consider these instances:

- I had access to a privileged system of primary, secondary, and tertiary education that allowed me to compete for a place at Oxford University for my doctoral studies.
- My early studies in theoretical physics and applied mathematics were a step removed from the discourse surrounding *apartheid* policy and practice that were more closely aligned with social, cultural, religious, economic, and political perspectives.
- After returning to South Africa in 1971, the influence of *apartheid* educational policies was more apparent. These imposed restrictions on who the universities could admit as students, or appoint as lecturers; as a result, the Wits teacher education programs largely catered to white schools.
- I was the beneficiary of a countrywide university system that confirmed me after three years in my position as a physics methods lecturer. Had there been a North American-style tenure system in place, the pressure to publish would have severely restricted the space I needed to explore a different career direction.
- In the early 1980s the worldwide growth in research in students' alternative conceptions led to suggestions that instruction should be geared to students' knowledge, beliefs, interests, etc. The ever-present danger in South Africa was that these suggestions could be co-opted as a justification of *apartheid* education.

In other words, the influences of the South African societal and political context on my professional career vary considerably from the immediate to the distant, and from the positive to the negative. In South Africa of the 1970s and 1980s the contrasts were particularly stark, and the edges of the divide quite jagged. Yet this facilitated clarity of vision of what was fundamentally important in education, and the realization that this vision was not confined to South Africa. Thus it also produced a corresponding ability to see the ways in which prior knowledge, and the social and cultural contexts that produced it operated in more nuanced, less confrontational settings elsewhere in the world.

On various occasions, events that I did not plan or anticipate have exerted a significant influence on my professional career. I was not involved in the establishment of the PEEP project at UBC or the creation of a methods position in physics at Wits in Johannesburg. I did not expect to participate in a conceptual change project with George Posner and Ken Strike when I accepted Joe Novak's invitation to spend a sabbatical at Cornell. Jane Kahle's invitation to a weeklong seminar came out of the blue. One way of characterizing these events is that I have been fortunate to be in the right place at the right time. Doors open and doors close, and if I am not there when they are open, I miss the opportunity of going through them. No doubt there have been many doors that were not open when I was around, or were open and I did not appreciate what opportunities they presented. The point here, however, is not to grieve over lost opportunities, but to recognize that there's more than simply being in the right place at the right time. In each case I responded to circumstances that resonated with my personal and professional interests and background. Unexpected events took place; they become serendipitous when I chose to get involved with them.

Another way of characterizing my professional career is in terms of the different communities with which I have been associated. In each of the instances that I have provided, there have been others with whom I have interacted, from whom I have learned, and without whom the outcomes of which I have spoken would not have been possible. The most important person has been Mariana. On multiple occasions, our professional interests have intertwined in productive and influential ways. Other communities have been of shorter duration, but important nevertheless. At one level, this is such an obvious conclusion that it hardly seems to be worth commenting on. At another level, however, my awareness of the significance of these various communities has increased considerably over the years to the point that I could not imagine writing this account without referencing them. There is, I suggest, a connection to the development of the epistemological frameworks of my work. My physics background gave me a view of the world that stressed the importance of unique, exact, confined, causal accounts of natural phenomena. Initially I did not see that this needed to be adjusted when I stopped focusing on atoms and nuclei, and turned my attention to a student learning physics. While I knew that there was a context in which people learned, the notion of an isolated system was a conceptual device for effectively ignoring it. In time, however, the limitations of trying to identify unique causes became apparent, and I came to appreciate the complexity of factors that influence learning, and their interactional nature. While there clearly is an individual, cognitive character to our learning, we are also social, cultural, affective creatures who respond to those around us in a variety of ways that are strongly influential of the questions we ask, the opinions we espouse, and the understandings we create. In other words, while this is not a paragraph I would have written thirty years ago, I can now affirm how essential community has been to my personal growth.

Peter Hewson
Department of Curriculum & Instruction
University of Wisconsin-Madison
University of Toronto

131

PENNY J. GILMER

AS A WOMAN BECOMING A CHEMIST, A BIOCHEMIST, AND A SCIENCE EDUCATOR

Writing about one's life in an autobiographical way can be a powerful way to learn and reflect on one's life and professional practice. I have read many autobiographies by scientists and science educators. I decided to write this chapter using an autobiographical point of view to address the principal elements in my classroom—collaboration and technology. I chose to do this to help me understand why collaboration and technology were so important to me as the two foci of my action research. I got the courage to write this autobiographical chapter after reading a science education autobiography by Tobin (2000). Here I compare two dominant themes, collaboration and technology, from within my biochemistry classroom, with the same themes in my own professional life. I focus on why collaboration and technology are important themes for me.

Reading Bruffee (1993) just before teaching the biochemistry course in a study of my classroom (Gilmer, 2004) greatly influenced my thinking. It was a revolutionary time for me. My life as a scientist had the authority of knowledge as its mantra, yet Bruffee's book challenged my beliefs. It made me realize that knowledge is tentative. "Knowledge is what is said—or perhaps what can be said—in some language, by members of some community, to other members of that community" (Bruffee, 1993, p. 142). I realize we construct our knowledge and come to agree as a community of scientists that a certain construction is our best understanding at a certain point in time. However, what I needed to accept is that these understandings are only constructions, not the absolute truth, or Truth. Bruffee emphasizes it is critical for students planning to become part of a specific community of learners to learn "the language in which community members construct the knowledge that is their common property. . . . The job of college and university teachers is to represent the knowledge communities of which they are members in a way that will most effectively re-acculturate potential new members" (p. 3).

Reading Bruffee encouraged me to attempt to develop a classroom in which the primary interactions would be among my students, with me being the "guide by the side," instead of the "sage on the stage." I wanted my students to learn primarily from each other rather than from me. I wanted them to learn that they could seek out knowledge, construct their ideas of what it means while connecting to their prior learning, and share those ideas with their peers and on the World Wide Web. I implemented such a design in a fifteen-week biochemistry course. I required the students to work in small collaborative groups. Each group developed ten web sites on a variety of topics of biochemistry. Through these activities, many of my stu-

K. Tobin, W.-M. Roth (Eds.), The Culture of Science Education, 133–145.

dents (but not all) learned what it means to work and learn collaboratively while using technology. In the process I have learned about learning.

COLLABORATING AS A SCIENTIST AND SCIENCE EDUCATOR

I reflect on how collaboration with other scientists and science educators influenced my own ideas on collaboration. As I reflect it helps me understand my own viewpoint, which, in turn, helps me to understand—and sometimes change—how I portray those ideas to my own students. I address critically my assumptions on collaboration.

Collaborating as an Undergraduate in Science

When I was a freshman in high school, there was a window of opportunity for me to get into science, with the Soviet satellite, Sputnik, circling our globe starting on October 4, 1957. Suddenly, our country realized that we were technologically and scientifically behind the Soviet Union in mathematics and science. Our country needed more young people (like me) to study science. Although I was female, there was a new opportunity for me to enter into the sciences because my country needed me.

At my women's college we had just five chemistry majors (just 1 percent of the students in my class), and each of us worked individually to make sense of chemistry. I always solved my own problem sets and wrote my own laboratory reports. Even in laboratory each of us worked individually. Since I am the sort of learner who can thrive in this sort of culture, this approach felt comfortable to me at the time. One of my assumptions for many years was that learning on one's own was how everyone learned best, because it happened to work for me. However, over time I realized this sort of competitive culture just happened to fit my learning style and that there are many other types of learners with their own preferences for learning. There may have been other people desiring to be chemistry majors who did not do so because they needed different tools, a community of learners, different rules, or a different division of labor (see Figure 1 through the lens of cultural-historical activity theory [Engeström & Miettinen, 1999]), which may have facilitated that process, thereby helping the other types of learners reach their objects and outcomes. Table 1 shows the resources that added coherence to my moving toward my objects and outcomes; in addition, situations that provided contradictions are listed as well.

At that time I did not realize that I had what Sewell (1992) calls "agency" (the power to act) to change the structure in which I was embedded. In my career I have felt that I have somewhat changed the structure within my department and in the profession of chemistry, by being a woman with different ideas than most of my colleagues in the male-dominated field of chemistry. However, had I understood activity theory and Sewell's theory of structure at that time, I may have acted differently to empower myself, thereby changing the structure more purposefully. One thing I have learned is that agency and structure are in dialectic tension, with

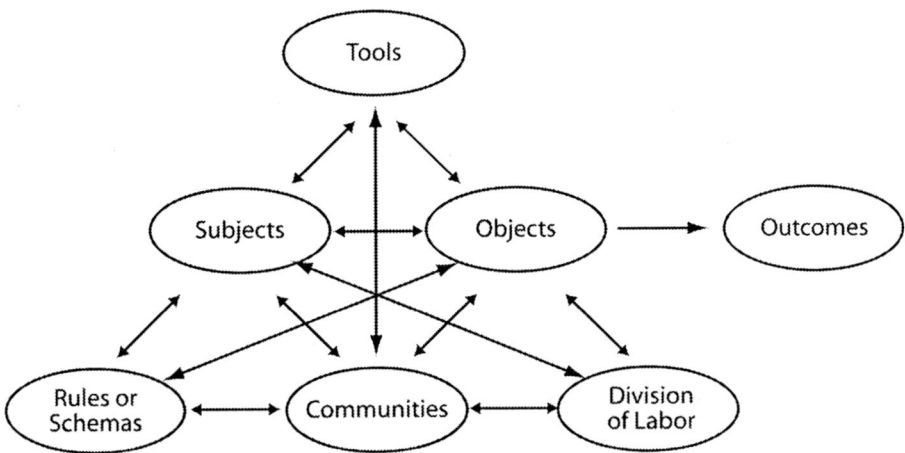

Figure 1: Cultural-historical activity theory diagram, which one can apply to any human activity system. One examines what adds coherence or contradictions to the subject moving toward his/her objects and on to outcomes.

each influencing the other. I have the power to change the structure but the structure also influences me and what actions I take.

Collaborating as a Graduate Student in Science

My first experience with professional collaboration was when I was in graduate school at the University of California, Berkeley, starting in 1968. My graduate research involved collaboration among several scientists (see "rules" in Table 1). In a sense I was the one who brought together the research of my two major professors. One of them, Esmond E. Snell, knew how to purify the transaminase enzyme and had done much of the pioneering research in vitamin B6-containing enzymes, and the other, Jack Kirsch, knew how to study mechanisms of enzyme-catalyzed chemical reactions using fast reaction kinetics.

I contributed to the collaboration through my knowledge of vitamin B6 chemistry (from my master's thesis at Bryn Mawr College) and from my perseverance to solve a problem, even when Jack Kirsch was on sabbatical in Germany for the year in which I collected most of my experimental data. In fact, that opportunity gave me confidence that I could solve my own scientific research problems—in those days it took two weeks to get a response from him—one week for my airmail letter to reach him and another week for his response to come back to me. I solved each problem by the time I heard back from him.

I do enjoy working at the interfaces between scientific fields. The interfaces in science seem like goldmines for research because one needs expertise in at least two areas to engage in the research. When I need to learn more, I have access to faculty members, graduate students and staff working in different fields (see

Table 1: Interactions of self in collaboration with scientists. Key features identified in the cultural-historical activity theory diagram (Figure 1).

Primary movement		
Subject	*Objects*	*Outcomes*
Self	Further scientific or educational research agendas Learn from others Work at interfaces of research areas Get research funding Publish findings	Become a professor of chemistry and biochemistry Do what I love to do (i.e., teach and do research) and be paid to do it
Coherences and/or contradictions		
Tools	*Rules or schemas*	
Scientific equipment Computers Chemicals and biochemicals	Encouraged to collaborate as a graduate student and postdoctoral fellow Discouraged from collaboration as an assistant professor (need to prove myself as independent) Encouraged to collaborate again as associate professor and professor	
Communities	*Division of labor*	
As graduate student and postdoctoral fellow: other students and faculty As associate professor and professor: other faculty and students from own institution and elsewhere	Difference in roles between faculty, staff, graduate students, and postdoctoral fellows	

"communities" and "division of labor" in Table 1). This is when collaboration can really make a difference. People with different expertise who have an interest in interdisciplinary research may decide to work together, to learn each other's language and ways of thinking, thereby developing the tools to move forward toward a new frontier. How well the collaboration works depends on the individuals (the "subjects"), the culture of the institution and the discipline (the "rules or schemas"), the scientific equipment and computers (the "tools") available to them, the division of responsibilities in the research (the "division of labor"), and the scientific researchers, technical staff, and graduate students (the "community") within the research enterprise.

Collaborating as a Postdoctoral Fellow in Science

Collaboration became my mantra by the time I finished my doctorate in biochemistry. For both of my postdoctoral fellowships I worked with two faculty members on each project at Stanford University. I worked at the interfaces between a traditional medical school discipline and chemistry. In the first project by working collaboratively with a physiologist, Peter Ramwell, and a physical chemist, Harden McConnell, I was able to apply biophysical techniques to study the interaction between a class of small but powerful bioregulator molecules called prostaglandins and the human red blood cell membrane. How I studied this interaction was very creative—I figured out how, through using the biconcave, discoid shape of the human red blood cell, to get the cells to orient as they passed through a small chamber between two very narrow parallel quartz plates in the magnetic field. By spin labeling the membrane lipids, I could monitor the motion of a membrane-embedded spin label, which reflected the deformability of the red blood cells in the presence and absence of prostaglandins.

I stayed for a second postdoctoral fellowship at Stanford University with Harden McConnell and an immunologist, Hugh McDevitt. I learned how to isolate the outer plasma membrane from leukemic T target cells using biophysical techniques to separate the proteins from the other cellular membranes. I used the purified plasma membranes to block the recognition (by competition) between a primed immune cytotoxic T cell and its specific tumor target cell. Using physical methods, I identified a number of the protein species on the surface of the tumor cell. However, there were both advantages and disadvantages to working within this collaborative environment. This research would have been much harder to do without the expertise available to me through collaboration. I would not have known how to address immunological recognition without the expertise of the immunologist and his research group, nor would I have had the expertise at that point in my career to know how to conduct the biophysical separations of various intracellular membranes and the plasma membrane without the advice of the physical chemist and his group. Therefore, the collaboration was critically important for success in the research.

Although I was able to contribute my understanding to important research questions that would have been unavailable to me without the collaboration, it was difficult at times to work within the culture of both a medical school and a chemistry department. The medical school culture was competitive with about a dozen graduate students and postdoctoral associates working within two small research laboratories. In our research group meetings in the medical school I felt we were pitted one against another, so instead of encouraging collaboration in the immunologist's group, it fostered competition. At one point my immunologist mentor told me to sweep some research results under the rug because he did not want to incorporate such results into his thinking. I remember feeling disillusioned with science in the medical school culture by his statement (and still do). Fortunately, it was different in the chemistry research laboratory. My chemistry mentor would never have said that comment about sweeping results under the rug. In fact, I remember my chemistry mentor saying that we were trying to find the Truth, not the truth based on pre-conceived notions. Still I bridged the two worlds of chemistry and the medical school that were physically just across the street from each other, but that were in many ways, so far apart culturally.

Collaborating as a Science Faculty Member

I have been a faculty member in the Department of Chemistry and Biochemistry, at Florida State University (FSU) for twenty-nine years.

Discouraged from Collaborating as an Assistant Professor

During my tenure-earning years, I was discouraged from collaborating with others in research. Their rationale was when it came time for my departmental and other university colleagues to vote on my promotion and tenure, it would be easier for them to know what I had accomplished in research if I were the only senior author. Also getting funding for my research through peer review was critical for demonstrating that the scientific community valued my ideas and methodologies.

My research program at FSU was a study of the immunochemistry of cell surface molecules embedded in mammalian cell membranes. At that time no one else on the faculty in my department or elsewhere in the university studied anything related to immunochemistry. There were a few biologists who studied membranes in living organisms, but none of them thought like a chemist. Therefore, I felt isolated. This feeling was partly due to the isolation of my research interests and also because I was the first tenured or tenure-earning woman in my department to conduct research, teaching and service like my all-male (about 25) colleagues. However, in my second year a new immunologist arrived as an assistant professor in the biological science department. His and my research groups met jointly for journal club once a week for about four years.

I did try to collaborate with another biological science colleague and his staff on the characterization of some mammalian membranes using spin labeling, a method I had learned during my postdoctoral fellowship. I taught the graduate student how to spin label the membrane samples and how to analyze the complex, slow motion of the spin label inserted in the membranes. However, when the graduate student wrote the manuscript on the research for publication, all that the major professor would agree to do was to thank me in a line in the acknowledgments for my technical assistance. I was furious for his not including me as a coauthor, and I told him to remove my name from the acknowledgments. I cannot imagine how he would think an assistant professor would spend precious pre-tenure time and effort, contributing only technical assistance, which certainly does not count towards promotion and tenure. I am sure that he would not have done this had I been male. The issue of authorship is a critical one in that, when rightly given, it confirms your intellectual property, but when denied refutes your intellectual contributions. This incident soured me from further collaborations with him and most other faculty until I was promoted and tenured.

Encouraged to Collaborate as a Tenured Faculty Member

In 1983, once I knew I would be promoted and tenured, I reconnected to collaboration through a graduate student, C. Deane Little. He wanted to work on a funded, collaborative project between FSU and Oak Ridge National Laboratory to isolate a microorganism that could biodegrade the solvent, trichloroethylene, which was pumped into the ground at Oak Ridge, Tennessee during the Manhattan Project during World War II. The plan was to use monoclonal antibodies as a tool to characterize the population of microorganisms. The Oak Ridge scientists with whom we collaborated were competent microbiologists and analytical chemists, and I contributed expertise in immunochemistry. This did result in a publication and the student's doctoral dissertation.

As I became more confident of myself as an associate professor I began to speak on how it was to be a woman in science. For instance, at a colloquium in mathematics and science education in May of 1991, I spoke of ten stepping-stones in my professional life. Concerning one of the stones, I said:

> [This institution] was lonely when I first came as [there were] very few women in the sciences. Margaret Menzel [a faculty member from Biological Sciences] was helpful. I struggled to get published and to get grants. It had always been easy to get papers published with my doctoral or postdoctoral advisors' names on it, but it was much harder on my own. I had trouble getting tenure, but I appealed and won. It was a hard time, but I put it behind me. Within a year I even had a grant for the [Undergraduate Research Participation-Faculty Professional Enhancement] program with the departmental chair [as co-Principal Investigator], who had not supported my tenure decision. Still I had a hole from which I had to pull myself out, but my earlier confidence in myself helped me.

I still remember that feeling of loneliness of being the only woman faculty member doing research for my beginning years as an assistant professor. I was the first woman who was on equal footing with the men because I did research. Research was (and is) the ingredient that constituted power in my department and at the university.

Collaborating with Science Educators

One difference in the culture between those of scientists and of science educators is scientists value *funded* scientific research while science educators value *any* educational research. Once I started to interact with science educators in 1991 (this was seven years after I was tenured), I felt the community-of-learners atmosphere that I experienced helped me to learn educational technology. Technology was a major contributor to my learning science education and collaborating with science educators. I have learned technology from my two co-major professors, Kenneth Tobin and Peter Charles Taylor. Each one influenced how I think about teaching, learning, and use of technology—especially when I teach my current classes.

My collaboration with my two co-major professors, in my quest for the second doctorate, required six and one-half years. The collaboration actually started two years before the start of the graduate program, when the three of us had dinner just before Taylor was about to return to Australia from his sabbatical while he worked

139

Table 2: My interactions in using technology with scientists and science educators. Key features identified in the cultural-historical activity theory diagram (Figure 1).

Primary movement		
Subjects	*Objects*	*Outcomes*
Self	Advance teaching and re-	Become a science education
Other scientists and sci-	search in science and sci-	researcher and improve my
ence educators	ence education	own teaching and learning

Coherences and/or contradictions	
Tools	*Rules or schemas*
High school & college: slide rule, French curve, logarithmic tables	Technology drives the world
College, graduate school, & postdoctoral years: early computers; hand-held calcula-	Those who know and utilize technology have power
tors	Technology without theory is wasteful of
Faculty member: progressively more	energy and output
complex computers, software, Internet,	
Palm pilots	
Communities	*Division of labor*
Two co-major professors in biochemistry	Learning technology from other graduate
Other scientists and science educators	students, other faculty, and in faculty work-
K–12 teachers	shops
Graduate students	My teaching others how to utilize technol-
Two co-major professors in science edu-	ogy and also my learning from my students
cation	

with Tobin. We set out in broad strokes the idea for an edited book on research in college science teaching, *Transforming Undergraduate Science Teaching: Social Constructivist Perspectives* (Taylor et al., 2002). I remember at the start I barely knew the difference between social constructivism and radical constructivism, but that is part of my journey. During this six-year time period, I have worked more closely with Tobin as my ideas and needs changed with time. I found my relationships comfortable with both Ken Tobin and Peter Taylor. I can say what I need to know and can ask for help, but also I can stand my ground when I have learned from my research, even if my ideas differ from their ideas.

USING TECHNOLOGY AS A TOOL

I choose technology as the second major common theme in my life and in my biochemistry classroom. Since the mid-1960s, technology increasingly became a powerful force in my career both as a scientist and as a science educator. It permeates my existence, my writing, my teaching, my creative energy, and my communication. Table 2 shows how the components in the activity diagram in Figure 1 focuses on how using technology mediated my objects and outcomes, both in science and in science education.

Using Presently Considered Primitive Technologies

There have been radical technological advancements now since my first days using technology starting in the mid-1950s (for juxtaposition of current tools with my original tools, see Figure 2). While I was in grammar school my father gave me my first technological tools as seen on the computer screen. When I went to graduate school at Bryn Mawr College, I used what we would now consider a primitive computer at Haverford College, where I did my scientific research for my master's degree in synthetic organic chemistry. This computer occupied an entire room, with heat-emanating machines containing whirling tapes and noisy printers. Even a few years later at the University of California, Berkeley, I had to punch each line of computer code onto a card and load them (in the right order) into a long box, which I would carry very carefully to the university computer center. At that time I was trying to separate mathematically two exponential curves with similar time constants, one decreasing and the other increasing in amplitude with time. I used a nonlinear program from the main computer facility. I had to enter my estimates of the amplitude and time constant for each exponential, and then the program minimized the error between the data and a theoretical fit, starting with the given estimates of the variables and then making slight variations in the four variables until the computer gave me the best fit.

Eventually, in our building at Berkeley, we got two computer terminals with a direct link to the computer center. It was a frustrating time as I was one of the first graduate students in the building to be using the new computer terminals in our building. The good part was there was no line, so I did not have to wait to use it. However, there was not much of a community of users with whom to converse, as most of my fellow students did more wet biochemistry and fewer calculations.

Figure 2. Technology now, technology then. Photograph of my current PowerBook G4 Mac computer displaying a photograph on the computer screen of the primitive technological tools I had used previously to learn science and mathematics up until the time I was a post-doctoral fellow: slide rules, French curve, and various types of graphing paper. My current Palm Pilot and HP calculator are in the background.

However, with persistence, I would get some results. It was agonizingly slow, however, that I wondered if I would get enough data at many concentrations of my reactants to calculate the kinetic rate constants.

My major professor for my first doctorate, Jack Kirsch, gave me a good idea. It was easier to change the reaction conditions than trying to separate computationally the relaxation curves. He noted that since one relaxation time was concentration-independent and the other was concentration-dependent, I might find concentrations of enzyme and substrate so that I could separate in time the two relaxation times from each other. I did this by increasing the concentrations of the chemical species to speed up the concentration-dependent process while maintaining the slower relaxation time of the concentration-independent process. I learned an important lesson that I could save valuable time and effort by paying careful attention to theory in the design of experiments. Because I was able to collect enough data using this improved protocol, I could use the theory of fast reaction kinetics and utilize computers to help me eliminate certain reaction schemes and obtain evidence for one scheme that was plausible that fit the experimental data. By paying attention to theory I could get to think about the more interesting aspects of science.

One of the important things I learned was that emphasizing a mathematical language enhanced my learning. In a competitive atmosphere like in my high school and undergraduate school, I tended to argue the mathematical and scientific points to myself. I became facile at developing the mental pathways to think concepts through to completion. I learned the mathematical language so I could think in scientific terms. Having the tools of mathematics gave me a sense of agency, despite the discouraging odds of becoming a woman in science. Had I used language within a community of learners, I might have gone further in science. I felt isolated, perhaps by being a woman in graduate school in the field of biochemistry (mainly dominated by males when I was there), or by my choice of the mathematical field of kinetics or by my personal idiosyncrasies (enjoying my own autonomy and pushing myself). I tended to do it alone so I did not learn as much from my fellow biochemistry graduate students and other colleagues had I been different or in a different situation or culture. Reflecting on this, I think that I felt this way because the community (very competitive colleagues) interacting with the rules or schemas (a culture that did not expect women to go into science, let alone into the physical sciences like my doctoral project in kinetics) did not support my needs as much as it could have. This contradiction of feeling out of synchronization with my chemistry and biochemistry colleagues despite being fully qualified as a biochemist still lives within me.

Connecting to Learning Through Technology

I remember in 1968 when I applied to graduate school at Berkeley for my doctorate in biochemistry that I wrote that I wanted to understand the biochemistry of memory and learning. There were not any courses I took that directly pertained to this topic—our graduate courses focused on using genetics and biochemistry to

142

understand DNA replication and transcription and ribonucleic acid (RNA) translation into protein at the time. However, while at Berkeley I taught an undergraduate seminar on (what little that was known at that time) the biochemistry of memory and learning (this was in an age near the end of the Vietnam War when students wanted relevance). I had about ten students in that seminar class. My library research on what was known on that topic whetted my appetite.

It is interesting that I come back now later in my career still wanting to understand learning. We still do not know the biochemical basis for memory and learning, but we have learned considerably about learning (Bransford et al., 1999). Using technology, especially the Internet, has brought us into the information age and, in the process, has allowed us to enhance the learning environment for many students. Through my participation in education research grants and my own action research for the second doctoral degree in science education, I have learned considerably about learning.

Technology has been an important component in my educational research. Initially, in my first education grants in the early- to mid-1990s the technology involved using word processing, electronic databases, and electronic communication. There was no World Wide Web yet with its many manifestations. Most middle school teachers with whom I worked had not used E-mail, and many had not experienced word processing. However, the teachers became immersed in technology, in part through a technology course that we offered to them but also by doing their assignments in my course, *Science, Technology and Society*, using their computers.

I started to make some strides in learning more technology by taking science education courses from Ken Tobin and Nancy Davis. Even after two classes with Tobin and one with Davis, I was comfortable with E-mail but was still "dragging my heels" in posting my comments on a course Web site. It took some time to learn to use the "tool," the electronic Web site to be part of a "community of learners." I did not feel that it was user friendly at first, but with time I came to appreciate it.

When preparing to do the action research in my classroom, I had worked with undergraduate students to create Web sites on the World Wide Web to provide what I called contextual learning opportunities for the students to focus and learn about autoimmune diseases and cancer. The first project was on lupus. Other diseases included Hodgkin's disease, non-Hodgkin's lymphoma, colon cancer, rheumatoid arthritis, and leukemia. I saw the opportunity for students to learn science by creating Web sites in courses that I taught, such as *Science, Technology and Society*. One collaborative group of undergraduate students taught themselves how to make a Web site on the topic, *Exploring Our Universe*. Their project had three foci, one for each student in the group to present in class. I was so impressed with the students' learning that I thought it would be a good idea to provide a similar learning opportunity for other students. Students used the language of science and constructed their ideas while sharing them with others.

A month later I taught a summer course in Miami, and my teaching assistant convinced me to encourage my students to use technology and the Internet. She helped our Miami students create their own Web pages. I saw it as an empowering experience. Meanwhile, I was learning to create my own Web-MC Web site in a series of workshops that the university offered for college faculty. I figured that having the students make Web sites related to particular aspects of biochemistry would be a good way to encourage my students to utilize the language of science in my biochemistry course.

In my *General Biochemistry I* class in which I did action research, ten collaborative groups each had to develop ten Web sites during the fifteen-week semester. It was tough for my students. I did not realize how much time each Web site project would take. Also I made them work within collaborative groups, which was hard because there was not a set laboratory time when they all could get together. Reflecting now I realize that I demanded too much of my students in how much I expected and how I assessed them. I had wanted to convince my biochemistry colleagues and myself that students could learn this way. I think that my students learned biochemistry, but it was not complete. I think that they learned some areas very well (probably better than you would learn in a traditional, lecture-based course), but the islands of deep learning often did not connect together to form a coherent view of biochemistry. Besides learning technology and biochemistry, I think many of them learned how to collaborate, but for some collaboration was a stumbling block.

CONCLUSIONS

In summary, my transition from a scientist to a science educator has been less of a metamorphosis and more a slow evolution over fifteen years since 1991 when I first started to work with Kenneth Tobin. There have been some punctuated spurts of growth, for me it has been a slow evolutionary process to change my thinking. It is quite a transition to change from a positivist to a social constructivist to an activity theorist. It has been a long journey. One catalyst for change has been writing the (second) doctoral thesis in science education. Writing my constructions helped me grow and learn—just what I wanted my students to do. I have found through this retrospective reflection that collaboration has been a major thread in my life and in my teaching. Use of the activity theory diagram (Figure 1) and my ideas on collaboration (Table 1) allowed me to see the coherences and contradictions as they relate to my teaching. The major contradiction in collaboration in science is through a "rule" that isolated me from collaborating as an assistant professor, especially in my first few years, thereby influencing the communities with whom I could work. I felt academically isolated for years! This influenced my own flow on the activity diagram, from subjects to objects to outcomes. It was not until I became a tenured associate professor that I felt I could collaborate with other scientists and science educators. It was seven more years after I became tenured that I became involved with the science education community. The academic science education community welcomed me as a scientist into their community of learners. Therefore, collaboration with educators enabled me to become truly a science educator.

I note the parallels between the collaborations that I experienced in my professional life as both a scientist and a science educator and the type of collaboration that I tried to promote in my biochemistry classroom. It was not until I was looking for themes in my teaching for my science education doctoral thesis that I realized the tension that exists between collaboration and autonomy. This is what has been called a paradox or the "joining of apparent opposites" (Palmer, 1998, p. 63). I tried to promote both collaboration and autonomy in my biochemistry classroom. Some students flourished in that environment, but some students could not live productively with that tension or paradox. Since recognizing the paradox between collaboration and autonomy, I highlight that this paradox exists and encourage each student to learn to be both an autonomous learner and a collaborative learner. One needs to take responsibility for one's own learning but contribute to and benefit from the learning of others. I continue to utilize technology in my teaching. However, I was a living contradiction for a long while, because although I had my students creating their own Web sites, it was not until fall of 2003 that I developed my own professional Web site.

Penny Gilmer
Florida State University
Tallahassee

THE EMERGENCE OF SCIENCE EDUCATION

The four science educators in collected together in this part B of the book represent three continents, Africa, Europe, and North America. Whereas science education was already an emergent discipline in the latter continent, the autobiographies of those from the other continents show a field in its emergence. In the following, we discuss the culture and history of science education exemplified in the four autobiographies in terms of the beginning of the discipline, the role of conceptual personae, dichotomies and contradictions, the role of scientists in the shaping of the field, and the barriers to women scientists and women in science.

THE BEGINNINGS AND MAKING OF A DISCIPLINE

Svein Sjøberg writes the first of two chapters featuring science education in Norway (Doris Jorde in part C is the other Norwegian science educator), which was greatly defined in and through his work. Prior to Sjøberg, there was no science education in Norway. Like Gallagher and Roth, Sjøberg was a scientist who had a passion for the inclusion of culture, ethics, and social issues in the study and enactment of science. Sjøberg was lured into science education mainly because an opportunity arose within his university to be involved in a laboratory for school science, which provided outreach services for the professional development of physics teachers. His switch to science education initiated a career that continues to this day and contained many milestones along the way, including his appointment as a professor in science education, the first such appointment in Nordic countries. Sjøberg's autobiography highlights power relationships within the science education community. He notes that his rapid rise to power within science education created some problems since his knowledge of schools, teaching, and learning was limited. Similar patterns have occurred throughout the world with scientists being chosen to lead national curriculum development and policy efforts based on an assumption that as leading thinkers in science they have the knowledge needed to guide the development of curriculum frameworks and advise on policy in science education. For example, in the United States, the National Research Council had oversight over the development of the *National Science Education Standards*, a document that is widely acknowledged as shaping curriculum development in many countries in the world. More recently Sjøberg has been involved in European projects associated with science and society and interdisciplinary research in Europe involving values, interests and perceptions related to science and technology.

Sjøberg's transition into science education is not unlike others at about the same time, who showed an interest in science education, just after the Sputnik launches galvanized action about the quality of science education, reasons for teaching science, and the recruitment to science of the best thinkers among the children of the time. Gilmer notes that she was a child who was synergized by the efforts to recruit talented youth into science. Hence, the structures associated with Sputnik were global and cross age. Universities provided resources to support the professional development of science teachers not only in the United States, but also in Norway (as described by Sjøberg) and Australia, as evidenced by Fensham's appointment in science education at Monash University and the creation of science and mathematics education programs within the physics program at the Western Australian Institute of Technology—in which Tobin enrolled to obtain his master's degree in physics (focusing on education). John de Laeter, a former science teacher and one of Australia's leading physicists, pioneered new roles for science departments in the professional development of teachers. His initiative was the forerunner to the development of the *Science and Mathematics Education Centre* where first John Deckers and then Barry Fraser were to provide leadership that would capture international attention.

When Sjøberg made the transition to science education there was no formal structure for science education in Norway and he looked to other countries for relevant expertise. Initially his role models were from England, notably at the University of Leeds, where Rosalind Driver was just returning from completing her doctoral studies with Jack Easley at the University of Illinois. Sjøberg's decision to study for a graduate degree at the University of Leeds opened the door for productive relationships with networks of scholars from England and beyond. In Part C of this book, we further explore this theme of science educators formed in one country, especially the USA, who then returned or moved to another country to mark the development of science education.

Sjøberg studied science education at the university of Leeds and created a social network that resulted in visits from leaders in the field, including Rosalind Driver, whose connections with Jack Easley in the United States and numerous colleagues throughout the world helped to shape the field of science education. Sjøberg mentions connections with influential science educators Edgar Jenkins and David Layton, who developed the journal *Studies in Science Education*, which was to influence graduate studies in science education throughout the world, especially outside of the United States. Outside of science education Sjøberg illustrates how scholars such as Basil Bernstein and Lawrence Stenhouse shaped his thinking.

CONCEPTUAL PERSONAE

Conceptual personae are figures that come to stand for a way of thinking, working, theorizing, or doing philosophy (Deleuze & Guattari, 1991/1994). The Socrates of Plato, the Dionysus of Nietzsche, and the Idiot of Nicholas of Cusa are typical conceptual personae. But historical persons themselves come to be conceptual personae in the sense that they embody and define a way of going about business. We come to speak about Platonism or Kantianism precisely because the historical person embodies a way of thinking and doing philosophy. In science education, Jean

Piaget is a prime example a *conceptual persona*, standing for a way of thinking, theorizing, and doing research: he is a subject on a par with his approach and way of thinking. As conceptual persona, Piaget, though seldom referenced today, has been a major force in science education until the influence of his work waned with the coming of sociocultural, cultural-historical, and discursive approaches.

Sjøberg applied Piaget's work to his science education and then pursued research that was similar to Driver's. Rather than working only with others' interpretations of Piaget, Sjøberg worked from original texts and focused on epistemological and philosophical frameworks rather than stage theory that was popularized throughout the world and reified in curriculum projects in the United States, Nuffield Projects in Great Britain, and the Australian Science Education Project. Inadvertently the widespread focus on stages of cognitive development focused on the deficits of pre-formal thinking and led to pervasive acceptance of hands on curricula with insufficient attention to critical issues of the nature of science and how children across the elementary, middle and high school years could meaningfully do, learn, and apply science.

Ken: In the early 1970s I was interested in Piaget as a developer of curriculum resources to support the teaching of science in grades 8–10. I read avidly in books and journals and came to know Piaget largely through scholars in the English-speaking world. When I began my research I was convinced that the students' capacities to reason formally would mediate their learning of science in grades 5–7 in Western Australian schools. I searched for a way to measure formal reasoning ability so that I could use it as a covariate in a quasi-experimental study. I came across the work of Dick Tisher, a science educator from Monash. He had adapted some Piaget-like formal reasoning tasks, administered as clinical interviews, to paper and pencil format. I used this measure in my research and, when I did my doctoral studies, I adapted a similar paper and pencil instrument developed by Tony Lawson, for use in my classroom research. My goal in developing the *Test of Logical Thinking* with Bill Capie was to create an instrument that would provide reliable and valid data to allow us to take into account differences in the formal reasoning ability of students as we explored their engagement in class. For a number of years we used the *Test of Logical Thinking*, which has been translated into many languages and has been widely cited, up until 2005. Other pencil and paper measures were developed, but usually for different purposes—to identify the stage of logical thinking and in many cases to tailor the curriculum to the stage of cognitive development of the learners. As Roth noted in Part A, after about a decade of pre-eminence in which researchers and curriculum developers used Piagetian stage theory, there was a shift toward constructivism and the dominance of science education by researchers such as Rosalind Driver in England and Joe Novak in the United States.

Michael: As many science educators and perhaps scientists, I was attracted to Piaget's logical approach to theorizing, for example, development of proportional reasoning, a topic which I studied during my dissertation. I am certain that my own background as a scientist was a crucial aspect for mediating my preference for Piaget's work, much as I was attracted to the structuralist approach Claude Levi-Strauss stands for in anthropology as compared to the narrative approaches that other anthropologist have come to represent. Equally interesting is the way in which many of us have turned away from Piaget, and in our turning, have changed the field of science education, which now supports very different concerns, topics, and theories.

Most science educators, despite the tremendous influence they have wielded on the field, do not become conceptual personae. However, their role and influence has to be acknowledged, especially because they are associated with major movements. For example, Easley, a collaborator of Robert Stake, was a pioneer of qualitative research and well known for his case studies in science education. Driver, however, was best known for her research on conceptual change in science education and became one of the leading international figures in science education, having an enormous influence on the field—especially through her program of research at Leeds, which became a magnet for science educators from around the world. Among the role models and collaborators listed by Sjøberg were Peter Fensham from Monash University, and Glen Aikenhead from the University of Saskatchewan (Canada). Sjøberg's collaboration with Fensham afforded a connection to Monash and its network of scholars, resulting in institutional collaboration between the University of Oslo and Monash University. Science and its relationships to society was the substantive focus for the networks in which Sjøberg participated. Accordingly, his primary professional affiliations were directed toward science technology and society and physics education—unlike the scholars included in the first section of this book, whose primary science education affiliations were with NARST. Sjøberg regards Joan Solomon from England and Glen Aikenhead from Canada as key mentors. Interestingly, Sjøberg highlighted a Portuguese initiative, *Ciencia Viva*, as an ambitious European project designed to promote the culture of science. He made specific reference to the efforts of a physics professor Jose Mariano Gago, who became a political leader in Portugal.

DICHOTOMIES AND INNER CONTRADICTIONS

As an emerging discipline, science education has become a gathering place for scholars from other fields. Cultural fields, however, are weakly bounded such that practices in one field may be reproduced in another (Sewell, 1999). This is especially the case in an emergent field, which comes to be defined, at least initially, by the practices of the root disciplines of its early practitioners. But when, as in the case of science education, these newcomers come from different root disciplines, the actions of different members may be at odds, even dichotomous, and lead to inner contradictions that the field as a whole has to work out. Science, psychology, and general pedagogy mediated contents and career trajectories especially in the early parts of science education history.

Like Sjøberg, Reinders Duit came to science education through the pathway of physics. Duit obtained his doctorate through studies of physics education, but within a framework of a degree in physics. However, from the outset Duit studied physics to become a science teacher whereas Sjøberg started out as a scientist and changed later because of a feeling that something was missing from science as he experienced it—the ethical, social aspects that became foci for his career in science education. As Duit explains, his approach to science education was always focused on the learner, incorporating German traditions while at the same time showing outside influences, which he describes as Anglo-Saxon. Duit was associated with a key European science education center, IPN—The Institute for Science Education at the University of Kiel, in Germany. The IPN was a magnet for collaborative activities in science education in much the same way that Curtin, Monash, and Leeds were institutional nodes in a network that supported science education's invisible college.

Duit alludes to an important dichotomy within science education—those with a primary allegiance to their discipline and those with a primary orientation toward teaching, learning and curriculum in a more generic sense. He notes that Edgar Jenkins had described a similar split in England. We have seen this throughout the world and it is not unusual for scientists coming into science education to affiliate with their subject matter professions such as chemistry educators associating with the American Chemical Society. Accordingly, science education can be seen in terms of somewhat disparate groups, including but certainly not limited to, physics education, chemistry education, biology education, and geology education. Such groups may have little collaboration with another group that identifies as science education and divides within this field into groups associated, for example, with teacher education, teaching and learning, curriculum, science technology and society, and equity issues. Each of these groups might have its own journal and as a result research within the groups might not easily diffuse outwards and social networks may not penetrate the boundaries. Perhaps an inevitable outcome of these groupings along traditional lines is for the members of the different groups to not be aware of or respect the scholarship of those from outside their own group. Duit addresses this in terms of the difficulty of having his scholarship recognized by gatekeepers who are grounded in the discipline—in his case physics. A tendency to fractionate science education can have deep implications for peer review in all the fields in which it operates, including earning jobs, publishing research, and winning grants.

Duit explains that his theoretical framework is grounded in the epistemology of constructivism, associated with the traditional German foci on *Bildung* and *Didaktik*, which he inherited. He embraced Piaget, Ausubel, and Kuhn and focused his research on conceptual change—with a primary concern for the ways in which individuals make sense of science, taking account of what they know and can do when they come to learning situations. He rejected deficit perspectives and endeavored through his instructional models to align curriculum with what students know and can do—with the goal of producing canonical understandings of science. When many science educators were arguing about which of the psychological models were most appropriate for science education it seems as if Duit was a bricoleur who took something from different models, even from Robert Gagné's behaviorist oriented learning hierarchies.

Among Duit's many collaborators were colleagues from the IPN, David Treagust from Australia, Shawn Glynn, a psychologist from the United States, and Michael Roth from Canada. This group is the tip of a very large iceberg. The conceptual change group became the dominant group in science education with worldwide networks involving powerful scholars and institutions that included universities, research centers, and professional organizations. For a quarter of a century, starting in the early 1980s the conceptual change group dominated science education and created a vast literature and support network. As the dominant group their impact in the literature and the structures of science education has become a mainstream ideology that has suppressed other frameworks. Interestingly, in the final part of his chapter Duit addresses the necessity to examine policy issues and the ways in which they can shape science education. Perhaps this is a sign of the willingness of the conceptual change group to consider theoretical frameworks that extend beyond psychology and include social and cultural phenomena more explicitly.

SCIENTISTS-BECOME-SCIENCE EDUCATORS

As new disciplines emerge, individuals who have received their training in some other field may shift into and populate them, with the result that the cultural aspects of the original fields are reproduced in the new disciplines. For example, during the early part of the 20th century, physicists, chemists, and mathematicians migrated into and defined theoretical ecology generally and population ecology more specifically (Kingsland, 1995). Similar trends also can be observed in science education. Thus, unlike Sjøberg and Duit, who were associated with physics but undertook studies of science education before doing their doctorates, Hewson and Gilmer both graduated with doctorates in science, Hewson in physics and Gilmer in biochemistry. Like Gilmer, Hewson completed his PhD in nuclear physics and his autobiography vividly describes his transition from science to science education. So much of what he describes seems at first glance to be serendipity. However, it is clear that social bonds involving individuals and institutions were present and supported his transitions.

As a nuclear physicist, Hewson became interested in science education through collaborations with physicists and science educators on the teaching of college physics. In Hewson's case a key collaborator was his spouse Mariana Hewson. While on a sabbatical at the University of British Columbia in Canada, Hewson got involved in research on the teaching and learning of college physics with Walter Boldt, a former student of Jack Easley's. The research introduced him to the uses of interaction analysis and Piagetian theories of learning.

From the beginning of the 1970s Hewson showed a strong curiosity about the learning of physics and when the opportunity came he applied for a position in science teacher education in South Africa. As Sjøberg had done, Hewson then set out to build his expertise in education and schools, first spending time in schools and in collaboration with teachers and eventually enrolling for a degree in science education—taking education courses alongside the students he taught science education.

Faced with an opportunity to learn during a sabbatical, Hewson did something quite novel in outlining his interests in an open letter to the members of the National Association for Research in Science Teaching (NARST). Already he was developing an interest in the work of David Ausubel, augmenting his prior interest in Jean Piaget. Hence, when Joe Novak invited Hewson to Cornell University he jumped at the chance. Hewson was then introduced to Novak's expansive network of scholars, including Ken Strike and George Posner. With them he was to write one of the most influential papers on the conceptual change perspective. Hewson was situated in a lively network focusing on meaningful learning in science from the perspective of conceptual change theory, of which he was to become one of the primary architects.

Joe Novak is undoubtedly one of the most influential science educators of the 1970s and 1980s, but, despite his tremendous influence on the field, is not a conceptual persona as Piaget has been. Novak, who was recognized by NARST with its 1990 award for Distinguished Contributions to Science Education through Research, had a profound impact on science education, pioneering the field of conceptual change theory through his extensive writing, doctoral students, post doctoral fellows and visiting collaborators. As well as being a leader in the advancement of conceptual change theory he also pioneered concept mapping in science education and expanded the theoretical frameworks being used beyond Piaget's stage theory.

Hewson identified the early 1980s as a period that was salient for his career as a science educator. During this time the conceptual change special interest group was developed within the American Educational Research Association, and Hewson joined the faculty at the University of Wisconsin. It was a time of change within science education. For example, in 1984 Tobin and Gallagher began their ethnographies of high school science classrooms in Australia and in the United States John Penick and Robert Yager initiated an agenda of hope, *The Search for Excellence*, in response to the gloom and doom era of *A Nation at Risk*.

When Mariana Hewson accepted a position in Ohio, the invisible college afforded Jane Kahle recruiting Peter Hewson to get involved in the large systemic reform projects she was undertaking in Ohio, beginning a collaboration that continued when Hewson obtained a large grant from the National Science Foundation to create an infrastructure to support research in science education in South Africa—a program that would involve several of the scholars in this book, including Kahle, Fraser, and Tobin. Clearly, the invisible college supported international projects through the interconnections among individuals, institutions, and individuals and institutions.

GENDER BARRIERS

There is evidence that gender has been an important resource for producing and reproducing inequities in science and science education. In a remarkably candid autobiography, Kathryn Scantlebury (2005) has described the difficulties a woman science educator might experience in a department of chemistry. Penny Gilmer describes a very similar situation, thereby articulating the barriers within our field at the institutional level to include women. It would not be surprising if these same forces opposing women like Scantlebury and Gilmer to succeed also mediated science learning, interest, achievement, and barriers at the high school and under-graduate levels.

From the outset Gilmer created social bonds in her graduate studies and then as an assistant professor. Surprisingly she experienced strong forces that encouraged individual scholarship over collaborative forms of scholarship, largely because of processes associated with promotion and tenure. Being a female in science appears to have been both an advantage and a disadvantage to Gilmer. Unlike many of her male colleagues Gilmer sought out collaborators and endeavored to produce science through collective activities in which her social bonds with others were explicit and strong. Her interdisciplinary preferences encouraged her to identify others from whom she could learn different perspectives, especially in graduate school and during postdoctoral experiences. The connections she made appear to have served her well and she endeavored to establish and maintain similar social networks when she was hired at Florida State University. Here she felt a sharp edge of competition as male colleagues made clear to her their preferences for individual forms of activity. Because they needed to make decisions on her promotion and tenure they wanted her scholarship to be clearly distinguished from others. This requirement was a strong force against collaboration as was an early attempt in which she contributed to a project intellectually but was denied co-authorship on the associated paper.

Although it is not appropriate to attribute cause to what happened to Gilmer, the fact that she was one woman among many males makes it a plausible assertion that Gilmer's gender was a salient factor in her being a scholar in an institution in which almost all of her colleagues were male. Notably, her preference for collaborative research was not shared by her male colleagues, some of whom held powerful positions as deans, heads of department—all of those who would vote on her promotion and tenure having higher rank and tenure. When her colleagues denied Gilmer promotion and tenure she was able to appeal to a broader university community that included females and non-scientists. At that point the decision of her colleagues was reversed.

In her post tenure years collaboration continued to be Gilmer's mantra, however, there was an ideology of individualism within science departments that discouraged her collaboration with colleagues. Although Gilmer doesn't address this issue in her chapter, Tobin was at Florida State University when deans and provosts discouraged scientists from collaborating with educators, arguing that if they had time to do this kind of work they were not fully committed to their research in science. Notwithstanding such structures, Gilmer reached out to science educators and in so doing swam against a strong current. Not content only to obtain external funding, Gilmer realized her limitations as an educator and set out to be as educated as the doctoral students on whose committees she served. Over the years she read dissertations, enrolled for and passed courses in science education, and then enrolled for a second doctorate. Just as she has done for her PhD in science, Gilmer sought the support of two doctoral advisors, Peter Taylor and Ken Tobin. Gilmer engaged seriously in science education, ensuring that she understood social and cultural theories to support her research.

As well as maintaining her ongoing interests in the ethical dimensions of science and her continuous involvement in the American Association for the Advancement of Science (AAAS), Gilmer became very active in the Association for Science Teacher Education (ASTE) and the National Association for Research in Science Teaching (NARST) and is currently president of the latter organization. She has remained in the Department of Chemistry and Biochemistry and has undertaken ongoing research and evaluation on the teaching and learning of college level chemistry.

Differences in achievement levels between boys and girls have been noted for many years; these differences have been attributed, in line with the work of Evelyn Fox Keller, to differences in characteristics (see Kahle, this volume). Sjøberg's gender equity studies occurred in Nordic countries where feminism was well established and in terms of international comparisons the Nordic countries were near to the top in terms of equity. However, as he pointed out in his chapter, there is a contradiction in that females failed to select to study subjects associated with mathematics, physical sciences, and technology. Sjøberg argues that a primary reason for their choices is that these disciplines have an image that is distant from historical, social, cultural, philosophical, and ethical issues. Throughout his career Sjøberg was an active participant in the Association for Gender and Science and Technology.

It could have been, however, that the sciences and science education, as disciplines, have contributed to the reproduction of a situation that has excluded women for millennia from the pursuit of initially natural philosophy and subsequently science. Such a thought might be dismissed a little too rapidly within our community; yet science educators have contributed to the reproduction of science teachers, who, in turn, have taught in ways that have maintained the gaps in achievement and retention rates in some science disciplines.

THE EMERGENCE OF JOURNALS FOR RESEARCH IN SCIENCE EDUCATION

An interesting sidebar on Sjøberg's biography is his affiliation with *Studies in Science Education*, a journal that is well known in Europe and Australia but not so widely accessed by scholars in North America. For us, this autobiographical comment raises the issue of scientific journals and how they shape a field and lead to its internal structuring and even geographical differentiation. For example, non-North American scholars may perceive it as too difficult to get their manuscripts through the peer review in the US-based *Journal of Research in Science Teaching* (JRST) and therefore send their materials preferentially to the *International Journal of Science Education* or to the newly created *International Journal of Science and Mathematics Education* based in Taiwan, because it also provides language support to non-English speaking authors.

The issue of the role of journals is not simple, because the different national organization of universities and the career trajectories individuals undergo mediate what scholars publish and in which journals they do. For example, JRST and *Science Education* not only have a high reputation but also make it into the top-ten list in terms of citations and impact ratings constructed by Thomson ISI, the main database sociologists of (social) science use for studies of invisible colleges, citation networks, the impact of certain theories and concepts, and so forth. In many North American universities, these journal rankings—together with rejection rates—are used to establish the "quality" of a journal and therefore the quality of the work a researcher publishes.

Other universities are less concerned with having their faculty members publish in highly ranked journals but rather count publications more indiscriminately but still require publications for the purposes of giving tenure. With the emergence of science education as a field, there was therefore a need to have more journal space available than was, for example, in the 1960s when our field began to grow following the launching of Sputnik and the race to attract more students into the technosciences. At that time, *Science Education* had already been in existence for more than half a century; School Science and Mathematics was founded in 1901, and *School Science Review* was founded in the UK in 1919.

Given the increasing number of science educators brought about by the creation of academic departments and professorial chairs—as indicated in the autobiographies—it is not surprising to find that many more science education journals have been founded after 1960 and to the present day. Throughout the 1980s there has been a rapid expansion of journals that publish research in science education, some international journals and others national in their scope. We cannot list all such journals here because to do so is beyond the purposes of this book. But in the English-speaking world, these include, among others, the *Journal of Computers in Mathematics and Science Teaching, Journal of Science Education and Technology, Science & Education, Research in Science Education, Research in Science & Technological Education*, and the *Journal of Science Teacher Education*. Most recently, our own *Cultural Studies of Science Education*, which takes an explicitly and decidedly cultural-historical perspective has been added to provide an alternative to the largely psychological orientation provided by the other journals in the field. Along with the increase in numbers were issues, including the selection of editors and reviewers, the genres of research that were published and encouraged, and the criteria that emerged to define scholarship in science education.

Ken: I vividly recall a conversation with my new colleagues in the curriculum branch. He was telling me how to get up to speed on the latest in curriculum development in science. I was a new hire in the curriculum branch of the education department and had just returned from England where I had taught some of the new Nuffield science programs in English high schools. "How do you spell that word?" "P-I-A-G-E-T." Within the next hour I was in the library poring over Volume 55 of a journal I had never before encountered—a journal with the name *Science Education*. Evidently it had been around for some time and right alongside of it were several volumes of the *Journal of Research in Science Teaching*. Before long I was into Volumes 1 and 2 of JRST, seeing for the first time the names of key science educators who had and would shape the field—Herb Their, Myron Atkin, David Butts, Milt Pella, Leo Klopfer, Fletcher Watson and Jim Rutherford, just to name those who contributed to the first volume of JRST and subsequently mediated my growth as a science educator. In volume 2 I found Jean Piaget and David Ausubel, in the same issue. This visit to the library was critical in my growth as a science educator because it was at this time I learned for the first time that people did research on the teaching and learning of science, and that scholars in the field were exploring issues that were central to my practice as a curriculum designer with a strong interest in the professional development of science teachers. Also evident in the current volumes of *Science Education* and JRST was the focus on producing curriculum materials and the hope that these would lead the revolution in science education.

My reading of research in science education was very focused on journals from the United States and the proceedings of the fledgling Australian Science Education Research Association. Indeed I was unaware of research journals in science education from other countries until 1983, when I began to publish in the *European Journal of Science Education*, edited at the time by Richard Kempa. Quickly this journal became my favorite and when I moved to the United States I became North American Editor of the journal, which was renamed the *International Journal of Science Education* in 1987. Established in 1979, the journal supplemented *Studies in Science Education* which, as was noted by Svein Sjøberg was established in 1974 as a journal that reviewed research in science education.

Michael: For me, the journals became perhaps the most crucial mediating element in my career as a researcher. I did my doctoral degree at an institution (in a department) where doing and publishing research was not valued—there are less than 10 publications listed in the *Thomson ISI-Web of Science* database involving science education researchers from the institution, and a considerable number of these have been written by graduate students. In this situation, reading journals allowed me to connect with the field as a whole because I attended my first conference after I had already graduated. In those days, I read every issue of JRST and *Science Education*—leading my dissertation advisor to joke that I was the only person she knew to read such journals as bedtime literature.

It probably comes as little surprise that these journals constituted the first outlets for my own work. I still remember how I propped up two JRST articles Anton Lawson, whose work was most similar to my own, had published and how I attempted to emulate his writing. Later, after having returned to the classroom and conducted interpretive research on learning I used several of your articles published in *Science Education*, which I attempted to emulate in its genre for writing up my own work. I only realized much later that in this way, I learned to write for my audience; and because my research had similarities with other published work, it embodied its concerns as well. Thus, I became a science education researcher mostly via my attempts to read and understand the published literature, and subsequently to write for the same journals.

Perhaps it was my training as a natural scientist that oriented me not only toward publishing but also to publishing in what some call "top-tier" journals. I learned about the *Thomson ISI-Web of Science* and the different factors they use to rank journals—including impact factor and total number of citations. Much like scientists, who orient their work toward making it publishable in those journals that are most read and most cited (as indicated by the impact factor), I began to orient toward journals in and outside education who had high-impact factors and were the leading journals in their field. This orientation therefore has mediated the content and process of my research as much as my interests, which do not exist but in a dialectical relationship with the collective interests of a discipline.

In recent years, an increasing number of journals is available online, and many journals have an online first policy, which means, articles are made available prior to their ultimate print version. There also have been a number of attempts to bring about online journals, including the *Electronic Journal of Science Education* and the *Electronic Journal in Science and Literacy Education*. For a variety of reasons these have been relatively unsuccessful, however, many of which have been articulated in a sociological analysis using a cultural-historical activity theoretic model (Roth, 2005b). Fundamentally, the tenure and promotion processes tend to favor "high-impact" and "first-tier" journals, so that many of the leading researchers have not sought to publish in the emerging electronic journals, which leads to their minimal impact on the field as a whole.

Part C

US-TRAINED SCIENCE EDUCATORS ABROAD

EXPORTING SCIENCE EDUCATION

Why would a person, interested in doing a degree in science education, enroll in a university of another foreign country? Since both of us did precisely that we have a personal interest in exploring this trend and placing it within a larger framework. The chapters written by science educators in different countries provide a context for this chapter, which examines the ways in which science education has developed in Taiwan, Colombia, Spain, Costa Rica, and Norway; we see these auto/biographies as examples of a pattern that numerous science educators have realized in their own career trajectories. Before looking at the themes addressed in these chapters we briefly address the most salient issues associated with our decisions to seek graduate education in the United States.

Ken: I had completed a master's degree in the physics department at the Western Australian Institute of Technology (to be renamed Curtin University). My research on wait time had captured my interest in educational research and I wanted to be a university science educator. At the time there were no doctorates in science education in my state and no support to study on the Australian east coast as Barry Fraser had done at Monash. Accordingly, I commenced a doctorate in psychometrics at Murdoch University, working with Barry McGaw, who had just returned from the US where he had studied with leaders in educational measurement. He advised me strongly to go to the US to study with leaders in science education. Hence, my reason for going to the United States to do my doctorate was to ensure that I could get well qualified in science education—a field that was rapidly emerging in Australia but which did not have well established and accessible doctoral programs. The opportunity to obtain a degree that specialized in science education rather than take a more generic degree in educational measurement was a primary motivation.

Michael: My own reasons as a Canadian to do a PhD in the US are probably very different than the reasons that drive most abroad. After my master's degree in physics, I stumbled into teaching because there was an economic downturn in the late 1970s and early 1980s. Even in my first few weeks of teaching science in a tiny middle school in an isolated village in northeastern Quebec, I knew I was hooked on teaching. Children must have been hooked too, for they returned to the school after dinner. The same happened when I taught in Lewisporte, Newfoundland, where students began returning to the school in the evenings to work and learn. One day, a superintendent asked me whether I would not want to have more impact on science education in the district; but for this to happen through a school board coordinator position, I needed an advanced degree in (science) education. I was burned out from going to

161

school, and told him so. He talked about the University of Southern Mississippi, which had a summer program so that I could do most of my course work during the summer and needed only one regular semester preceding or following a summer term to complete my two-semester residency requirement. So I went for one summer term, found a professor to work with, and beginning with the subsequent summer term, went to complete my degree as a fulltime student. After completing my degree, I stayed for another year in the US but then, not in the least because of my political tendencies—fully supporting, for example, a medical system funded by the collectivity and available to all its citizens—I returned to Canada. Nevertheless, my orientation has been to the American educational organizations and journals, which I feel have a more international orientation than our Canadian journals and organizations. I believe that among my colleagues there are many who, in different ways of my being, recognize this orientation toward the American education culture rather than toward the Canadian one.

Each writer in this section identifies role models who made a big difference to their progression within the field. Fortunately for Jorde she had the only female professor in plant pathology as her boss and hence she had a role model from whom she could gain insights into how to succeed as a woman in science. Until Jorde came to Norway as a colleague for Sjøberg she had mainly women as role models. Notable among these were Jane Kahle, who provided her with advice on how to cope with family issues, especially childcare issues, while earning tenure. Later, during a sabbatical leave Marcia Linn and her colleagues at Berkeley served as role models for Jorde, who redirected her life trajectory into the field of science education as a result of what she learned. Rosalind Driver also was a role model for Jorde, who visited Leeds, thereby reinforcing the social bonds between Leeds and Oslo that were initiated by Sjøberg.

Chao-Ti Hsiung's chapter exemplifies the role of the home environment in building dispositions toward learning science. Even though she appears to have grown up in a somewhat frugal economic environment—an aspect she shares with Michael Roth—she developed habits of mind that are typically associated with science, a willingness to be persistent, curiosity, and a capacity to see science in everyday activities—such as mixing milk and baking bread. From what she writes it is possible that both parents structured her scientific ways of thinking and being.

Hsiung describes how Russell Yeany went beyond what a science educator might do by assisting foreign students to settle in by building bookcases for their use and also assisting them to become culturally aware by involving them in sporting events that might not otherwise capture the attention of graduate students or seem important to them at the time. For Hsiung it was an international cycling event—for Tobin it was a triple-header baseball game in Atlanta and annual softball matches between the faculty and graduate students in science education. Though foreign students may be focused on getting a degree and returning to their home, the shared emotions of events such as these no doubt contribute to the creation of social bonds that serve the field well in the future. In fact, it is precisely by means of such shared events that Scandinavian workplace design projects have brought together the quite different groups of computer software engineers and the future users of the software. Because they have spent a lot of time not only working together but also engaging in after-work activities, these heterogeneous design groups gelled to produce software designs exceeding anything that each group could have produced in terms of specifications for the requirements of the user system.

Espinet's chapter begins with a shocking insight into education within a dictatorship in which she was not officially allowed to use her native language to aid her learning. Her story of her elementary education and the pride she has for the efforts of her libratory teachers is a reminder of the political dimensions of education and the ways in which oppressed people can find resources for power. It seems such a long journey from the oppression of Catalonians in Spain to minority groups elsewhere in the world and yet the parallels are quite distinct. For example, research in urban schools in the United States suggests that all students, but especially immigrants, should be allowed to use their stocks of knowledge to make sense of science in ways that involve hybridization as they make sense of classroom experiences in terms of their own cultural resources. What might emerge in the process is a creolized science, which reflects the cultural resources from their homeland and their experiences with science. (We do recognize the contradiction in all of this when international tests are constructed in terms of a hegemonic concept of science where everyone is required to know the same concepts in the same way—a lot of work remains to be done in the science education community to deal with this and other contradictions.) Paramount is having the autonomy to use their cultural capital as a foundation for learning. Of course this is what Espinet did, even though the uses of the Catalan language were not sanctioned. Given her father's involvement in mathematics and engineering it is possible that he father structured her interests in science.

Lilia Reyes got involved in educating adults from her neighborhood, showing a strong value for educating not just individuals but in fact the community as a whole. When she was in teacher training she continued this practice by offering programs to enhance the science literacy of employees at the university—workers who might not normally gain a science education beyond their basic schooling.

For a considerable number of science educators beginning in rural setting is a first experience with teaching; it may also be that growing up in rural settings (Hsiung, Alfaro, Roth) mediates interest in science and the opportunities it provides. Gilberto Alfaro's chapter provides a stunning contrast to what might apply in many countries—especially in the 2000s. His descriptions of elementary education are consistent with those that might be written by many Australians or Canadians (for example). We (Ken, Michael) both have experienced sparsely populated rural towns where students had to travel some distance by bus to attend high school or even to live away from home because of the distances involved—in Michael's case, students had to leave the village after ninth grade to complete their high school 1,000 miles away; appropriately preparing the students so that they would not drop out and return the village is quite a challenge indeed. However, several factors are salient in Alfaro's chapter. First, the transportation available between home and school was horse. Second, attending high school was not taken for granted. Third, he stayed for six years with the family of his elementary school teacher. A fourth point, one not addressed by Alfaro, is provided by Tobin who, in the early 1990s, accompanied Alfaro on a visit to elementary schools a short distance from San José. Conditions were very different than might be experienced in countries such as Australia, England, and the US. The classrooms were neat, tidy, and clean but had limited access to electricity and no glass in the windows. The context in which Alfaro was educated was profoundly different than was experienced by Tobin—even though he too experienced rural education. Alfaro's high school science teacher was a role model who appears to have instilled in him a spirit of inquiry and a love of science. Perhaps because one teacher taught all sciences, Alfaro learned the value of learning with others and teaching to learn—valuable lessons for a future teacher, teacher educator, and researcher.

It turns out that Roth, too, taught science under minimal conditions: a basement where the ceiling was only 5 feet in some places, no water, a few sparsely lighting bulbs, no working fire extinguisher—and this was in the early 1980s in Canada. But there was a kit that came with the *Integrated Physical Science* textbook that largely featured experiments, and together with the students, Roth extended and added to these experiments to bring about a two-year science course that students actually liked. Rural education often means having to be creative to bring about interesting lessons; or rather, being creative might mean that students become excited because they come to see and experience science (and other subjects) in ways that had not imagined before.

CHAO-TI HSIUNG

A TAIWANESE JOURNEY INTO SCIENCE AND SCIENCE EDUCATION

I worked as a teacher of agronomy, meteorology and fertilizer studies at an advanced vocational agricultural school for one year and as a teacher of biology at a senior high school for fourteen years. While teaching biology at high school, I received a government scholarship to study abroad for a master's degree. Then, with the support of a University of Georgia research fellowship, I got my doctorate in three years before returning home and joining the faculty in a normal institute known in Taiwan for its training of primary school teachers, which just had founded a department of science education. That was seventeen years ago.

DEPARTURE OF MY JOURNEY: FAMILY AND SOCIETY

Thanks to my parents, I always did well in school; both parents introduced me to science-related things. My mother, who liked to be involved in our education, took us to observe how chickens hatched. We looked through the eggshell into the embryo using a lamp. Then she made us raise the chickens; and I studied how to combine rice bran with rice so the birds better enjoyed the feed. All the eggs became the main dish of our lunch at school. When the chickens grew old and died, we cried and built a tomb for them. I remember once my elder sister wrote an essay in memory of an old hen, which moved her schoolmates to tears. We raised dogs and cats; and from time to time we saw wall geckos creeping along the roof ridge and window edges. Such experiences assisted my biology learning at school, since it was never hard for me to understand and memorize things like structures.

When he was young, my father dreamed to grow flowers and grapes in the garden and orchard. The grapes we grew bloomed and bore fruit; the purple clusters looked so much like the pictures in the book. On holidays, father and mother took us mountain climbing. On several occasions they would say a certain plant or animal or ore stone looked the same as those in the mainland hometown. On the seventh day of the seventh month of the Chinese lunar calendar, when the figures in China's fairy tale—the Herd-boy and his lover Weaving-girl, the youngest daughter of the Jade Emperor in the heaven—are to perform their once-a-year rendezvous in the middle of the Milky Way, which has separated the couple, father would show us the seven stars that shape the big dipper. He also pointed to the stars Altair and Vega, which are known as the stars of the Herd-boy and the Weaving-girl in China, telling us about the Milky Way.

A former physics major, father loved to discuss relativity theory with me. I was not very interested. Later on, when he found I favored biology and nature, father

K. Tobin, W.-M. Roth (Eds.), The Culture of Science Education, 165–174.
© 2007 Sense Publishers. All rights reserved.

encouraged me to read Zhuang Zi, an ancient Chinese thinker known for his philosophy of nature. Father held that it was not enough to merely memorize biological structures and functions, but it was important to internalize meanings contained in nature. However, I was occupied with going to an institution of higher learning, and had little time to digest father's words. Now I have realized that back then, father already perceived that contemporary science education should not only value knowledge, skills, and exploration for learning, but also provide a profound understanding of why man should coexist with nature.

After six-years of primary education, a student had to pass highly competitive examinations to enter a high school; the entrance rate was not high. It is hardly imaginable now that an eleven- or twelve-year old child should get up as early as 5:30 A.M. every morning, climb down Bagua Mountain to go to school, and not return home until 8:30 P.M. when it was already dark. Those students whose families could afford it would have milk and small bread ordered from the school, while my mother, to save money, would send along milk she made for us. That milk was made of milk powder and water, which saved me the trouble to understand the solute, solvent, and concentration in my chemistry lessons. The homemade steamed bread allowed me to understand fermentation and breathing, as I knew what kind of flour to use, how much water to add, and how to maintain the temperature of yeast by wrapping it.

As for science education at primary school, I recall how a middle-aged woman teacher in my fourth grade taught us about solar eclipse, lunar eclipse, and the rotation of the earth. She showed us all the natural phenomena with a globe and light bulb, and draw graphics on the blackboard. Then I began to take notes of the key points of what the teacher said in class, and the teacher specially recommended my study method to the class. In fifth grade, we had a new teacher, fresh from college. She soon got married and gave birth. So my class lagged behind all other classes in every team event. Sixth grade was our last year in primary school and under the pressure of the high school entrance examinations some students continued to have some supplementary studies at the teacher's home after 8:30 P.M. We did all the four fundamental operations of arithmetic, recited the dictionary of proverbs, and recited the plosives. The high school entrance examinations usually covered three subjects: Chinese, arithmetic, and science (social and natural). But in that year of our graduation, the adults said they should alleviate the examination pressure on children and withheld the examination on science, since it merely required rote learning. That ended our class of science, and we spent almost the whole year of sixth grade on arithmetic and Chinese.

At high school, I remember only the science teacher in seventh grade. He intended to take us to do some special research, for things like science exhibitions; and I felt as if he was a scientist. Other teachers of chemistry and physics all graduated from this junior high school and later went to the best universities in Taipei and became the best students there. So they all were our idols. Throughout my high school years, aside from the experiments required by the curriculum, the teachers did some demonstrations. Sometimes we were asked to experience chemistry lessons to be taught at senior high school, for instance, we were asked to observe the burning of candles in the first class, and note the phenomena: the more we wrote the better. Sometimes we were allowed to see chemicals of various colors in the laboratory, just to prove what the books said was true. We heard of new textbooks at the time, such as new mathematics and new chemistry. These "new" things were probably imported.

It was chemistry that impressed me most. Everyone was saying how important the experiments were. Yet we could not remember the results on the tests. One classmate and I discovered that there was a manual in our library that included all the test results. So everyday we took some time to copy the results into our textbooks. As a result, at the most important uniform examinations, I ticked the answers directly without even bothering to calculate: I got high scores. For the college entrance examinations, I scored the first place for the agronomy department. I might have chosen this department because in Taiwan, we value going to prestigious universities and practical specialties. Many farming households were cultivating mushroom and asparagus, which were quite profitable, and the growers could have a chance to work at the Agriculture Committee and get picked as a member of the agricultural technology corps, with handsome salaries. Half of my class consisted of overseas Chinese students from Hong Kong, Myanmar, Malaysia and Singapore. Most of them went into agriculture following graduation. In contrast, few students from Taiwan went to work in a field relevant to what they studied.

BECOMING A SCIENCE TEACHER

Prior to my graduation, a professor recommended me to work as a research assistant in an agricultural institute. Yet the institute director turned me down, saying it was inconvenient to have a woman work with them, as it was dangerous to collect and study medicinal herbs in mountains. He did not bother asking whether I was really afraid of going into the mountains to pick medicinal herbs for study. I sighed at my misfortune; based on the number of strokes in my Chinese name, I was supposed to have a high-ranking position, at least had I been a man. But I became lucky. One day I was walking down a street at Taiwan University and bumped into the head of the agricultural department. He told me that some school needed a graduate from the department to teach meteorology. I said I had never earned a credit in education. The department head said, "You'll learn it while working!"

I went to a rural vocational school to teach meteorology although I knew little more than cold and warm front. I also taught agronomy and fertilizer studies, both subjects I took in college for just one semester. Once on the teaching job in school, I found what was taught in college was actually urbanites' concept of agriculture. The students of the vocational school all came from farming families and their experiences made me wonder what I could really teach them. How could I teach them how to carry manure on their shoulders?

Once I was asked to give a presentation on typhoon at the school's weekly rally, so I searched materials from schoolmates at the meteorological bureau and from book markets in Taipei, drafted my speech at home and rehearsed dozens of times. All the faculty members and students remarked that I had clear enunciation after I finished the presentation, but it seems nobody had paid attention to the content. When the school planned to buy a barometer, I was given some order catalogues I had not seen before, and the equipment chief of the school asked me to choose a model together with him. I actually just followed his lead, because when I did experiments I used whatever instruments were available in the lab at school. I never realized they might differ in quality.

After one year, I was transferred to a senior high school near my home to teach biology. This time I had only one subject for seven classes; each class meeting three periods a week. The work looked easy, but the school attracted the best boys in Taiwan. So the principal told me to study hard so as not to be overwhelmed by the students. These students indeed were good at learning. I could never simply concentrate on the knowledge of textbooks. More than half of my teaching used supplementary materials. Influenced by the students, I had to read the bestsellers of the time, like *Silent Spring* and *Brave New World*. This allowed me to discuss various possibilities in scientific reasoning with them. Several years later, a junior student of a medical college returned to school and thanked me for my teaching, saying the physiology classes in the junior year were all conducted in English making it hard for him to comprehend, yet unexpectedly, the biology notes he took in my class in his twelfth grade were very helpful. Every vacation, while preparing lessons, I read relevant books to supplement my teaching materials for my students. I would also gather test papers from Japan, Russia, and the United States, translate them into Chinese, and transform them into daily or monthly test papers for my students. I became well known for my ability to predict the examination questions and prepare students for them.

In recent years I have been concerned with the specialty of educators coaching science practice, and in several interviews and school observations I often recalled the process of how I became a teacher. While at the advanced vocational agricultural school and in my first two years at the senior high school, I was regarded as a teacher on probation. Since I never earned a credit in education, I was not a formal teacher. Yet my salary was not much different from that of a formal teacher, and I was not concerned. Later on, the school paid for me to take an eight-week course to earn the required credits in education during the summer vacations.

When I took the summer courses, I felt the studies of psychology were very practical, with many cases true to life. But the classes of teaching methodology on biology did not impress me much, except that the professor seemed to explain the general biology curriculum with some abstract graphics. Since I had taught the subject and had my methods to help students comprehend the general structure, I did not gain substantially from the lessons.

Four years into my teaching, I received another stipend to enroll in advanced studies at a prestigious university, which offered special training programs for junior and senior high school teachers. I took summer courses for four consecutive years. In the very hot Taiwan summer, the only one standing fan in the classroom blew toward the instructor alone out of our respect for the teacher. In the fourth and last summer course, there were moments of great excitement: I saw an electronic microscope and used a Geiger counter. This gave me the feeling that the invention of these instruments enabled humans to perceive natural phenomena indiscernible to human eyes and the journey to science.

GRADUATE SCHOOL ABROAD

In 1979, education authorities of Taiwan decided to change their policies on awarding government scholarships. They decided to fund mid-career primary and high school teachers to spend one year abroad for advanced studies in administration. I chose to study at the University of Georgia (UGA), then renowned for its science education program.

During the year at UGA, I found it rather a pleasure to study in the USA, as the tools you wanted to use were always available to you, and the library was open for long hours. Also after class I could borrow the teaching materials or books the teacher used. The theory of mastery learning, in which I was very interested at the time, was applied in the veterinary department. With the permission of the instructor, I studied the teaching methodology in the small laboratory of about a dozen people. On the lab table were bulls' legs for dissection, and the assistant professor just gave oral instructions. As there was a learning center, we went to the center to borrow cases of slides to prepare our lessons in advance. Wearing earphones, we operated the projector, guided by instructions. This helped us a lot in our studies. The experiments virtually integrated what we learned in books with real objects. In case we failed to score 80 percent on an examination, we would study and then get tested again.

In the four years after I returned to Taiwan and continued teaching in my senior high school, I never stopped thinking of going on to do a PhD. In 1985 I gave up my position as a biology teacher and became a student again. This turned out to be a wise choice. A friend of mine once remarked that it is a reward to oneself if one could spend a few years in a lifetime to experience exotic cultures. What I did not expect was that my pursuit of a doctoral degree enabled me to meet science educators from various countries in the world. This altered my career, shifting from an emphasis on teaching content—pondering about how to share with others philosophies of science education and transform them into a research program.

Once again I came to UGA. On the faculty of its department of science education were Dr. Yeany, Dr. Okey, Dr. Butts and Dr. Capie, who were active in the science education community, and visiting professor Dr. Kenneth Tobin, also a graduate of this department.

There were eight doctoral students in the year I entered the university. But I most associated with Mariona Espinet from Spain, Zurida Ismail from Malaysia, Antonio Bettencourt from Portugal, and Hsiao-Lin Tuan from Taiwan. The only American whom we saw more than others was Nancy Davis from South Carolina. We often had discussions and spent every Thursday evening together. At the time the theory of constructivism prevailed, and I even put up the fallacy of "destruction first, construction next" when I saw the school going in for large-scale construction. In my first year of doctoral studies, I could always feel Dr. Tobin around us. I read his article on wait time before I took notice of the time between questioning and answering and the time for thinking, and it dawned on me that science education should use such scientific methods of measurement to point out problems in teaching.

Dr. Tobin also guided us in getting to know qualitative research and classroom observation. He also was our instructor of statistics. The studies in the first semester of our first year were rather quantitative, but the exploration of constructivism and qualitative research in the second semester made a deep impression on me. I understood his desire to discover really new things through research. When he taught us program evaluation, he even agreed that we did an internal evaluation of our department. It was said that some professors were unhappy about it!

My tutor was Dr. Yeany, the head of the department who was smart and kind. He later became the dean of education. He was very busy, but he could go straight to the heart of an issue that puzzled us for a long time and offer solutions to it. He also helped us structure a pile of materials in clear-cut outlines. What touched me most was the way he got along with people. He took us to the small studio in his home and made bookshelves we needed in our office. One Sunday afternoon, while I was buried in my paper, he came to the department and said to the American students that they should take me to town to watch world championship bicycling race. I still see the exciting scenes of the racing first-class cyclists! Dr. Yeany also took us to Atlanta to work as judges for a science fair. I was worried that my views might not be in conformity with the local U.S. culture, but he wished to hear views of teachers from different cultures on the children's works. He all the more encouraged us to take part in international conferences such as those organized by the National Association for Research in Science Teaching or American Educational Research Association. We could get our travel expenses covered if our papers were accepted. Dr. Yeany's style inspired me to be concerned with graduate students later in my work, after I returned to my country.

BECOMING A SCIENCE TEACHER EDUCATOR

My return to Taiwan was timely. Previously the students who trained to become primary school teachers were junior high school graduates, who had to study at normal colleges for five years before going to teach. Their schooling was similar to that of sophomores in the US. Just when I returned, the system changed considerably. The future teachers were senior high school graduates who had passed the uniform entrance examinations for higher learning institutions and were full time university students. By then there were only five departments in my college, one of which was the Department of Mathematics and Science Education. Chen Jingtan, the dean, was a doctor in chemistry and concerned with science education. Many assistant professors or lecturers who only had a master's degree received financial support from the National Science Committee to pursue PhD degrees abroad. As a result, there were six instructors with a PhD in science education, which made quite a strong faculty in Taiwan's teacher training institutions. My institution therefore had a considerable impact on science education in Taiwan.

In the sixteen years since I returned to Taiwan, I experienced two so-called reforms on primary and high school curricula. In 1993, the subject of science was changed to nature. Literally, the former focuses on the transmission of knowledge, while the latter particularly emphasizes the nature of science and inquiry. Such an approach, however, is hard to realize concretely. We helped with compiling the curricula and videotapes of teaching, and we gave talks and ran workshops on the theme everywhere. But Taiwan's political atmosphere became tricky in 1998, when the education authorities turned out a uniform curriculum for the nine years of compulsory schooling, and the orientation of education was considerably affected by politics. Up to then Taiwan's teachers had been trained in normal schools, but this practice was considered to constitute a "monopoly." Accordingly, the education authorities gave permission to many other institutions of higher learning to offer education courses. This soon led to a saturation of human resources in teaching. Now it is as difficult to get a formal teaching job as it is to win the lottery.

Some people describe Taiwan's current policy on teacher training as being as changeable as the moon from the first to the fifteenth day of the lunar month, with its shape always changing. It is hard for people to get readjusted to such frequent changes, yet government officials say the only thing that is not changed in this world is change. Perhaps the twenty-first century is an era in which you cannot do things following a rule!

My main work at school is to give instructions on teaching methodology of nature science to the junior students and supervise the senior students during their education internship. As a mentor, I advise them from the first year to attend various presentations sponsored by science departments of the neighboring Taiwan University. Because students enrolled from 1988 to 1998 were naturally assigned to work as primary school teachers at various counties and cities upon graduation, I would make arrangements for them to observe programs in different primary schools.

I was once in charge of the curriculum and teaching research institute and mathematics and science education institute, as well as the master's program in teaching designed for mid-career teachers. The latter offers good opportunities for teachers to improve their competencies. It even requires a thesis for receiving a degree; hence there now is widespread acceptance in Taiwan that teachers are researchers. The research subjects of these teachers ranged from various myths among the students, education strategies, and consolidation of the curricula to solutions to problems. I appreciate that teachers really can apply what they studied to their classroom teaching. But later there were too many candidates and most of them intended to finish their studies in two years.

Over the past two or three years, a lot of changes have occurred. My institution has changed and no longer specializes in teacher training. In the summer of 2005 it was renamed the University of Education, along with other normal schools. This is like the entry into the World Trade Organization—it only gives you some nominal delight. As a university, it has to be restructured—many research centers are eliminated in the process. I am not a decision maker and I do not understand why some entities that have outside funding also are cancelled including the environmental education center that promoted biodiversity and sustainable development. Added to the school were a toy and game research institute and departments of information science and digital data. At the same time graduate students' offices from the mathematics and science education institute were withdrawn to provide space for department heads' offices and classrooms.

The utmost tragedy was to transform the Department of Mathematics and Science Education into the Department of Applied Sciences. There were actually no data to suggest that graduates of this department had limited prospects for employment. On the contrary, compared with other departments, many graduates of this department have become outstanding primary school teachers. Many years ago, there were few students registered in the departments of Korean Language or Veterinary Science, but people with vision insisted that these specialties should survive. Over the years, they became hot specialties in Taiwan. The situation regarding mathematics and science education is similar. We have a very strong faculty of science educators and there still are possibilities for a bright future of the department and its graduates.

KEY RESEARCH FOCI

To receive tenure as a professor, I had to engage in a variety of activities that were valued institutionally: offer advice or give presentations at various schools or research institutions and conduct research. My personal research has been related to my administrative work. Initially I was chief of the computer center. I therefore took the students to work on computer-assisted instruction related to primary school teaching materials. Then I became chair of my department, becoming involved with the research training of educators. Since we entered the era of online education, I have tried to set up and apply websites to remote education. In recent years I have been involved in policy studies sponsored by the science education section of the National Science Committee, specifically on concepts concerning biology for junior and senior high school students. Following recommendations made by David Treagust, Taiwan established a database on the development of students' science concepts. Recently, education authorities have made plans for teaching appraisal and grading of teachers; some administrators even intend to disqualify training institutions by means of research. Some of the research facilities are taking up a mentoring system of internship. I have participated in the research on school-based science educators' training programs.

My research is to observe the work of mentor teachers on the subject of nature in primary schools, but in reality the mentor teachers for those interns are mainly homeroom teachers, usually teaching mathematics or Chinese. Those who teach nature at primary schools are subject teachers and are seldom designated to work as mentors for intern teachers. This research compelled me to ponder the question of whether the mentor educators at primary school are born with the capacity to mentor or are trained to do so in university courses. Is their capacity based on accumulative experiences so that latecomers can follow their example? Are some courses necessary to steer them to develop in a more correct orientation and more efficient way? I chose to study curriculum designed for the school-based mentor teacher approach because of my personal experience. I have always held that there is a gap between higher learning institutions and primary schools, since everyone has problems to solve in his own sphere. Currently I am studying a case where a principal runs a curriculum and program for intern teachers. This principal intends to take a whole-school-based mentor teacher approach and attempt to enact the best possible mentor teacher curriculum and program. My research will explore what happens and the extent to which the school benefits from the program.

REVIEW OF KEY ACCOMPLISHMENTS

In December 2005, I was invited as a keynote speaker by the Hong Kong Academy of Education. They invited me probably because my colleagues and I had written some books on methodology of teaching nature. We had also translated two books, which have been used as reference books by some teacher education institutions in Hong Kong and China's mainland. They also seem to serve as a bridge for the academic exchanges between the education circles in Taiwan and China's mainland. Because they spotted my name on these works, they also invited me to work as an external examiner of their nature curricula. Therefore, it is necessary to produce some potentially useful books. As I assumed that most participants were practical workers in reality, I introduced the primary school teaching model I developed with a group of teachers over the past three years, and developed an approach using Robert Sternberg's theory of intelligence to verify that my model could successfully help improve the children's intelligence as well as their opportunities to succeed in life. I was satisfied with the completion of this teaching resource. Yet I am still uncertain of its ability to improve intelligence in just six months of training; although the test results show there is improvement.

Personally I feel gratified that I spent one year at Florida State University as a visiting scholar. Urged by Dr. Tobin, I conducted classroom observations at the nearby laboratory school and wrote a report of *Hook's Science Classroom* based on the experience, in which I used classroom observation and interpretive research to study the interaction between 21 eighth graders of physics and chemistry and Mr. Hook (the teacher), centered on the belief in science classroom learning environment and teaching from a constructivist perspective. In the course of my study, students participated in two topics: electricity and alternative energy sources. The unit itself was centered on challenging questions, with the aid of rich study resources, cooperation among students, and strategies like alternative assessment that best display students' strengths. The teacher played the role of mediator rather than controller to assist the students in constructing learning that was significant to the individuals involved, based on their current knowledge and experience. In the process of learning, the students controlled 67 percent of the time. Mr. Hook fully respected the students' rights to control their learning; while he stressed that they should be responsible for it. Based on four-months of intensive classroom observation and interviews, my research provided insights into the enacted curriculum, student learning, teacher-student interactions, and performance assessment of individuals as the curriculum was enacted.

Another paper worth mentioning is "A study of creating a distance supervision hot line." The purpose of that study was to create a means of using the Internet to collaborate with new teachers and help them to understand and solve problems that arose in their first year of elementary school teaching. The results of this study provided an alternative way for teacher educators to think about how to utilize information technology to improve teaching supervision. Our conception and website design later became a model for many universities to coach their intern students. I prefer to design curricula and teaching materials and, on campus, I also involve graduate students in creating a comprehensive illustrated handbook of plants for teachers plus a companion CD disk.

Chao-Ti Hsiung
National Taipei University of Education
Taipei, Taiwan

GILBERTO ALFARO VARELA

RISING TO THE TOP

Science Education in Costa Rica

I was born in a small rural district in Costa Rica on January 6, 1954. I am the second in a family of thirteen. Big families like mine were very common in the rural areas of the country until the 1970s. My parents attended school until fourth grade and they did not have a strong conviction about the possibilities and importance for their children attending school beyond elementary level. I attended elementary school with my brother, who is a year older than I am. There are a few images in my mind about the experiences in elementary school: playing football, reading some things, visiting some places around the school, participating in some cultural activities, cultivating some vegetables. When I finished elementary school, it was the recommendation of my elementary school teacher that influenced my parents to give me permission to continue into secondary school. Now I understand that it may not had been an easy decision for them, because I was just twelve years old and I had to leave my home to go on. For traditional families, children were supposed to live with their parents to help on the farm until they got married; and that was not the case with me. The town where I studied was far away from home. In order for me to reach that town, I had to ride a horse for about four hours. Given the distance, I lived with the family of my elementary school teacher, his mother and his sisters and one brother. I lived with them for about six years, while I was studying in secondary school and in the university. After these years of living with them I became like a member of the family. Nowadays I am "uncle Gilberto" for the children of those who were like my brothers in that period of my life.

In some respects secondary school was a hard time, getting used to a different style of life, plus economical restrictions. But very early on I decided that I needed to continue if I wanted to make a difference. My teachers generally were supportive, but I had very good relationships with the science teacher and the librarian. Over the course of five years, I took all the courses in science with the same teacher. The same teacher in that secondary school taught courses in general sciences, chemistry and biology. She is, until today, a very active person, with a very open mind, and with the idea that students are able to learn if they work hard. One day she also expressed to me the belief that her students had to go beyond the point she had been able to reach. Until this day I have vivid images about the excursions organized by the science teacher, the way she used to explain in the classes, plus some of the hands-on activities she designed for us to learn more and in a different way. In those days I used to spend much of my free time in the library, reading books that were not on the lists of the recommended texts we had to read as re-

K. Tobin, W.-M. Roth (Eds.), The Culture of Science Education, 175–183.

quirements for the courses. I just wanted to read. In the meantime, I helped other students with their homework.

By tenth grade I decided that I wanted to go onto university, and then I realized that I needed to devote even more efforts to my academic preparation. Teachers encouraged high academic preparation and the recommendation was to study in groups and to help one another. The way I could help others was being advanced in mathematics, chemistry, and biology; working in the groups, I learned the subjects by explaining to the others. I took that challenge and got enough expertise in these areas. At the end, I was the second in my group in mathematics in the national examination, I succeeded in the admission exam to go to the university and I decided to go into the chemistry area.

One day in secondary school, I had a conversation with a teacher, who worked in the same elementary school that I had attended previously. He said that he could see me one day being a good teacher. I did not have an idea about being a teacher; perhaps there were some prejudices in my mind. When I was in the university as a student of chemistry, I had the feeling that something was absent in my preparation. I enjoyed laboratories, chemical theories in the classes but I did not see myself all the time dealing only with that. At the time I was struggling with the future and I took the chance to obtain professional advice in the university. They recommended that I take some courses in education, because they saw a lot of potential for me in that area. At the beginning, it was a little bit scary, like fulfilling the idea that the elementary school teacher had put to me. But I took seriously the recommendation.

It was exciting when I started. I began to connect my academic preparation in chemistry with learning theories, philosophical ideas about education, historical perspectives of education, and school praxis. I became so involved that I really fell in love with education. I saw myself as a teacher. I wanted to work with students. I spent time in schools observing classes and talking with teachers. I knew it then: this was my field. I started working in a secondary school as an assistant librarian. Books were surrounding me again and there were many opportunities to interact with students. I also made some money to buy books in my fields, chemistry and education—my personal library started to grow. In the science teacher preparation program there were some efforts to develop an orientation toward hands on activities in science education, plus the inclusion of some of the new technologies (video, audio), in which I opted to participate. I received the highest possible grade in the practicum. Years later my supervisor told me that such a grade was not well accepted by some of the professors, because I was so young; some of them also did not believe that a young student could teach chemistry so well while enjoying it so much. During these experiences I learned an expression from my supervisor, a young energetic woman convinced that being a teacher was something worthy to live: "Go to the field, work hard, fight for your ideas and try to influence others, but be aware that, when people get to adjust to your ideas, it is time for you to move away, because the possibility to promote change has gone."

GETTING A STABLE JOB AND BECOMING A REAL TEACHER

Once I finished the chemistry teacher preparation program in the university, I also received a degree as a secondary school chemistry teacher. My certifications allowed me to garner a position as a chemistry teacher in two secondary schools (Colegio Nocturno de Naranjo and Liceo Francisco J. Orlich), about 10 kilometers apart. The first one was for adults and the second one was for young students. I came to these secondary schools with the clear idea that I needed to know and understand students' perspectives in order to make adjustments to the programs and to my way of approaching teaching. I had my first professional confrontation with the authorities of one of the schools, because of the idea that before the students arrived I had to have the program that I was planning to develop with them. At the time, the *CHEM Study* program was used in Costa Rica. I had prior experience of using the *CHEM Study* in my practicum in an urban high school, a context that allowed me to develop such a program. The high schools in which I taught did not have laboratories, nor books and other materials for students. Even though the experiences in both schools were very good opportunities for me to learn, I focus in the following on the experiences of the adult high school.

The adult high school was a relatively new school, running for about five years when I arrived. I was part of the first generation of teachers who arrived after the founders. Chemistry did not have a good reputation in the school community. In the beginning, even the school director was not clear about the importance of distinguishing a general science teacher from a chemistry teacher. After my clarification of the differences, he was proud of the "new acquisition" for the school. He introduced me to the school community, giving special emphasis to my special degree in chemistry. The first reaction of the students was not good. Some of them did not want to participate in my class. If the previous chemistry teacher in the school, who was a general sciences teacher, generated such a negative attitude towards chemistry, a new teacher who is prepared only to teach chemistry will create even more problems. After some negotiations I convinced them to try to explore the different ways of learning that I suggested. I made adjustments to the program and even when the school did not have laboratories, I introduced hands on experiences, some explorations in the community about the importance of chemistry, short experiments to be done at home and in the process, little by little chemical reactions came, nomenclature emerged, symbolic representations were on the blackboard, the notebooks and in the exams. My hope was that they became in some way, valuable instruments for thinking. A common language started to be used to explain chemical processes, familiarity with chemistry emerged and a respectful relationship was established with students, and the image of chemistry started to change in the school. The second year, I was not only teaching chemistry in tenth grade but in eleventh grade as well. All the students wanted to study chemistry. That was a challenge, but in a way I was pleased to see so many of them develop so much interest.

For the next eight years, I had the chance to be involved in many academic activities besides my chemistry classes. I was part of the evaluation committee of the school, the director's advisory board, and the delegation for the school in the teacher union conferences. At the same time, I participated in a project in which the pedagogical assumptions of Paulo Freire were developed. This was a project coordinated by the National University with schools in different parts of the country. In that project, I not only taught in the school but I studied at the university toward a Bachelor's degree in the pedagogy of communication. The idea was to integrate the contributions of different disciplines as a way of helping students in the construction of solutions to problems that teachers and students found in the community. It was necessary to work with an open mind, coordinate among all teachers, and establish connections among the disciplines. But the most important aspect was to be open to the questions that students had and to the solutions that they proposed. It was a valuable time to learn how to learn to be a teacher and how to be a teacher without being afraid of not knowing. Deep discussions were common among groups in the school, expositions of the students were part of life, and the documents generated by students were part of the products that I enjoyed reading. Nowadays I realize that at the time I read many books on education (theory, history) and ideas that came from Costa Rican educators who worked at the beginning of the twentieth century. Those were well-prepared educators who had studied abroad or who were well connected with the mainstream in the USA and Europe.

During that period of time I attended meetings once a month in which chemistry teachers discussed academic topics related to the programs, shared hands on experiences that could be developed in classes, shared ideas on evaluation, planning and teaching that helped others to see alternative ways of teaching. It was like having a special course, learning how to become a better teacher.

My experiences in that first school were overwhelmingly positive. It was there that I learned about the concept of a school community, the need for a good relationship with students as a way of supporting them, the need to work hard to maintain good teaching, and the compromise with education as a profession. I still have vivid memories of my students, my colleagues, and administrators—even of those with whom I had conflicts. The discussions nevertheless created conditions for putting into practice ideas that I wanted to promote.

BECOMING A TEACHER IN THE UNIVERSITY

While I was working in the secondary school and studying the courses on pedagogy of communication, I attended courses in general sciences and chemistry. I also enrolled in some courses in geology and some special courses on chemistry teaching. The last ones were courses that analyzed the content of the program that we, as chemistry teachers, were supposed to teach in secondary school. The courses provided opportunities to better understand the content of general and organic chemistry but at the same time teaching methodology was included in the discussions: "What to know and how it could be taught." During my participation in those courses I met colleagues who were working in different institutions and with whom I had the opportunity to talk about my perceptions of science education. By the end of the 1970s and the beginning of the 1980s many general science teachers in Costa Rica were teaching chemistry. I was in a better position than others, because my major was in chemistry. In 1980, based on the results of my participation in these courses, I was invited to come to the university to teach an introductory science course. The idea of the introductory science course was to expose students to hands-on activities in the laboratory and systematic organization of the information and discussions in order for them to facilitate their learning of physics, chemistry, and biology. This was an interesting way to get involved in teaching at the university level, because it was like a transition stage for the students and for me. I was used to teaching students at the end of secondary school and now I came to work with those students who just began university studies. The course was organized by the College of Natural Sciences and was part of a broader program for new students. We had to coordinate with professors of mathematics, language, and logic.

After my experience in the College of Natural Sciences I was invited to come into the School of Chemistry as a part time assistant. I was still working as a chemistry teacher in secondary school. The idea was to collaborate in an in-service chemistry teacher program and to support chemistry teachers as they implemented new methodological ideas in their classes. I was involved in the follow-up part of a program organized by the schools of chemistry of the public universities of the country and the Ministry of Education. The intent of the program was to improve the quality of chemistry teaching. That job allowed me to visit almost all the secondary schools of the country and to identify different kinds of problems that teachers faced while they implemented new methodological ideas. In the organization of the supportive strategies for teachers that program developed research on chemistry learning and teaching. I got involved in these research projects, sometimes doing research, at other times being part of research on chemistry teaching. In those days, science education was focusing attention on the ideas of Jean Piaget and some of his followers as the main referent for research on teaching and learning. I got really involved in reading and trying to understand his ideas. I thought they could help me better understand my double role as chemistry teacher and supportive advisor of chemistry teachers. On our team, there also were psychologists, educators, chemists, mathematicians, and philosophers. The idea was that we needed to understand not only the concepts of chemistry but also the ways the concepts are constructed by scientists and students and the ways in which such concepts fit with each other and with the whole structure of the discipline. If we were able to understand that then we could create approaches for teachers to teach chemistry in a better way. Today I can say that being involved with that interdisciplinary group was a marvelous experience: learning was central, tolerance for other disciplines was necessary, and openness was a requirement. All the participants in that group generated a deep understanding of the complexities of teaching, the responsibility with the disciplines, and the right to learn that all teachers have.

Gradually I expanded these experiences and, as part of the collaboration of the School of Chemistry, I was able to get involved with colleagues from other areas and assist them to create similar programs in mathematics, physics, general sciences, language, and literature. Accordingly, I increased my time working in the

179

university and little by little I reduced my permanence in the secondary school. In 1984, I ended my contract as a high school teacher. During the two preceding years, while still working in secondary school and in the university, I participated in a formal academic program to get a Bachelor's degree in chemistry teaching. The intention of that program was to improve our understanding of chemistry, the epistemology of science, the psychological theories of learning, and the educational approaches to teaching. It was a program for in-service chemistry teachers and we all learned in different ways. Students and teachers had to learn together, because the organization of the program did not contain traditional courses. Courses were organized to allow professionals from different disciplines to work together with students in order for them to construct approaches that contributed to them developing a better understanding of chemistry and how to teach it.

A DREAM COME TRUE

As part of the program of chemistry teaching, as I mentioned before, I was involved in several research projects. By 1984 I attended a seminar on qualitative research that made an impact on me in the way I thought about research. At the beginning I was influenced by the statistical approach to do research. But I realized in that seminar that educational problems were more complex than what previous research projects tended to show. In that seminar I met Martha Montero-Sieburth, a professor from Harvard University, who encouraged me to apply for a scholarship to study in the school of education at her institution. What appeared to be a dream became reality two years later when I actually received a scholarship from the Latin American Scholarship Program of American Universities. I had the chance to study for a master's degree in education, combining curriculum, teaching, research, and clinical supervision. Here I had the opportunity to deepen my understanding of Piaget's theories of learning and moral development and the application of Freire's contribution to understand curriculum among many other things. That scholarship was an opportunity for me to open my mind toward other aspects of teaching, research and to myself. I realized about many potentials that we have and the need to share with others in order for us to grow.

Being at Harvard changed my life—shared by my wife and two sons at that time. Science education was present in all my contributions to the program. I always made connections with the area of science in the courses I studied in curriculum, teaching and learning, supervision, and research. Even in a moral development course I had a chance to see the moral responsibilities that teachers have when they promote one way or another for students to learn. I came across very influential professors who reinforced my ways of thinking. Eleanor Duckworth is a professor with whom I have been in touch since then. She is a Piagetian with a sensitive way of approaching his ideas; her thinking tremendously influenced how I began to see Piaget and his contributions to development and learning.

When I finished the master's degree, during the commencement ceremony, the dean of the school of education made a statement that had a powerful impact on me: "Nothing of what you have learned at Harvard will be good enough to solve the problems on which you will be working in the future. Wherever you go, you will have to identify the problems and construct your own solutions. Just remember that you made good academic friends here." From that moment onward I understood that recipes do not exist in education, nor do standard solutions, but good communication with friends is a good way to approach the construction of solutions to educational problems.

After the year at Harvard, I returned to my position at the National University to work with chemistry teachers and in a science teacher preparation program. I became involved in new research projects with teachers and students and I started to include more sensitive ways to approach reality when we talked about chemistry teaching. I also got involved in science education experiences beyond chemistry, including K–9 school teachers, trying to get them to think in different ways about science teaching and learning. The science coordinators in the national Ministry of Education invited me to participate with them in seminars, short courses or just some presentations to draw the attention of science teachers to specific aspects that we needed to take into consideration to facilitate learning and understanding for children in the school. Based on the experiences that I had with different levels of the school system, invitations to participate in conferences came, and my first international consulting experience took place. In 1989 I was invited to conduct a seminar on science education in the Superior Normal School of Tegucigalpa, Honduras. That seminar was financed by a program of the California State University in Chico that supported the improvement of science education in Honduras. There were science educators in Honduras with whom I keep in touch; this allowed me to become part of a broader science community.

In 1989 I met Kenneth Tobin, who visited my university. We worked together in the organization of a seminar on constructivism and science education. That was a provocative occasion for me to familiarize myself with the perspective and language of constructivism. I think that there were many common ideas perhaps expressed with different words, but we had interesting ideas and experiences to share. At the end of his visit I received an invitation to attend the international conference on History and Philosophy of Science and Science Teaching that took place in Tallahassee, Florida, at the end of that year. It was an opportunity for me to meet or just to see part of the international science education community, plus an opportunity to pursue the idea of getting involved in a doctoral program in science education.

I was accepted as a doctoral student in science education at Florida State University and the acceptance note indicated that I had an assistantship that would support me while being there. The National University provided economic support to take my family with me (wife and four boys). I spent three years in the program, Kenneth Tobin was my major professor, and I participated as an assistant researcher in different projects that were developed in the science education program. We accomplished a lot in our collaborative work. The times were exciting. In classrooms and the hallways, there were many deep discussions focusing on constructivism. The years at Florida State also allowed me to connect to other science educators while participating in national and regional conferences. Many involved in the program had direct and indirect impact on my academic growth. As I had planned, three year later I returned to my university with a dissertation on the social construction of professionalism in Costa Rica and a PhD diploma in my backpack.

Back in Costa Rica I became involved in different science education-related projects in which an understanding of constructivism was required. Elementary school teachers were dealing with constructivism, secondary school chemistry teachers wanted to enter into the discussion, and university chemistry professors were interested in this new epistemology as a basis for doing research in their classrooms. Personally, I wanted to do research on teaching and learning in my own classroom and, with some colleagues, I organized a project in our general chemistry courses. Many things emerged from those activities and projects, including the following foci: issues of language and learning in science, mathematical bases required to construct and understand science explanations, environmental perspectives required to teach chemistry, pedagogical content knowledge, and the organization of learning environments.

The period between 1993 and 1996 included intensive collaborative research in classrooms with science teachers from different grade levels. I had to teach courses on chemistry and on research in science education and I had to be involved in the organization and development of some research projects with students as partial requirements for graduation. At the same time, I was partially involved in the development of some evaluative processes in the university. My expanding knowledge of the university and a broader perspective on teaching, learning, curriculum and research allowed me to contribute meaningfully in these activities.

OTHER EXPERIENCES

Over the years, I increasingly got myself involved in projects on institutional evaluation, accreditation, and quality management. These involved networks—collaborative projects with universities from the Netherlands, Spain, Portugal, Chile, Colombia, Costa Rica, and Germany. My participation was a base for me to participate in projects within Central America. For example, I became the coordinator of a newly created regional evaluation committee that was charged with the development of a regional system of evaluation of universities. The system still is in place today after almost seven years.

Based on the connections I began to establish with other science educators, I was invited in 1999 to participate in a course on university staff development that played a powerful impact on my comprehension of the complexities of the university and the ways they can be improved. In my university I have been moved from one place to another where authorities considered that I could better help to improve the quality of different actions in the university. I was the general coordinator of an internal evaluation process within the university focusing on teaching, research, extension and administration (1996–1998), coordinator of evaluation (1998–2002), and director of the international cooperation office (2002–2005). Even when I was involved in all these activities I kept a foot in the field of science education, not through teaching but as part of committees to support student research projects in science education. I also have been part of the academic team of a doctoral program in education in the University of Costa Rica, in charge of the courses on educational research.

Gilberto Alfaro Varela
Professor of Education
University of Costa Rica

MARIONA ESPINET

BECOMING A SCIENCE EDUCATION RESEARCHER IN SPAIN—EXPERIENCES AND TENSIONS

This chapter is part of a courageous book edited by risk-taking scholars. I understood the book as an opportunity to give voice to a wide diversity of science educators and science education researchers worldwide. The approach I have chosen has been to isolate the most important tensions I felt along the process of becoming a science educator and researcher in my country. Although writing this chapter has been a very rewarding task, I also feel uneasy. The fear of not being able to represent science education in my country and to take all the relevant people and institutions that have contributed to my professional life into account has pervaded all my writing. I am uneasy about all the missing and the overemphasis that might appear in the present chapter and also about the possibility that I might have hurt someone in building this personal view. It has never been my intention although tensions always reveal problems as a consequence of social relationships. The story told in this chapter is not about my research but about my perception of what made my own science education research and practice possible. What can be found in this chapter is a particular life within a particular context that can only be illuminated by the comparative efforts of the editors: Good luck in this work!

BEING A STUDENT IN PRIMARY SCHOOL: EDUCATIONAL INNOVATIONS IN DIFFICULT POLITICAL TIMES

I was born in 1956 at the peak of the Spanish dictatorship and started my basic education within a school (Talitha School) led by key educators in Catalonia at that time. My primary education teachers were also teacher educators who dared to undertake a new teacher education model within the dictatorship when the teacher education curriculum was tightly controlled by the fascist central government. Very young and courageous primary teachers who believed in social, political, and educational change undertook new didactical approaches for science education. As a student I was very much aware of what the school was doing and felt very comfortable and happy. As an example, Teresa Codina, the director of the school and an outstanding teacher, left Talitha School when democracy started and moved to the most deprived areas in urban Barcelona to teach gipsy children. Some of the teachers also participated in the development of the leading Catalan teacher association, Rosa Sensat. This association aimed at providing teachers with the necessary resources for continuous educational innovation and also at becoming a social movement that enforces quality in public education. Later on, when I became a science educator at the Universitat Autònoma de Barcelona, I met several of my

K. Tobin, W.-M. Roth (Eds.), The Culture of Science Education, 185–195.

former primary teachers with whom I have been working until the present days. As an example of the courageous actions that could be observed in my school, I remember one of the difficult moments I experienced in primary education. Our language, Catalan, was forbidden in schools and other public places. When inspectors from the central government came to supervise the school, both teachers and students knew what to do: hide all materials written in Catalan (children's work, posters, books), sit in our chairs silently and pretend to read a Spanish text book. The message for me was clear, education was something important, challenging, and worth risk taking. Later in 1989 I dedicated my dissertation to the educators who contributed to change education in Catalonia, Spain.

BEING A YOUNG SCIENTIST: SCIENCE RESEARCH AS A TOOL FOR EDUCATIONAL AND SOCIAL GROWTH

I arrived at the chemistry school of the Universitat de Barcelona, the larger public university in Catalonia, in 1974 one year before the dictator Francisco Franco died. I can say that my university education was shaped by the Spanish transition from a dictatorship to a democracy, one of most internationally valued political transitions for its inclusiveness and peace. Many intellectuals, professionals, or workers who had already worked towards the failure of the dictatorship continued to participate in politics, such as my chemistry master's thesis advisor, Heribert Barrera, and others started to collaborate, such as my father, Josep Espinet, a mathematician and civil engineer, who became member of the Catalan parliament.

By that time it took five years of chemistry studies before one could begin a master's or doctorate degree and thus be able to really "taste" research. The times of strong political engagements were slowly fading and I had the chance to be able to attend classes regularly and without interruptions. What struck me the most during those years was the lack of intellectual engagements in laboratory activities and regular classes. I constantly asked laboratory instructors about when we were going to pose and answers questions? The answer from the laboratory instructors always was the same: next year. My decision was clear; I had to move from the Universitat de Barcelona to the new second public university in Catalonia, the Universitat Autònoma de Barcelona (UAB), where university teachers were younger, more dedicated, and more open to new teaching and learning approaches.

My inclination to theoretical thinking moved me to choose the area of Inorganic Chemistry where I met two of my most influential role models in science: Heribert Barrera, one of the motors of the UAB Chemistry School among others, and Mercè Izquierdo, a doctoral student who later became my science education companion for the time being, and the motor of the science education department at UAB, among others. Heribert Barrera was an older chemist and a very active member of the political party Esquerra Republicana de Catalunya. During the dictatorship (the nineteen forties and fifties), he had to leave the country for political reasons and had the opportunity to undertake chemistry research in France and the USA. The newly created UAB had very little financial resources and research equipment was lacking. When I chose Heribert Barrera as my master's thesis advisor I knew I was choosing a committed citizen, someone who was very much aware of the Spanish scientific research limitations and also someone who had his life split in two. On the one side, there was his political dedication as a president of the Catalan parliament, and on the other, he was a university professor teaching courses and undertaking research. His views and personal example on the role of research in developing citizen's culture influenced my appreciation and value of research as a tool for education.

The topic of my master's thesis was to explore the possibilities of the ligand cyclohexane-ditiol to build complex compounds with heavy metals such as chrome or mercury. As chemists know, this is not an easy project since the cyclohexane-ditiol is one of the most stinking organic compounds (two sulfur atoms directly bounded to one carbon!), the resulting products were very far from being a crystal (they were brown and with a muddy aspect), and mercury compounds need a lot of care for handling. Heribert Barrera knew the limitations of Spanish chemistry research but he valued the engagement into research even if the topics were not the most salient or urgent ones. In fact, at one time he told me, "We cannot afford anything else but the research topics that foreign leading chemistry researchers throw into the sink."

It took me almost two years to isolate a rather stable compound! This was possible thanks to Carles Bayón and other colleagues in the inorganic chemistry department who helped me through while they were becoming associate professors. Their closeness, continuous help, and role modeling in the course of two years helped me understand the difficulties of being a science researcher day and night. They acted as role models of a new generation of scientists who understood that research could be only undertaken through effort, discipline, team work, and updating. Carles Bayón belonged to a group of young science researchers who raised the level of the Spanish university research by opening up to new experiences that came from being in contact with researchers from other countries with long scientific research traditions like USA, France, or Germany. He traveled abroad to the USA and it was him who encouraged me to travel abroad if I wanted to become a good science education researcher in my own country.

BEING A NOVEL SCIENCE TEACHER AND SCIENCE EDUCATION RESEARCHER: THE EXPERIENCE OF A "COPERNICAN REVOLUTION"

I had the chance to start teaching physics and chemistry in high school during the last years of my university education. Mercè Izquierdo, a doctoral student in inorganic chemistry and also a high school science teacher in the private Sant Gregori School, asked me to collaborate since she wanted some time off to finish her dissertation. In Spain it was possible to become a high school teacher in a private school without any specific training in didactics or pedagogy. To teach in a public school there was the need to obtain a certificate after a very short and inadequate preparatory course that I also took. We spent several years together sharing our inorganic chemistry research and high school science teaching. Mercè had been teaching in this school for many years and she was developing an approach to chemistry education based on the centrality of chemistry phenomena. Partly inspired by U.S. curriculum innovations of the sixties like the influential *Chemical Bond Approach* chemistry course, she strongly believed that chemistry could only be understood if students moved away from formalism to chemistry phenomena with the mediating role of language. Mercè acted as a teaching advisor and helped me see and directly experience an investigative approach to science teaching by working hard, and trying out new ways of teaching. Actually my first publication in science education was written jointly with Mercè and dealt with the experimental approach to high school chemistry that we were developing.

However I have to admit that these years as a novel science teacher were very difficult for me. The need to write this article has pushed me to reread some pages of my personal diary as a novel teacher that I found by chance the last time I was moving. Starting to teach for the first time put me in a state of considerable turmoil and high levels of anxiety. It was difficult for me to cope with the amount of uncertainty involved in science teaching: How can we know how to do it? How can we do it better? Is research into science education a tool to teach science better? Anger against the Spanish system of becoming a science teacher was pervading my daily professional life. No science education department was established in Spanish universities before the eighties. The only option was to start a new five-year undergraduate programme in pedagogy without any specific focus on science education. The key at that time came from another completely different field: the Piagetian genetic epistemology.

By chance I had the opportunity to attend a seminar on genetic epistemology and operational pedagogy given by Montserrat Moreno and Genoveva Sastre two highly respected Piagetian psychologists, research fellows in Geneva and Paris and directors of the Institute of Psychological Research Applied to Education in Barcelona (IMIPAE). These psychologists succeeded to apply genetic psychology to education (operational pedagogy) and were running an experimental school guided by this pedagogy. This seminar had a tremendous impact on my vision of science education and I compared this change to a "Copernican revolution" in science education in that the focus of reflection moved from science content to the science learner. I spent two fascinating years reading Piaget, reading the works of the researchers in the institute, and collaborating in Piagetian research on science learning at the IMIPAE. A new world was opening to me, one in which teaching science and investigating science learning was possible. I was twenty-five and decided to devote my professional life to science education and science education research.

BEING A DOCTORAL STUDENT IN SCIENCE EDUCATION:
THE QUANTITATIVE VERSUS QUALITATIVE DEBATE IN SCIENCE EDUCATION

In 1982 the Spanish democracy was only five years old and for the first time in the last forty-five years the Spanish socialist party (PSOE) won the elections. Education was a priority for that government and the effects were quickly noticed at all levels. Promoting scholarships and research at higher education was one important priority and social and financial institutions offered collaboration by promoting scholarships among other actions. Science education was included as a new university research field and science education departments were starting to operate by 1984. The problem for science education was how to build, in a short time, science education research teams composed by science teacher educators from teachers colleges and also by young science education researchers. The Catalonian bank *La Caixa* offered a very interesting and ambitious program jointly with the Fulbright Commission to send a considerable number of young postgraduate Catalan students abroad to obtain master's and doctoral degrees in the USA. At the same time the Catalan government, and more specifically the education department through Mercè Izquierdo, who at that time moved to the Catalan educational administration, offered special scholarships to science teacher educators to get part time doctoral degrees in collaboration with the King's College at the University of London.

I applied and received a La Caixa-Fulbright scholarship to obtain a doctoral degree in science education in the USA. The problem, however, was to choose a science education department that was big, active, and influential enough to help me socialize into the community. I wanted to avoid the isolation experienced within big, multidisciplinary university departments in which many science education professors or small groups were inserted at that time. In September 1984, I started as a doctoral student at the science education department in the School of Education at the University of Georgia (UGA), Athens at the time when the department head Russell Yeany was editor of the *Journal of Research in Science Teaching* (JRST). Making this decision had important consequences in later stages of my career that I was not totally aware of at that time. This decision detached me from the European science education tradition and my recently acquired background and moved me towards the U.S. tradition by being exposed to different and conflicting science education research foci and approaches.

The science education department gave me the opportunity to meet doctoral students who have continued to be active science education researchers such as Steve Oliver, Elisabeth Charron, Nancy Davis, Hsiao-Lin Tuan, Chao-Ti Hsiung, Antonio Bettencourt (who died several years ago). Russell Yeany became my doctoral advisor. He was a very bright scholar, trained in the purest quantitative tradition in science education research. His sensitivity, intelligence, and openness to the new helped him initiate the difficult but necessary transition of *JRST* from being a journal only supporting quantitative science education research to accepting other more qualitative research approaches. Other well-known and respected university scholars from outside the science education department shaped the turn to qualitative research approaches within our community. For instance, Judith Goetz, a professor in the social science education department at UGA and a leading scholar in educational qualitative research through her seminal 1984 book *Ethnography and Qualitative Design in Educational Research,* became a member of my doctoral committee and helped several of us to cope with the literature and the assumptions underlying qualitative educational research.

Ken Tobin, who at that time was a visiting professor from Australia, exerted tremendous impact on my views on the problems and methods of science education research. He was moving towards qualitative approaches in his research on classroom research and teacher change and managed to involve all doctoral students in intense teamwork on science education research. His personal commitment to educational improvement, his involvement with educational phenomena, his high productivity, and his positive leadership acted as referents for my later work in the field. In fact, my first publications in science education on science teachers peer coaching where written in co-authorship with him and other doctoral students and represented an important learning experience. I still remember one science classroom observation study we did jointly with Russell Yeany, Ken Tobin, and several doctoral students whose aim was to use qualitative and quantitative approaches to investigate teaching strategies of the best science teacher of the state. To do this, two teams were built. The one led by Russell Yeany conducted quantitative observations and the second team led by Ken Tobin, in which I was also participating, conducted qualitative observations. Communication between the two teams was not allowed until the end of the study during the presentation of the work at the annual meeting of the National Association for Research in Science Teaching (1986). Although it was certainly a risk taking experience all participants worked with high emotional involvement and mutual respect avoiding the good/bad polarization characterizing some of the quantitative/qualitative debates happening at that time. A publication in *Science Education* emerged out of this work that reflects part of the experience and knowledge gained through it (Tobin, Espinet, Byrd, & Adams, 1988).

While the European science education research community was leaning towards students' learning through the development of the pre-conceptions research program, the U.S. science education research community was more interested in teaching and the teacher through the development of the teacher thinking and teacher change research program, among others. I then experienced the need for choice between my former interests in students' learning and my later interests in teacher thinking and change. I chose the science teacher as a research focus, a decision that contributed to increase the difficulties of adjustment when returning to my country.

While being a doctoral student in the science education department at UGA I acutely felt the tension arising from the confrontation of quantitative and qualitative methodologies. This unique experience forced me to reflect on the importance of assumptions when choosing a methodological approach to science education research and equipped me to get deeply involved in the teaching of doctoral courses in educational research methodology during the later stages of my career. In addition, it gave me theoretical arguments and depth to defend the importance of action research in science teacher development and change. At present I am still working with this approach when undertaking innovations and research in science or environmental education.

STARTING AS A SCIENCE EDUCATION UNIVERSITY PROFESSOR: THE BUILDING OF A NEW IDENTITY

The need to find roots and to start a family, and the personal commitment to contribute to the improvement of science education in my country drew me back home right after my dissertation defence in December of 1989. In 1985 a new science and mathematics education department was created at the Universitat Autònoma de Barcelona (Departament de Didàctica de la Matemàtica i de les Ciències Experimentals) that gave me the opportunity to work as a science teacher educator and researcher. I had gained a Spanish postdoctoral scholarship that would support me throughout the professional adjustment to this new created community. I went back to UAB totally convinced that I was landing in the best possible science and mathematics education department in Spain.

Strong science and mathematics educators where leading the department such as Mercè Izquierdo, Neus Sanmartí, Rosa Maria Pujol, Roser Pintó, Carmen Azcarate, Josep Maria Fortuny, Jordi Deulofeu, and David Barba, among others. Their professional profile was similar: All of them were former very good science or mathematics teachers for a long period of time; before becoming science or mathematics educators because of their outstanding and dedicated work as teachers. While being science and mathematics educators at the Escola de Mestres Sant Cugat, a non-university institution, they had to undertake a professional adjustment. The Spanish education system accepted to transform the Escuelas de Magisterio (teachers colleges or normal schools) into university institutions, thus encouraging teacher educators to obtain doctoral degrees to become university professors. The fact that the dissertations of my colleagues mostly concentrated on students' pre-conceptions and were directed by professors from the King's College of London—for example, Paul Black and Rod Watson—determined the research interests of the UAB science and mathematics education community during the 1990s. The problem for me was that neither my research interests nor my previous research training in the USA were similar to those espoused by my colleagues.

I arrived at the time when a new doctoral program in didactics of science and mathematics education was launched at UAB in 1990. This program was well accepted by the best secondary science and mathematics teachers in Catalonia, who saw it as an opportunity for professional growth. The pre-conceptions topic of research was very attractive for them since it gave room for the development of their subject-oriented identity and also for the development of classroom teaching strategies that would promote students' understanding in different scientific domains. The reflective teacher as a focus of research was not appealing to them and it continued to be so until the arrival at the doctoral program of Latin American science teacher educators interested in teacher training contexts.

In addition to the efforts to consolidate a science education research community the political context was encouraging all university teacher educators to get deeply involved into the development of a new educational reform. For the first time in many decades university experts had been called to contribute to the philosophy of an educational reform, to participate in the design and implementation of intensive teacher training, and to collaborate in science textbook writing. It was a reform inspired by a constructivist approach to teaching and learning that highlighted the role of the teacher as a contextual curriculum planner and aimed at introducing changes at both the content and the pedagogical level. It was certainly a very demanding educational reform for the teachers!.

This opportunity to have educational impact could not be lost and most of us got involved in one way or another into such intensive professional activities during the 1990s. As an example, I spent a lot of time during the mid nineties writing primary science textbooks from a constructivist perspective for the publishing company *Bruño,* whose writing team was led by Joan Pagès and secondary science textbooks for the publishing company *Baula,* whose writing team was led by Rosa Maria Pujol. Other university colleagues were doing the same for *Barcanova, Casals,* or *Vicens Vives* publishing companies. The books we wrote then are out of print today and most of the textbook companies have changed their business strategy by backing old educational approaches and hiring low cost writers without authorship. Nowadays business companies no longer trust educational innovations and innovators. When I check the CV of most of us during the nineties I can see that this decision affected our research productivity, which has had inevitable impact on our possibilities for promotion. In some ways it could be interpreted as if university standards never considered the leadership role we had in the development of Spanish science education as a consequence of a strong and unique social demand.

BEING A SCIENCE EDUCATION UNIVERSITY PROFESSOR: MULTIPLICITY OF ROLES, MULTIPLICITY OF CHALLENGES

The nineties were difficult times for university hiring and UAB was a rather young community with few retirements on the horizon. I took me ten years after the completion of my doctoral degree in the USA to obtain a tenure track position in the Departament de Didàctica de la Matemàtica i de les Ciències Experimentals at the Universitat Autònoma de Barcelona. Science education was not a very competitive area, so I did not experience promotion as a very painful process. It was not the same in other more consolidated areas, such as pure sciences, economics, or humanities, where hiring committees often showed disrespect towards the high numbers of potential candidates. In contrast with such areas, Spanish science education, today, is still having difficulties attracting highly qualified candidates into the field. In one more year this context will change tremendously since, for the first time ever, Spain is going to have a university master's-level program for the training of future secondary teachers, and infant and primary teachers will need graduate degrees to become certified. This new context will provide a rich field from which to attract good candidates to postgraduate science education research programs.

International Recognition of Spanish Science Education Research Journals and Conferences

Becoming a stable member of a department meant to me the development of new professional roles as well as consolidating old ones. The department held the editorship of *Enseñanza de las Ciencias*, the most influential Spanish and Latin American science research journal and today the number one Spanish education journal. Members of our department such as Neus Sanmartí and later on Carmen Azcarate and Mercè Izquierdo acted as editors of the journal for many years. I became section editor of this journal and also a permanent member of the organizing and scientific committee of the major science education research event in Spain, the *International Congress on Didactics of Science Research* (Congreso Internacional de Investigación en Didáctica de las Ciencias). My participation in this type of professional activities gave me the opportunity to see the field from many angles and to contact many different scholars. The most important difficulty that our journal faces in the near future is to become an international refereed and referenced journal within a world where only English based research journals reach this position. I think that the international rules for the establishment of quality research standards need to change so that they become more culturally sensitive. If this was so, Spanish science educational researchers would be better equipped for promotion since the most basic criteria for getting funded and also promoted is the number of international publications in high-impact journals, which usually are published in English.

The Framing of Science Teacher Education within a European Context

Another role I had to fulfil was the becoming of a science teacher educator both at the pre-service and in-service levels. This represented an important challenge for me since I felt weak in it, and most of my colleagues were really good teacher educators and very well integrated within the local and national educational community. The strongest value held by most of the teacher educators at our university was the coherence between what to teach to student teachers and how to teach it in university courses and in real schools. I came to share the value that teacher educators used innovative approaches in their university teaching supported by innovations in schools so that theory and practice would benefit from each other. Thus, I became deeply involved in an area that was rather new, infant science teacher education, to develop collaborative science and environmental education innovative projects in schools, jointly with teachers, student teachers, and doctoral students. Spanish teacher education is going to be shaped in the near future by the consolidation of the European space for higher education. As a consequence of this process, teachers will need to develop new competencies for international partnerships. In fact, during the last several years I have started to work collaboratively with European science and environmental teacher educators to build a frame for conducting multinational science and environmental in-service teacher education programmes using the resources provided by the European Community such as the *Socrates* program.

Science Education Research as a Valuable University Field

Spanish university professors, at some time in their career, need to become involved in university management. For me, this new role arrived much earlier than expected and four years after tenure I became department head. Although I was not very happy with this new management position, I approached it as a temporary service rather than a permanent privilege. One of the most important challenges I faced was to struggle for the recognition of our area within the university community. Being department head certainly detached me from science education research consolidation but it also gave me perspective and tools to confront the future of my own science education research.

Consolidation of Two Research Groups: Language in Science Education and Environmental Education

Being a university science education researcher implies leading or participating in research groups constituted by individuals with different academic status and levels of involvement but with the same or similar research interests. In addition, these groups are characterized by hidden power relationships that have productivity consequences such as number of doctoral students, number of projects funded, and politics of authorship for the publications emerging from the group. I decided to participate in two of the research groups ascribed at the department. One of them, *Language in Science Education* (LIEC), led by Neus Sanmarti and Mercè Izquierdo, had gone through different phases following different research interests. The initial focus, pre-conceptions supported by a constructivist approach to teaching and learning, moved to evaluation as regulation when embracing social constructivism and the Russian social psychology, and finally to language in science education as a consequence of realizing the importance of language for modeling. Despite these changes, there was a common and ultimate goal: building an idea of school science as a scientific activity where there are some parallels with formal science but also important differences. My most recent contribution to this field deals with the study of science teacher discourse and the multimodal nature of communication inspired by the social semiotic approaches developed by Gunther Kress and Nancy van Leuven at the Institute of Education in London.

Several other Spanish and Brazilian science education research groups (M. Jiménez Aleixandre and Marina Castells in Spain, and Isabel Martins and Eduardo Mortimer in Brazil) share similar interests in language and common projects are underway and close to producing published products. Although the specific focus, frameworks, and methods are not the same there is a common will for collaboration and a shared belief that convergence is possible. To the extent possible we are making important efforts to find common foci and assumptions—to build worthwhile international partnerships in science education research. I see the challenge for the future of any type of international partnerships as one of building collaborations that do not fade away but highlight the particularities of science education research traditions.

The second research group at the Departament de Didàctica de la Matemàtica i de les Ciències Experimentals was a younger group named *Complex*, which is led by Rosa Maria Pujol and deals with environmental education. Most of the group members come from the interuniversity doctoral program on environmental education. In fact, several academics from our department have been very active in environmental education since the seventies. This interest has been kept alive until today when the area has been given university status and recognition. The approach chosen to frame the work of the group has been the complexity paradigm inspired by the work of several Latin American philosophers and also by Edgar Morin. The fact that several members were also participating in the LIEC group forced us to analyze the ways in which the complexity paradigm would affect our view on school science. These reflections have been collected within a monograph on complexity published in the Spanish journal *Investigación en la escuela*.

I have been very active in developing international partnerships on environmental education for teacher education. As an example I participated in European networks such as SEED Comenius III Network on *Environmental Education for School Development* led by Johannes Tschapka from Austria or Latin-American networks such as ACES Alfa Network *Ambientalización de los Estudios Superiores* (The greening of higher education) led by Anna Maria Geli from Spain. These networks have been a useful platform for reflecting on the difficulties of innovating and researching science and environmental teacher education. My relationship and collaboration with Michaela Mayer, an Italian science educator, a specialist on evaluation for school change in environmental education, and a member of *Environmental and School Initiatives* (ENSI) has been very rewarding. She has introduced me into the world of international partnerships with European teachers and has given me the opportunity to reconceptualize and reinforce action research as a tool for science and environmental teacher change.

I consider that science education research in Spain is in good health. The department I worked in has played a leading role in the consolidation of Spanish science education and might continue to do so if we all face the new challenges that our university system, national regulations and international globalization pose to the community. A good balance between local impact and international participation would be necessary for the improvement of science education at all levels.

Mariona Espinet
Universitat Autònoma de Barcelona
Barcelona, Spain

LILIA REYES-HERRERA

SCIENCE EDUCATION IN COLOMBIA

Possibilities and Challenges

Still we have a hidden country to discover in the middle of the disaster, a secret Colombia that does not fit into the molds that we had forged with our historical errors.
So it is not, then, amazing that we begin to see the apotheosis of the artistic creativity
of Colombians and realize the good health of the country definitely conscious of who
we are and what we can do. (Gabriel García Márquez)

Gabriel Garcia Márquez captured the state of Colombian society quite well in his
speech occasioned by the bicentennial celebration of the Universidad de Antioquia.
There are contradictions in our society that also can be felt in the field of science
education. This field incorporates many dilemmas and intellectual challenges and I
hope that the following reflections on my autobiography can illuminate the tensions associated with traditional and alternative ways of thinking about science
education in this new century. My reflections are grounded mainly in a research
program that began ten years ago, with my doctoral studies in science education at
Florida State University and continue with research and teaching in the doctoral
program in education and the biology department at the Universidad Pedagógica
Nacional (UPN) in Bogotá, Colombia. As we (teachers, professors, researchers,
and students) do research on our pedagogical practice and better understand our
beliefs, actions, expectations and contexts, we construct and adjust our identities.
In the process we can construct theoretical frameworks that enlighten our present
and future actions, allowing us to be agents—leading to peace and harmony between the environment and ourselves, and becoming effective leaders in transforming and improving society.

Science educators have the goal of lifelong transformation, becoming dynamic,
autonomous and responsible beings, responding to the challenges of using their
own science literacy to improve the quality of life. Colombia is no exception. Because of its responsibility as an institution for the education of educators, UPN has
educated, and has educated the educators of, generations of Colombians. The Universidad Pedagógica Nacional attempts to serve the entire Colombian population
and does not exclude students for economic, political, or religious reasons. UPN is
more than a place to train teachers to teach, since it accepts the broad goal of educating teachers as human beings—preparing teachers to embrace a Colombian and
Latin-American humanism; having the knowledge and courage to speak out to
convince Colombians they have the right to a better future. For 33 years I worked
at UPN—teaching, learning and researching—contributing maximum intellectual
effort and maintaining a high ethical stance. As a university science educator I seek

K. Tobin, W.-M. Roth (Eds.), The Culture of Science Education, 197–205.
© *2007 Sense Publishers. All rights reserved.*

to meet my goals by doing what I know best, which is to educate educators. I provide ideas for and insights into social thinking, looking for explanations in regards to societal concerns.

MY PARTICIPATION IN SCIENCE AND SCIENCE EDUCATION

The first experience I had as a teacher was in an adult education school in our neighborhood. My three sisters and I went with some of our friends (young students following different career paths at the Universidad Nacional de Colombia) to volunteer our time to benefit people living in our neighborhood. With the help of local community authorities we opened an informal night school in a building that functions as a kindergarten school during the day. We offered lessons in math, language, science, and ethics. About fifty adults participated—they were happy and receptive to our classes and we were pleased to foster the well-being of our neighbors. We were committed to the school and taught there for about two years, until we stopped our involvement when our family relocated to another part of the city. Unfortunately the evening school closed soon after we left. However, I built on these experiences in developing a similar center at the UPN.

During the last year of my studies to become a biology teacher I volunteered to be part of a team that founded an adult education center. The center offered primary education for UPN employees and their relatives, who could not read or write—some being unable to sign their names when they received their salary checks. The students included gardeners, mechanics, guards, cooks at the restaurant, and cleaners. The university employed me and my roles included secretary, coordinator and teacher. With the director, I did everything needed for the adult education center to run.

The classes taught at the center provided memorable experiences for the participants, who enjoyed seeing adult students wanting to learn, participating in ways that supported their learning, being taught by enthusiastic teachers. Most of the teachers were younger than their students, enrolled at the UPN, in training, and under the supervision of university professors. The university authorities granted two hours of compensated time for the workers who attended the center. Many of the advanced student-workers obtained a high school diploma and some were promoted to better jobs inside the university or began professional careers.

In my view, the center program was successful because the students were adults and were treated as such. The experience for the prospective teachers was also meaningful and significant because teaching adults was a challenge and they needed to identify solutions for the problems posed in class. Through their participation in the center's activities the prospective teachers learned to prepare their classes in ways that took account of what the worker-students knew and could do; it was on this knowledge as a basis that they were able to add further understandings.

In 1976 I earned a second bachelors degree at the Universidad Nacional—this one in biology with a research thesis in plant physiology. During the 1980s I taught general and plant biology to students at UPN in a more or less traditional way and supervised the teaching of prospective teachers in public elementary and high schools. Simultaneously I studied for my master's degree in agricultural sciences doing research in plant physiology. My research was in science and I had no interest in education research. I prepared for the classes I taught by studying the most recent books—to lecture effectively and evaluate my biology students. I was not overly concerned with their individual understandings because I considered my chief teaching responsibility was to provide students with the most up-to-date information. This, I thought, was a sufficient criterion for effective teaching, which I regarded as a routine. I thought that if improvements were to be made in my teaching effectiveness it would probably involve my uses of education technology.

It was not until the 1990s that I began to think of the possible value of education research. At about that time I attended an education conference and heard some Spanish guest professors talk about constructivism. But I was not quite convinced by what they had to say. My reluctance to consider constructivism may have reflected the relative chaos that characterized most traditional institutions in Colombia. In the name of constructivism students were empowered to criticize their teachers and ignored "traditional" sources of authority. Many teachers lost their jobs and their confidence in teaching because the importance of the information they knew was questioned and their pedagogical practices were criticized severely without offering them alternatives. In Colombia there was no tradition of critical pedagogy and most professors and teachers taught in much the same way they were taught by their professors. Approaches to teaching and learning were rarely critiqued and there was widespread acceptance of traditional ways of setting up and maintaining classroom environments.

My attitudes were about to change. In 1992 I began my doctoral studies in science education at Florida State University. Those doctoral years were the most significant in my pedagogical career and in my life because my professors challenged my traditional ways of thinking—gradually I began to understand and accept critical and social constructivism. My reading of one book in particular shaped my learning and thinking about education: *The Practice of Constructivism in Science Education* (Tobin, 1993). According to Kenneth Tobin, who was teaching at Florida State University at the time, science does not exist as an independent body of knowledge separated from the knower but is considered as a set of comprehensions about events and scientific phenomena that are socially negotiated. In this way, the emphasis in amount of content knowledge in the teaching of science is replaced by understandings that make students able to reach their goals in specific contexts. During my doctoral years I learned about the importance of doing, learning, and teaching how to do education research. I was taught by example, with their teaching, research, and attitudes that there are possibilities when science educators reflect deeply about their professional practices. There are possibilities even when it is difficult to introduce structural changes in the actions and beliefs of professional science educators,

From my major professor Kenneth Tobin I learned that "At the base of this (educational practice) are the referents used by educators to identify problems, give operational meaning to them, develop a plan to solve them, formulate solutions and reform policies and procedures." This key idea was central in my research program and due to that I began to explore teachers' beliefs about the nature of science, the nature of learning and teaching.

CONSTRUCTING PEDAGOGICAL KNOWLEDGE THROUGH RESEARCH

In 1996, after finishing my research and obtaining my doctoral degree, I returned to Bogotá and my "alma mater," the Universidad Pedagógica Nacional. I was the first woman in Colombia with a PhD in science education. At UPN I began to teach, write research proposals, and encourage colleagues to collaborate with me in the study of teachers' beliefs. My first research project was an exploration of science teachers' practices and beliefs—to characterize their mainly traditional ways of thinking about and doing education, and to reflect with them about new possibilities for teaching and learning to improve students' understandings of science.

The purpose of my research is to explore, reflect on, analyze, and learn about the relationships between science teacher actions and their philosophical, epistemological, sociological, and disciplinary referents. This research was done in collaborative projects where students and university professors collaborated with public school students and teachers. The talking and thinking together allowed the participants to analyze and transform their own pedagogical practices for the purpose of constructing a more dynamic community of learners. At the same time these research projects created opportunities to develop collective responsibility for science teaching and learning. Interactions among the participants generated short and reflective discourses, which also helped to validate their knowledge and practices.

In the years since my dissertation, I frequently have attended national and international teaching and learning meetings. In one of the recent congresses of the National Association for Research in Science Teaching, I presented a paper, which advocated a more significant science teacher education in Colombia in which teachers undertook rigorous and systematic reflection on their professional practice. These processes in which teachers reflect on their pedagogical knowledge and their understanding of becoming good educators constitute powerful instruments for change; they empower teachers to make appropriate decisions about what they know and how they teach. Research involving secondary school teachers and university professors provides a range of opportunities to enrich learning and teaching.

The teachers and professors participating in the research consider themselves as equal academic partners in a process of dialogue and content negotiation. This negotiated content is very powerful in the transformation of teachers' beliefs and educational practices. But in spite of the many reflective dialogues of this type we have observed only limited serious commitments to pedagogical improvement. However, I do acknowledge that some of the teachers made concerted efforts and showed great willingness to change their traditional ways of thinking about education and adjusted themselves to new visions about science teaching and learning.

By means of this participative action research, different generations of teachers have studied their own pedagogical beliefs and their professional actions, and have considered the possibilities of change for the purpose of improving the learning of science of their students, especially in those public schools that predominantly serve vulnerable populations. The analysis of teacher beliefs and the exploration of different alternatives to teach and learn made this productive and satisfying labor, which was very significant for the participants in the research and probably others who like them seek better ways for their students to learn.

LEARNING ORGANIZATIONS AND TEACHER EDUCATION

Never was knowledge more important than today. Not only the countries on the northern hemisphere but also those like Colombia are characterized by an increasing shift to a knowledge society. Every day there are more activity systems that require increasingly sophisticated knowledge—as is the case of learning to be a teacher. In general, organizations that continually adapt to the unfolding experiences as they arise are productive places for ongoing learning—enabling participants to experience, observe, and benefit from meaningful changes as they occur. Hence, the institutions responsible for teacher education cannot continue to enact programs in traditional ways and ignore what we learn from our research. Our research points to the need for teacher education institutions to redesign their programs and teaching models to adapt to and align with the values, expectations and needs of students.

If teachers are sufficiently qualified, they can participate in more intelligent and competent ways in the important roles that teachers have in society—the education of its children and youth, who constitute the future of a country. Well-qualified teachers can offer a quality education because it is well known that good teaching facilitates learning. This is why universities, as institutions that educate future citizens, have a responsibility to be agents of change. To be proactive is a desirable characteristic of a learning organization like the school. Institutions that manage knowledge in an efficient way are clear about pedagogical knowledge construction and they know how to use research results to teach teachers in individual and collective ways.

Teachers play an important role as models of what knowing and learning can be. They are knowledge workers whose mission requires them to be open to dialogue and open to problem solving with certainty. The type of knowledge that teachers promote in their classes structures possibilities for students to expand their agency and learn. It is impossible for teachers to know everything about facts, concepts and methodologies required for their actions, but it is important that they have specific knowledge that allows them to function in competent ways. For learning organizations to be successful in teacher preparation, all the participants need to be qualified enough so each one can contribute in an intelligent and competent manner in the duties required.

A SCIENCE EDUCATION DOCTORAL PROGRAM:
CREATION, DEATH, AND REVIVAL

One of the most significant experiences that shaped my approach to doing research and teaching was the doctoral program in science education at UPN, the first doctoral program at the university that until recently was a single purpose teaching institution involved with teacher education.

At the beginning of the 1990s the Universidad de Antioquia, the Universidad del Valle and the Universidad Pedagógica Nacional began conversations about offering a doctoral program in Education in Colombia. The Universidad Nacional and Universidad Industrial de Santander supported this idea and, at the end of 1991, the five directors of these public universities signed an inter-university accord to develop this program. After some years of talking and academic discussions the proposal for the inter-institutional doctoral program was submitted, and in 1996 the Ministry of Education gave its approval.

At UPN, in the faculty of Science and Technology, there were only six professors with a doctoral degree. I collaborated with two colleagues in chemistry, one in animal physiology, one in theology and another in educational technology to create the area of science education within the doctoral program. We were assigned half of the time to the doctoral program and the other half to teach in other graduate or undergraduate programs. This degree had a basic common curriculum structure and a joint set of regulations. The curricular structure had three main parts: theoretical foundations, research and thesis, and integration and complementation activities. The duration of the program was planned to be three years of full time study (though studying full time was not convenient for many doctoral students in Colombia). The science education program contained five research foci: the nature of science, actions and beliefs; evaluation in science; scientific concept formation; problem solving; and technology and education. Cohorts of doctoral students were admitted in 1997 and 1998. Students and professors engaged in research and academic seminars focused on theoretical foundations and other scholarly activities appropriate for the doctoral program. Because the research was highly contextualized, most publications were written in Spanish and published in national journals.

In November 2003, the first doctoral student in science education at UPN, Gerardo Andrés Perafán, received his degree with a laureate thesis. I was very proud of him because he was my student and I co-directed his doctoral dissertation. Since then, four more students have graduated, three are completing requirements for graduation and six are on leave from the program. Perafán's dissertation presents evidence of the epistemological diversity that constitutes the thought of some physics teachers and calls for the need to discuss and favor the epistemological polyphony attendant to teachers' subjectivity. The contemporary developments in the didactics of science show that it is pertinent to attain new meanings about the nature and functions of teachers' professional knowledge.

In 2001 the National Ministry of Education, office of evaluation and accreditation reviewed the doctoral program. The evaluators were not science educators and had backgrounds in physics and chemistry. They did not recommend the accreditation of the program even though there was solid research activity in each focus area, three of them having external funding. The sole negative point the evaluators highlighted was insufficient publications in internationally recognized journals. Accordingly, we were not permitted to admit any more students to the doctoral program. Presently, the six professors associated with the program, including me, continue to do research, publish, attend national and international meetings in science education, and conduct doctoral seminars for the students who have yet to complete their degree.

During 2004 and 2005 some professors who had recently obtained their doctoral degrees assisted in reformulating the program. This time the Universidad Pedagógica Nacional collaborated with the Universidad del Valle and the Universidad Distrital to design a doctoral program in education. The result was a plan for an inter-institutional doctoral program in education, not only in science education, but to include mathematics education; history of education, pedagogy and comparative education; education, culture and development; philosophy and the teaching of philosophy and other areas associated with active research programs in the participating universities. Finally, in December 29, 2005, after new evaluations, accreditation visits and completing all the requirements, the program obtained approval from the Ministry of Education to begin registration and admission. The signs are very promising as we experience the revival of a doctoral program in science education in Colombia.

TRANSFORMING SCIENCE EDUCATION: OPPORTUNITIES AND CHALLENGES

The Universidad Pedagógica Nacional recently celebrated its fiftieth anniversary. The slogan selected for this celebration is "History with Future." This seems to be a proper time to reflect about the past, the present and the future and it is in the context of this occasion that I proposed a book, the product of my one-year sabbatical leave entitled "Opportunities and Challenges for Science Education Transformation." This book recognizes that education is a field that involves a complex process of permanent interaction among economic, political, social and cultural forces. The ideas in this book contribute in a significant manner to the empowerment of our institutions and to the formulation and development of an educational intelligence. This book highlights the complex realities of educational development specifically in the area of science based on my experience as a professor, science education researcher and my lived experiences with prospective and practicing teachers.

The presentation of new theoretical frameworks will contribute to structuring other possible models of learning and pedagogical practices for the enhancement of theoretical understanding and cultural change. Although this book intends to maintain its original purpose of being a fountain of theoretical resources for professors, teachers, education researchers, and students, it also intends to provide relevant and up-to-date information about problems in and of science education. Furthermore, it intends to contribute to the production of student teachers and practicing teachers' professional knowledge. I have the conviction that when we open ourselves to new experiences and when we learn to look back to critique traditional scholastic systems, we will find new ways of appreciating education.

It is evident that when we compare the Colombian educational system with those in other societies, we can understand better that there are reasonable alternatives. We can learn from other societies and at the same time we can offer our knowledge. One of the main contradictions we face here in Colombia is related to research about teachers' own professional practice: schools and practitioners have not always considered it a useful resource. On the other hand, the curricular development in teacher preparation programs demonstrate that sound pedagogical knowledge and practice is not always taken into consideration on the part of the majority of teachers. In this context the questions for discussion are: What are the possibilities of development in science education? Which are the epistemologies associated with teachers' actions? How can we relate teachers' educational beliefs to favorable education policies? What epistemologies are at the base of educational practices for social change? Which relationships between science, research and society could favor greater theoretical and practical knowledge of science teacher education?

To understand teachers' professional knowledge we need to consider different epistemologies, not the ones of common knowledge, nor the ones of academic knowledge. New definitions are needed that move away from those traditional dichotomies and consider the possibilities of multiple epistemologies in the construction of the professional knowledge of teachers. We should recognize the importance of knowing teachers' epistemological beliefs to understand teaching practices.

THE ROAD AHEAD

A key interest of mine is to explore the relationships between scientific literacy of teachers and their beliefs, roles, goals and the context for a better world. In this area of scholarship I employ a conceptualization of scientific literacy that includes holistic perspectives seeking better life conditions for the planet and humanity. Scientific literacy is a different way to understand the world with a more responsible awareness that implies social, political, economical, ethical, and environmental consciousness. I believe that scientific literacy could be a way to address some of the fundamental educational problems in Colombia. It is possible to say that the option for making deep structural changes in education are those in which teachers' beliefs, roles, goals, and context about learning constitute referents of public consciousness.

Scientific literacy proposals go further than local and national frontiers to allow students to be world citizens. A scientifically literate student is able to make independent judgments about scientific and technological issues. This context recognizes that knowledge as a resource is a function of the particular needs of each society. What is true and has worth depends on the society in which it exists; each society therefore has different needs for thinking about its future and for realizing the actions that concretize the envisioned futures. For this reason it is possible to say that there is no absolute knowledge that can be considered truth for all people, under all circumstances, and in all cultures. Each people, while looking outwards to the solutions others have found, have to find the referents that are resources in guiding their development. The identification of philosophical referents for their actions and interactions and the social forces that constrain their educational actions have implications not only for the people as a whole but also for the teachers who educate future citizens. Teachers, through critical reflection about their own scientific literacy practices, will provide their students with more coherent education in harmony with the desired societal principles and values.

The twenty-first century is a time for progressive development of all human knowledge, specifically in science and technology. Even though contemporary society has multiple forms of information, very little is known about essential learning that can make the difference in our lives not only as individuals but also as society. This is a time for critical and social education, a time for ethics and human values development. In Colombian society as well as in many other societies around the world, education is not a neutral or independent system. On the contrary, education should be deeply involved in the formation of an educated public and should be against any kind of discrimination. To think seriously in education is to think seriously in the human capital of a nation. To think seriously in teacher education is to form teachers capable of leadership in their communities (Roth & Barton, 2004).

Some key questions that help us to think seriously about education are: (a) How to address scientific literacy in efforts to promote justice and social transformation inside neglected communities? and (b) What is the role of school science in the disequilibrium of power relationships in social groups? Scientific literacy is a social construction structured by social interests and practices. Accordingly, answers to questions such as these require acknowledgment of an expanded view of scientific literacy within a social group, and acceptance of multiple understandings, interests and goals of the individuals and the collective.

Lilia Reyes-Herrera
Biology Department, Doctoral Program in Education (Science Education)
Universidad Pedagógica Nacional, Bogotá, Colombia

DORIS JORDE

A BOX OF CHOCOLATES

On Making Choices to Become a Science Educator

> My momma always said, "Life was like a box of choco-
> lates. You never know what you're gonna get."
>
> (Forrest Gump)

ON BECOMING A SCIENCE EDUCATOR

I am not sure I ever really knew what I wanted to be when I was young. What I definitely know is that I never intended to study science and become a scientist. In reality, I think there are a number of career paths that would have been equally exciting and interesting for me to pursue. So how did I become a scientist and how did I become a science educator? The story of my career path into the field of science education is much like the box of chocolates Forrest Gump refers to; so many choices along the way.

On Choosing to Study Science

Reflecting back on my career as a science educator has been an extremely reward-ing process in which I have come to realize that certain people and events have played a major role in directing my career choices. My story is not filled with dreams of becoming a scientist or a science teacher. In fact, I think I could have been good at a number of different types of careers. My journey in becoming a science educator is one that points to the importance of chance as well as taking chances.

I suppose it would be realistic to say that I am a late bloomer when it comes to science. Neither my parents nor any of my five siblings had any interest in science. Like my older siblings, I seemed to be on my way to a career in the social sciences or teaching when a required college science course happened to spark an interest in me. A great female professor in anatomy and physiology was all it took to change my direction and become engaged in biology. I have often thought about this pro-fessor and her methods of teaching since she had such an important impact on me. I now know that she helped me to find my niche by choosing science. She chal-lenged me to work hard but more important, she allowed me to see myself in the role of a scientist through her own commitment to science. She provided me with a role model for action.

K. Tobin, W.-M. Roth (Eds.), The Culture of Science Education, 207–217.

It took a while for me to find my direction in biology. I moved between ideas of becoming a high school science teacher to eventually choosing an undergraduate major in Microbiology. I remember my amazement with how microscopy opened up this totally unknown world of microbes. After graduation I took a job at an agricultural field station in central California, run by the University of California. I was hired to work with bacterial infections in plants of all things! Previous plans for teaching or becoming a social worker were now replaced with scientific laboratory work and field experiments. These were great times since there was so much to learn and I was truly a novice in my field.

My "boss" at the field station just happened to be the only female professor in plant pathology in all of California at that time. Beth Teviotdale was another role model in my life—recognizing my potential and eventually inspiring me to move on to advanced studies at UC Berkeley in plant pathology. Would I have done this without her encouragement? Probably not! Deep into a master's degree on *Agrobacterium Tumefacians* I came to two very important realizations: the first that Petri dishes do not talk back, and the second, that plant pathology was a very male dominated field not yet accepting the presence and importance of women. I struggled at this point in my education to make sense of my decisions. How had I arrived at such a study, so far away from my interest in people and communication?

I suppose it is times like these when a new piece of chocolate is the best remedy for helping to make future decisions. I started searching for a new pathway able to combine my interest in science together with a pressing need to be engaged with people.

Doctoral Studies

I consider myself lucky to have attended UC Berkeley in the late 1970s to early 1980s when my search for a new direction led me to a PhD program called SESAME—Search for Excellence in Science and Mathematics Education. This innovative program was designed to be interdisciplinary, allowing students and professors to work together from multiple faculties including science, mathematics, engineering, education and psychology. The four years I spent in SESAME were the starting point for my career as a science educator. These were exciting times to study science education since many ideas were new and unexplored. The Sputnik era was still an influence on science teaching in schools, but the emphasis had changed from a concentration on teaching the elite to engaging all children in science. Psychology was meeting science teaching as Robert Karplus introduced us to the learning cycle, recognizing the importance of the interaction between the individual learner and the curriculum. Jean Piaget provided the theoretical framework leading to the *Science Curriculum Improvement Study* (SCIS) and later projects like *Great Explorations in Math and Science* (GEMS) and *Full Option Science System* (FOSS) from the Lawrence Hall of Science (LHS) at UC Berkeley. Curriculum development was at a peak in the United States as innovative programs were introduced into all levels of the education system aimed at engaging all students in science and not just those choosing science as a career path.

Watson "Mac" Laetsch (at that time director of the Lawrence Hall of Science and later Vice Chancellor of UC Berkeley) became my PhD advisor. Mac led a group of students interested in informal learning environments, including museums, zoos and other types of out of school activities. Researchers/lecturers in the SESAME program included Fred Reif, Marcia Linn, Lawrence Lowery, Elizabeth Stage, and Mark St. John. In addition to schools, our research arenas included the Lawrence Hall of Science and the Exploratorium as we explored learning in informal and formal settings. In my own doctoral work I chose to look at the dynamics of science teaching at a high school, exploring the methodologies of educational anthropology and ethnography in this setting. I was very much alone with my methodologies as I worked on my dissertation, trying to apply them to traditional classroom settings. Little did I know that less than 100 miles away at Stanford a group of students led by Lee Shulman were exploring ideas of pedagogical content knowledge as they also were observing teachers in classrooms!

Looking back at my four years as a doctoral student in the SESAME program, I now realize how fortunate I was to be a part of such a dynamic and creative environment where new connections in research and development were forcing established ideas in science and math teaching to change. It was during this period of my academic development that I understood the importance of creativity and even intuition in research and development. Good ideas come from people who have a solid theoretical background and at the same time see new connections for the application of theory.

ON BECOMING A FIRST-GENERATION SCIENCE EDUCATOR

Another unusual set of unpredictable events in my life placed me in Norway at the end of my doctoral education. I met a Norwegian student at Berkeley, married him and together made the decision to live in Oslo, Norway. I arrived in Norway with a fresh PhD in science education and absolutely no understanding of the Norwegian language. I remember thinking that nothing could be more difficult than completing a PhD . . . well I was wrong. Learning a language at the age of thirty (and continuing to learn it throughout my life) is the most difficult thing I have ever done!

Whereas the US had been working with Science education as a discipline for a number of years, Norway was just formally beginning in 1984 when I arrived. My colleague Svein Sjøberg (also in this volume) had the first doctorate in science education in Norway and was asked to lead the development of a new center for science education. Svein recognized the importance of international connections at this center and took a chance by employing a new PhD from Berkeley. For the second time in my young career, I was a part of a first generation team of science educators—there were no others before us to lead the way. Not only were we the first, we were also the only institution building up science education as a research field in Norway.

Our center was not large, having only 2.5 fulltime academic positions. I mention this because at a very early stage in my career in Norway, I realized the importance of working at all levels in the educational system. Norway has a population of about 4.5 million people. When I moved to Norway we were the only two people with doctorates in science education. We were involved at the classroom level in the development of science curriculum. We were critically looking at recruitment issues and motivational differences between boys and girls in science. At the same time we were involved at the national level in the development of the national science framework. Our efforts were needed at the school level but also at the political level to influence and stress the importance of science in the national curriculum. Had I stayed in the United States, I would never have experienced the many facets of science education as quickly as in Norway where I was literally thrown into the fire. Science education was a discipline in the making and I was fortunate enough to be a part of this unique experience.

First generation science educators have, out of interest and necessity, a much broader research and development arena than those following afterwards who are much more able to specialize. From 1984 until 2005 our Norwegian center at the University of Oslo built up a master's degree program and later a doctoral program in science and mathematics education. Our graduates are now found in almost all teacher education colleges and the other universities throughout Norway. Recently we registered a third generation of doctoral students—a very proud accomplishment.

MY RESEARCH CAREER

My early research projects were centered on science teaching at the primary level—something that we found to be almost non-existent in classrooms in Norway. Building on the ideas of constructivism and the role of the individual in constructing her own knowledge, we experimented with new ways of designing teaching sequences in science. I considered myself an "outsider" to the Norwegian school system for a number of years—an important and useful place to be when conducting classroom observations. Informants are very eager to talk to outsiders and this became my advantage as a researcher. I saw things that my Norwegian colleagues were taking for granted and was able to compare my observations with schooling in California. The life of the classroom was very different in Norway with children being given much more responsibility than is customary in U.S. classrooms.

In 1988 I had the privilege of attending a British Council course titled "Learning in Science: Issues for Research and Practice" at the Centre for Studies in Science and Mathematics Education at the University of Leeds. The director of the center, Rosalind Driver, was the leader of the course that assembled people from around the world to discuss ideas related to the teaching and learning of science. This event, and particularly Rosalind Driver, had a huge impact on my evolution as a science educator. For two weeks we worked on diverse topics including: "whose science?"; perspectives on learning; constructivist perspectives on learning and teaching; promoting learning in the classroom; and relating curriculum development to children's learning. Our multinational perspectives were constantly challenged by a dedicated group of speakers including Wynne Harlen, David Layton,

Robin Millar, Philip Scott, Joan Solomon, Susan Carey, Jon Ogborn, John Gilbert, and Richard Gunstone. Looking back at this course almost twenty years later, I realize what an extraordinary event I took part in. The Children's Learning in Science Project (CLIS) was underway with the development of learning activities that took into account children's previous experiences and ideas (often called alternative ideas or conceptions). The ability to diagnose these alternative conceptions changed the role of the teacher and the curriculum so that existing ideas could be challenged and hopefully also moved towards a more scientific understanding of scientific concepts.

Rosalind Driver was a remarkable woman and science educator having a major impact on science education. *The Pupil as Scientist* (Driver, 1983) and *Children's Ideas in Science* (Driver, Guesne, & Tiberghien, 1985) together with the article, "A constructivist approach to curriculum development in science" (Driver & Oldham, 1986) were important early contributions to my thinking on curriculum development. Rosalind Driver passed away in 1997. Her dedication to science education was an inspiration to many teachers and researchers throughout the world. She was, and continues to be, an important role model in my life.

Throughout my career as a science educator I have worked with ideas of inclusion in science and science education. Early research projects were directed at improving the situation for primary science teaching and especially for including girls in science. Using observation and video, we looked carefully at how alternative conceptions could be revealed and challenged in the primary classroom. We accepted that children had naïve ideas in science and that the role of the teacher and the curriculum was to help them explore their ideas and improve on them. We always knew that a good science curriculum was also a must for teachers if any science was to be taught since teachers' own background in science was greatly lacking. This tends to be the case in most countries where primary teachers are educated as generalists and yet the curriculum often assumes that they are subject specialists. So in the end, ideas of inclusion in science were just as much directed towards teachers as they were towards their pupils. In the book titled *Gender, Science and Mathematics* (Parker, Rennie, & Fraser, 1996) I wrote a chapter with Anne Lea titled "Sharing Science: Primary Science for Both Teachers and Pupils," in which we describe the challenges of creating science curriculum for the primary classroom.

Working with my colleague Anne Lea, we produced many different types of curriculum materials during the 1990s based on activity-based ideas of teaching science. We learned how important introduction activities were for leveling the playing field—allowing all children the opportunity to have a meaning about a topic because they also were provided some experience with the topic. We learned how to produce materials that could be used by primary teachers so that they were not threatened by the science content. It was, however, at this period in my research career that I came to the interesting observation that no matter how hard one tries to inspire teachers and pupils, there will always be those who just do not enjoy science or have a passion for learning science. I realize this may sound like a simple message, but it took me many years to learn! I remember thinking that if I could create an exciting and motivating curriculum, I could literally convert teachers and pupils into loving science. Though my goals for curriculum development remain high, I have come to realize that at best there are a few marginalized students and teachers that may become inspired and that perhaps this is good enough.

I have not been the type of researcher who has concentrated on one theme or topic throughout my career. Partially this is due to the fact that we are so few science educators in Norway and there is so much to do; but more so, it reflects my restlessness or even eagerness to move into new directions. Science museums have always been an important part of the interest I have in communicating science. Working with doctoral students Ellen Henriksen and Merethe Frøyland (e.g., Henriksen & Jorde, 2001) we have looked at how these centers might play a larger role in classroom science teaching. Though I am not as involved in these activities today, I continue to see the need to expand the types of out of school activities we provide in our culture so that science becomes a more integrated part of how we relate to our world. Science centers are wonderful places for expanding the four walls of the classroom since they are able to present topics in a way quite different from usual science lessons.

Continuing on with my interest in the science classroom, I had the unique opportunity to participate in the SMSO (The Survey of Mathematics and Science Opportunities) study from 1993–1995 together with colleagues from Japan, USA, France, Spain, and Switzerland. This project was designed to perform research in a subset of countries planning TIMSS participation to develop and validate a comprehensive battery of survey instruments addressing student, teacher and school factors that would help understand and explain cross-national student achievement differences in mathematics and the sciences. Members of the SMSO research team worked to develop shared understandings, meanings and conceptual frameworks through discussions of terms, concepts and research methods. The group felt the need to publish their findings, resulting in the book, *Characterizing Pedagogical Flow. An Investigation of Mathematics and Science Teaching in Six Countries* (Schmidt, Jorde et al., 1996). Working comparatively as we did in this project and subsequent TIMSS and PISA projects certainly makes one realize the importance of looking to other countries for ideas and critical reflection of practice in one's own country. I continue to be a supporter of international comparative studies, both qualitative and quantitative as they provide a data base of information allowing us to analyse, question and improve our own practice in teaching science. I only wish we had many more doctoral students around the world using this type of data in their studies!

CAREER AND FAMILY

Before moving on to my current activities and action plans for science education in Europe I thought it might be interesting to reflect on how career and family have been integrated throughout my journey. Young researchers are often torn between the demands of starting their career and those of family life—both of which often start at the same time. I bring up this particular topic because I am proud that I can look back and know that I managed both—but not without a great support system.

Jane Butler Kahle (also in this volume) was a guest lecturer at our center in 1987. She and her husband Floyd Nordland stayed in Oslo for three months during which time they shared their wisdom and experience with our new center for science education. Jane was particularly interested in matters of science for all and female inclusion in science—topics also prioritized at the center. Without knowing it at the time, Jane served as a very important role model for me. I had one small child and another one on the way. I had career ambitions and at the same time wanted to be the greatest mom in the world. Jane recognized my duel roles and gave me some of the best advice I have ever received. She said, "Doris, if you are able to keep your head above the water during the next ten years and not sink, then you are going to be a great success!" I have not only repeated these words endless times to myself, I now also share them with my doctoral students finding themselves in similar situations.

Norway is a fantastic place to combine family life and career since the Norwegian society decided many years ago that everyone was needed and desired in the workforce. In order to make that happen, child care needed to be abundant and of excellent quality. My support system included my husband, a great day care facility at the University of Oslo, an understanding workplace (Svein Sjøberg has children the same age as mine) and family and friends who supported my career. As it turns out, I never came near to sinking and as my children grew up, I became a much more accomplished swimmer. Now in my fifties I realize that I have experienced the best of both worlds and hope for the same for my younger colleagues—both men and women.

RECENT RESEARCH AND INTERESTS

Another turn of events and once more my career interests shifted. I was granted a sabbatical leave during the 1998–99 school year and decided to return to UC Berkeley with my entire family (my children were then twelve and fourteen). I planned to work at the Lawrence Hall of Science, picking up my interest with informal science. However, while at NARST in 1998, I attended a session given by Marcia Linn and the WISE (Web-based Integrated Science Environment) group. I was so impressed with their project of web-based curriculum development that I asked to work with them for my year in Berkeley. I have never regretted that decision.

I immediately became a member of the WISE group, learning about the strategies and challenges for designing web-based curriculum materials. I was invited to become involved with curriculum design as well as classroom trials of projects (Clark & Jorde, 2004). Working closely with Marcia Linn, Jim Slotta, and Douglas Clark as well as other members of the WISE team allowed me the opportunity to start a similar project in Norway (Viten) upon my return. Today, six years later, Viten has become a national project for web-based science curriculum, we have translated our materials to several other languages and we participate in E.U. projects connected to innovation and design of web-based environments. Viten has produced a number of master's degrees and recently my first doctoral student working in the area of ICT and learning, Sonja Mork, completed her dissertation while others are on the way. Viten is leading the way in Norway as an exemplary web-based curriculum project with new projects taking us into research and development domains that did not even exist a few years ago. The Viten project team has recently won two very important prizes for communication of science in Norway and as project leader, I am especially proud of this accomplishment.

Up-rooting an entire family for one year is by no means an easy task. Looking back, however, I would definitely say that the effort was well worth the extra work it took to make such a year happen. A sabbatical year provides the opportunity to take a step back and reflect on past accomplishments while at the same time providing impetus for new ideas and projects. I have had an extremely creative period of research and development after returning from Berkeley working with ICT and learning. I will no doubt continue to be involved with web-based curriculum development for years to come as we explore the possibilities it provides for presenting science in a modern environment for students.

Interest in information technology has also entered the classroom in other ways as we have initiated a "high-tech" video study that we call PISA+. Starting with Norwegian results from the PISA study, we became interested in entering classrooms to help us interpret PISA results. The PISA+ video study is looking at science, mathematics and language teaching at the junior high school level, targeting observations of teachers and pupils as lessons occur. We have three cameras (one whole class, one teacher and one student group) and four sound channels of information for hours and hours of teaching lessons in multiple schools. Data are recorded and stored immediately on a hard drive, allowing access for immediate playback for student and teacher interviews. The results from this study will help us to better understand the actors in the classroom and how the curriculum is implemented and received and understood. The original video study from the 1995 TIMSS (involving the US, Japan, and Germany) has now grown to be a much accepted method of conducting classroom research studies.

I am not a highly technical person myself, but do enjoy using technology in my research and development activities. I mention this because I think it is a way of staying connected to the younger generation. We know that information technology is quickly becoming an integrated part of our lives—and how silly it would be to not use this to our advantage in science education research!

NARST AND ESERA

Throughout my professional career I have been involved with NARST (National Association for Research in Science Teaching) and ESERA (European Science Educational Research Association). NARST has been an important link back to my roots as it has allowed me to follow the development of science education in the United States. ESERA is a newer organization formed at a European conference on Research in Science Education held in 1995 at Leeds, England. Our bi-annual conferences attract participants from throughout the world. ESERA organizes a bi-annual summer school for doctoral students in Europe. Sixty students and about forty supervisors (science educators) will meet for one week in Portugal in 2006 to discuss work in progress for these graduate students. The summer schools and conferences are important channels for communicating science education research in Europe. In 2003, at our bi-annual conference held in Noordwijkerhout, The Netherlands, I was elected President of the organization. I thereby became the first female president of ESERA and make that statement with pride. ESERA and NARST officially have established links for exchange of sessions at their meetings, bridging the gap across the Atlantic Ocean. Though research traditions in the two organizations are very much alike, there are small recognizable differences one sees when looking at the content of the meetings. ESERA presentations (especially those from Southern and Eastern Europe) tend to be a bit more connected to the subject disciplines in science; something that may be due to the fact that ESERA is a younger organization. At any rate, it is interesting to note that the newly elected president of NARST, Jonathan Osborne, is from Europe and the current president of ESERA is originally from the US!

NEW DIRECTIONS FOR SCIENCE EDUCATION IN EUROPE

Recent conferences taking place in the EU as well as the OECD indicate that Europe is in great need of more scientists in the coming years. Databases of information on student recruitment to the sciences show the same trend in most countries—there are not enough young people choosing science, mathematics and technology (SMT) as career paths. Not only are numbers declining in high school science classes, they are also declining at many levels of higher education. Whereas the number of women and other minorities has risen in many countries, it still is below that of men except in the life sciences. Science educators are now being asked to help solve this problem as politicians and industrial concerns recognize the role of formal and informal science throughout the cultural fiber of a country. At the same time, Europe is in a process of standardizing higher educational programs through what has come to be known as the Bologna process. All of these factors are important to mention since they point to a common agenda for Europe in student exchange in higher education as well as research programs, strategies and agendas.

Crucial questions are being asked about the image of science, math and technology (SMT) amongst the public, career information provided to students in SMT, the status of the curriculum in SMT, the status of teacher education in SMT and finally the role of gender and other minorities in SMT. It is apparent that if a negative trend of recruiting to SMT is to be turned around, countries must be willing to invest at all levels of the school system as well as in the informal environments where science is presented (science centers, zoos, television). Large sums of money from the EU will soon be channeled into the science and society component of the seventh framework program (a program for funding research in Europe).

I used to be a science educator working in Norway. Today, I am a Norwegian science educator working together with colleagues throughout Europe and the US.

My interests are directed within Norway but always with an eye towards broader applications beyond country borders.

In many ways I feel as though we are entering a new sputnik age where science education is once again on the political agenda. I think that Europeans will be looking for common solutions for solving the problems of recruitment—and this can be a good thing. The EU projects I am currently involved in stress the importance of shared competencies within Europe as we learn from each other—comparing strategies for science education in a broader perspective. I am not afraid of creating global ideas of science teaching as long as local adaptations are possible. Certainly the environmental challenges we face at the global level cry out for thinking globally as we act locally!

MAKING WAY FOR A NEW GENERATION

In 2000 I was awarded the title of Professor in Science Education—an international evaluation placing value on publications as well as contribution to the field of study. The committee commented on my commitment to my students as evidenced through numerous papers written together with them. As we strive to get ahead in our careers, it is not always evident that joint publications are a positive thing for a personal evaluation. I mention this because it is an important strategy I take into my career as a researcher. I took a chance with joint publications and in the end, was rewarded for my efforts.

At some point in a career we move from being inspired by role models to actually becoming a role model for others; realizing that our experience is worth passing on. The greatest satisfaction of my job at the University of Oslo comes from advising doctoral and master's degree students and forming research groups around them. What a privilege it is to be drawn into new areas of research as students expand the definition of science education. When looking back at a successful career, one realizes that the associations we have with our graduate students is often the impetus for expanding our research horizons.

The responsibility of mentoring a doctoral student, however, is a process that challenges most of the science educators I know. During the doctoral period it is important to help students realize their potential by listening to and challenging their ideas. They need us as discussion partners and supporters yet, at the same time, expect our critical reflection of their work. It is clear to me that graduate students should and need to be a part of a larger research group so that they can experience first hand the challenges involved with the research profession. The relationship that develops between advisor and student becomes very complex since we often become close friends with our students—a situation that often confuses our roles. As I become more experienced in my role of advisor, I understand the importance of providing support and experience while eventually allowing my students the freedom they need to shape their own career paths.

When our students become our colleagues, they continue to need our support—as mentors and role models. In 2001 I decided to put together an anthology of chapters demonstrating the creativity and breadth of science education research in

Norway (Jorde & Bungum, 2003). The main reason for the book was to celebrate Svein Sjøberg's 60th birthday in 2003 (a person who easily could be called the father of science education in Norway). Svein clearly has had an impact on all of the authors involved with the book, many of whom were just starting out on their research careers. I chose Berit Bungum to work with as coeditor for the simple reason that she was an up and coming science educator and this opportunity would no doubt influence the rest of her career. Today Berit is one of the editors of a new Nordic journal on science education (NorDiNa) and she has recently taken a top position in science education in Trondheim. Sometimes my schemes work!

IN CONCLUSION

Throughout this chapter I have tried to relate my career journey by mentioning those people and events that have had an impact on my life. I know how important role models have been for me—even though there have been few to aspire to! As I complete this chapter, I am still the only female professor in science education in the Nordic countries, which indicates to me that my role as a mentor continues to be important. And I realize that I define my success as a science educator not only through my own accomplishments, but also through the accomplishments of my younger colleagues and students who are not necessarily picking the same pieces of chocolate in order to realize their goals in science education.

Doris Jorde
University of Oslo
Norway

INTERNATIONAL FACES OF SCIENCE EDUCATION

In section B, our main theme has been the strong affiliation many science educators have had with the natural sciences prior to becoming science educators. This pattern is repeated in this section, where each author has had a strong affiliation with science. More so, all had completed science education degrees in the US; and all became major driving forces of science education in some other country. Jorde, Hsiung, and Espinet did their initial degrees in science and turned to science education whereas Alfaro and Reyes initially trained as science teachers before strengthening their backgrounds in science. Espinet, like Hsiung, became a teacher without initial teacher education. Fortunately for Espinet, she had a close colleague with whom she could interact as she taught—akin to coteaching—learning at the elbow of a senior colleague. After completion of the degree, four of the authors returned to the home country, but Jorde followed her spouse to his home country Norway and then became a shaping force of science education there. Through their autobiographies a number of emergent themes arise for further explication in this epilogue, themes that are further explicated in the remainder of the book.

GENDER AND SCIENCE EDUCATION

In Parts A and B issues of gender have arisen in relation to the career pathways of Jane Butler Kahle and Penny Gilmer, for example. On the one hand, Kahle was in the right place at the right time and gender appears to have worked affirmatively for her to take advantage of the breaks when they arose in her career pathway. On the other hand Gilmer faced obstacles that were apparently gender related and increased the difficulty of her succeeding professionally as a scientist and perhaps setting the stage for a move into science education. In this section Jorde and Espinet draw attention to the vexing issue faced by many female assistant professors when career and family goals come into conflict. Untenured women professors may require particular forms of support to afford their career goals. Many of the female scholars who contributed to this book are mothers and have had to contend with the contradictions associated with the temporal overlap between the years in which tenure is earned and those in which children are born and raised. Supportive colleagues, families, and structures within the university are essential components of solutions to problems that face female scholars to a greater extent than males when families are produced and raised—problems that also permeate gendered boundaries and can also be deleterious to male scholars with young families. But all too often, as we know from years of experience in academia, department heads and deans irrespective of their gender, use the uncertainties that come with untenured status as resources to exploit especially new female scholars. When Espinet returned from completing her doctorate in the United States she took a postdoctoral fellowship, gave birth, and readjusted to being a science educator in Europe. Whether or not Espinet was held back by her family commitments, it took her a decade to secure a tenure-earning position in science education, a field she describes as undersubscribed.

Throughout the chapter written by Jorde gender issues arise when she entered a male-dominated field of plant pathology; but she had the good fortune to have a female boss. In contrast Chao-Ti Hsiung completed a degree in agricultural science and was excluded from that field when she sought a position—informed that it was not convenient for a woman to work in the area. Interestingly, teaching is often described as "women's work" and it is a coincidence that Hsiung was hired to teach meteorology, even though she had no teaching qualifications and scant background in the subject area she was to teach. It was felt that she could learn the subject matter by teaching it—learning to teach by teaching. Learning to teach out of field by teaching out of field is an experience shared by many science educators because the scope of science education is so broad and many teachers with a degree in a given area of science are well out of field in relation to what they have to teach (e.g., a major in physics having to teach biology). Hence, even those science educators who were scientists may be called on to teach science topics that are quite distant from their areas of specialty.

Of course, gender is not necessarily an issue for all women, and some auto/-biographies written by female scholars do not exhibit gender as an issue. Similarly, male science educators often are oblivious to the different treatment their female counterpart receive or experience as receiving. Thus, gender is not a salient issue in most auto/biographies written by male science educators. It is precisely because of these differences in experience that the science education community as a whole has to engage with the topic of gender as a structure that mediates career trajectories, research interests, and approaches in the field. In a recently edited book on the use of auto/biography as method in education, readers will find an interesting auto/biography of a female science educator working in a chemistry department that articulates many of the tensions that women experience in academia but that are not salient to most (perhaps all) of their male counterparts (Scantlebury, 1995). We return to the issue of gender in part D, where it is the main theme.

COLLABORATION, ALLIANCE, AND GLOBAL SCIENCE EDUCATION

Science education in the US has mediated to a considerable extent the developments of the field in other countries. Because of the migration of science educators and science curricula across national boundaries (e.g., see part E), internationalization of science education has required a lingua franca, which, because of a number of historical factors, has been English. But the language of communicative exchange is not value neutral, and genres for reporting science education research are different in different national contexts—one of us (Roth) has worked with many German and French science educators, assisting them in publishing in Anglo-Saxon journals and found in the process that ways of articulating issues, making points, organizing texts differ across cultures and languages. This means, English is not just a lingua franca but also a means for implementing a worldview, and therefore, a means of dominance and hegemony. By wanting to publish in English-language journals, members of other cultures and speakers of other languages submit to and reproduce this hegemony (and all its ideologies). Resisting this development, bringing about national science education movements, and emphasizing publications in languages other than English constitutes a form of resistance to this hegemony.

We therefore have to ask the question, "To what extent is science education a global field and to what extent are national and regional issues of such salience that they counter trends toward globalization of science education?" Within the executive committee of the National Association for Research in Science Teaching (NARST) there have been several efforts to promote the international faces of the organization with resolutions to change the name of the organization, hold international meetings in Europe and promote native language sessions at the annual meeting. One such effort involved planning for the annual meeting of NARST to be held in the Netherlands. Barry Fraser was a central figure in planning the meeting and plans were well advanced when resistance from leading science educators in Europe urged NARST not to proceed. Some of the leading science educators in the world, including Rosalind Driver, argued that Europe should first be given time to organize its own *European Science Education Association* and schedule its first meeting. Other efforts for NARST to be more inclusive floundered because of the dominance in the organization of the English language. For example, while Tobin was NARST president the organization promoted sessions in which Spanish was used and translation was offered for English speakers. This innovation was expanded in a subsequent annual meeting when poster sessions were in Chinese—catering to a large contingent of NARST members whose native language was either Mandarin or Cantonese. However, resistance from within the organization persuaded subsequent presidents-elect (those who plan the annual meeting) to discontinue the practice.

Language and geography appear salient in forging alliances, although some authors of the chapters in this part of the book have forged relationships despite being in different parts of the world and speaking different languages. There is a pattern of reaching out while looking in. Sjøberg and Jorde created alliances with centers throughout the world but focused their efforts on science education in Nordic countries, a regional group of nations with a sufficient number of cultural issues in common and geographical proximity. Similarly, Alfaro has been primarily concerned with science education in Costa Rica and Central America, although he has been involved in collaborative projects in South America, Europe, and the United States. Hsiung has also participated in collaborations with scholars in Europe, Asia, Australasia and North America.

When Espinet returned to Spain from the United States she quickly felt the differences in her own science education and the more Anglo-Saxon approach adopted by her colleagues—their graduate studies being directed by scholars such as Paul Black. The conceptual change approach to research, which was dominant in global science education, had a foothold in Spain and it was to be some time before her studies of reflective practice in teacher education would be well received by her colleagues. Since obtaining her doctorate most of Espinet's professional efforts have focused on science education in Spain and Europe. Even so, she has consolidated the social bonds she created while in the United States. Espinet has invited both of us to participate in Spanish conferences in science education as well as other well known international scholars such as Jay Lemke and Richard White.

Globalization, the requirement of a lingua franca for engaging in communicative exchanges, and the issue of hegemony are not to be taken lightly in the science education community. They require as much sensitivity as gender issues. But in a world dominated by journals appearing in English and featuring Anglo-Saxon editors and reviewers, and the requirements even universities in non-English speaking countries impose on their faculty to publish in journals with high-impact factors (as per the ISI ranking) reproduces the cultural-historical hegemony of the Anglo-Saxon culture and language. We return to these issues in part E, Science Education Around the World.

INTERNATIONAL COMPETITIVENESS

An interesting aspect in the field of science education constitute international comparisons, which create opportunities for those who conduct the comparisons but may constitute constraints to the developments of the field within the countries where there are different local needs. For example, in Canada the needs of students from the various First Nations require very different approaches to curriculum than in countries such as the Netherlands where there are no North American aboriginals; the needs of Maori students may be quite different from those of students attending school in Finland (Finns, Russian immigrants). Conducting comparisons requires flattening or eliminating cultural differences. This is a consideration especially important in the light of the fact that precise translation is impossible because the semantic shifts and differences involved even in rendering the same conceptual content in different sentences of the same language (Derrida, 1998). But comparison requires the assumption that the differences are due to differences in ability or teaching *everything else being equal*.

Despite some evidence for globalization of science education the connections between science and the military and the economy have fuelled international competitiveness. For example, in part B Sjøberg referred to his role in international assessments of achievement in science education. Jorde also mentions the *Third International Mathematics and Science Study* (TIMSS) and the *Programme for International Student Assessment* (PISA). In many regards the international studies of attainment in mathematics and science have been the epitome of international competitiveness in science education. In a project conceptualized in Germany at UNESCO's Institute for Education, a number of leading educational researchers planned a pilot study in 1959–60 involving the comparisons of science achievement (among other subjects) of 13-year olds in 12 countries. The group of scholars, that included well-known icons such as Benjamin Bloom, Robert Thorndike, and Thorsten Husén, were to form the *International Association for the Evaluation of Educational Achievement* (IEA). Subsequent highly influential international studies undertaken by the IEA included The *First International Science Study* involving 19 countries with data being collected from children aged 10 and 14 and in their terminal year of high school. The *Second International Science Study* replicated the first study with data collected in 24 countries in 1983–84. The IEA, in conjunction with Boston College, was involved in the original TIMSS with data collected in 1994–95 and then follow-up studies referred to as TIMSS-R and TIMSS-2003.

Barry McGaw was a key figure in the development of assessments to examine science literacy beyond the classroom in 49 countries. The target group for this study constituted 15-year olds and samples were collected in 2000, 2003, and 2006. The Paris-based *Organization for Economic Cooperation and Development* (OECD) organized the research, which was part of PISA, while McGaw was the director of the education directorate.

One key bottom line in relation to the international studies of achievement in science is the polarizing effect it has had in many countries because they are said to have lagged behind Asian countries. Accordingly, in countries such as the United States a high percentage of the resources to support curriculum development, professional development of science teachers and research have been targeted toward redressing the unfavorable ranking of the United States in the international studies. Of course a funding priority also has been for additional international studies. The same has happened in Germany, a country that has ranked low despite its educational system focused on providing appropriate instruction to students of different ability, streaming academically oriented students differently than those who would instead receive high-quality vocational or technical training. The international comparisons have oriented German educators away from the high-quality workers they produce, workers that are welcomed and searched for in countries such as Canada, to focus now on making students competitive on tests that show little other than doing well on tests and—as shown by ethnomathematical research (e.g., Lave, 1988)—are unrelated to successful practices people deploy in their everyday worlds on the job, in the streets (markets, betting), at home, or while shopping.

PUBLISHING IN HIGH-IMPACT JOURNALS

As we discuss in earlier chapters, there has been a tendency for the number of journals in science education to increase and along with that a system has evolved to sort out the most prestigious journals. Many countries now have paid attention to the impact rating of journals and in the case of books most weight may be given to the university presses—especially those at Ivy League universities. As well intended as this might be, there are obvious shortcomings. Take for example Reyes' description of how the doctoral program she had created was shut down on the recommendations of a group of scientists without a background in science education. She notes that their sole stated criterion to support a decision was a lack of productivity of the faculty in terms of them publishing in high-impact journals. Of

223

course there might have been other unstated reasons for failing to recommend continuance, but consider the written recommendation. Why should scholars from Spanish speaking countries be expected to publish in English speaking journals? The expectation that they can and will is probably unrealistic and not all that useful. With considerable support some foreign scholars will coauthor with their advisors during graduate school, but the number diminishes when the scholar returns to his or her home country—as has been the case for all scholars featured in this section. Unless a researcher grew up with English as a language that was learned alongside of the native language, the chances of reaching fluency and being able to fully express ideas in English is diminished. Not only that, to reach scholars facing problems that are similar to the ones they explored in their research it makes sense to publish in journals that are accessed by scholars within the nation in which the study was undertaken. Hence, it is encouraging to see national science education journals in South Africa, Taiwan, Australasia, Spain, England, and Norway. It seems silly that evaluation panels, such as the one in Colombia, would use data like publications in high-impact journals as a criterion of scholarship.

The other side of the coin is an increasing internationalization of areas such as science education. For example, during the weeks that we complete the writing of this book, the European Union has expanded the number of participant nations to 27. To communicate, apply for, and distribute funding across the European states, a small number of languages have to function as linguae francae, including for the publication of research. There are therefore tensions between a necessary homogenization and reduction of Babel to a few languages all the while addressing the particular needs of linguistic and cultural differences.

It is not surprising that there would be efforts to promote high-impact journals written in languages other than English and that as Espinet reports, *Enseñanza de las Ciencias* is a high-impact journal for Spanish speaking scholars. In many respects a field is defined in terms of those who contribute to a journal, by reading it and publishing in it. Hence, a community of Spanish speakers has a strong affiliation with this journal and a series of Spanish speaking conferences.

The issue of where to publish books also is contentious. Both of us have a strong association with Sense Publishing, the publisher for this book. Our rationale for publishing with Sense includes accessibility. We want our books to be read and if that is to be the case they should be accessible throughout the world and affordable. The idea that Ivy League presses publish the best books is simply erroneous. We know of departments within Ivy League universities making a payment to the university press to ensure that, if it would publish a book written by one of its junior faculty, a minimum number of sales would be guaranteed. Similarly, we know of senior administrators requesting a university press to publish a book authored by an assistant professor struggling to make tenure.

When we proposed a new journal in science education to Springer Verlag, we did so in light of the fact that there already were about seven journals whose purpose was to publish research in science education. Part of our argument was that they were pretty much the same as one another. That is, the history and culture of science education led to a homogenization of language, genre, and psychology-oriented approaches and produced a lack of alternatives open to scholars interested in cultural-historically evolved differences within and across nation states. The goal of the journal we created, *Cultural Studies of Science Education*, though it continues to be published in English, is to be sensitive to the local contingencies that come with cultural and linguistic diversity. It is therefore not surprising that in its second year, *Cultural Studies of Science Education* will feature a special double-issue focusing on globalization (internationalization) of science education.

TRENDS IN THEORY AND RESEARCH METHOD

The years during which science education became international, as part of the increasing movement of students between countries, theories and research methods in use underwent considerable changes. The major research-driving theories changed from behaviorism to cognitivism to Piagetian and then radical constructivism only to be replaced by sociocultural, cultural-historical, and sociological theories. In regard to research method, the years of increasing internationalization fell together with the emergence of qualitative-interpretive approaches to the extent that much of research today employs interpretive approaches.

The Rise and Fall of Constructivism

During the 1970s and 1980s, Piaget's theory predominated in the field. Thus, Espinet, for example, also was influenced strongly by Piaget, attending a seminar on genetic epistemology led by experts in Piagetian psychology. On her road to becoming a science educator Espinet raised questions about her ways of teaching and brought together a preference for investigative forms of learning and dispositions towards theorizing learning. Alfaro's graduate education was undertaken in the United States at Harvard and Florida State University. At Harvard he was deeply influenced by Eleanor Duckworth, who had worked with Piaget—a role model with whom Alfaro is still in touch. It is interesting to note that pedagogy in Costa Rica was mediated by the pedagogical ideas of Paulo Freire as well as Piaget.

When Espinet and Hsiung were together studying for their doctorates in science education at the University of Georgia Tobin was there as a visiting professor. He had been searching for ways to make sense of teaching and learning science and especially learning to teach science. At the same time Ernst von Glasersfeld was in the psychology department and was collaborating with mathematics educators where constructivism was already well established among many of the leading research groups in the United States. Glasersfeld had a strong impact on Tobin's thinking, especially about the role of metaphor in making sense of experience and the epistemological and ontological aspects of radical constructivism. As Tobin grew to understand radical constructivism and used it in his research, many of the graduate students took classes with Glasersfeld and began to make sense of their research using constructivism as a referent. Within science education there were ripple effects and many of the conceptual change researchers began to use constructivism claiming that they now had a name for what they had done and thought all along.

Glasersfeld, however, sent shock waves through the field when he pronounced that those who only attended to the salience of prior knowledge to learning, without also thinking about the epistemological aspects of knowing, were trivial constructivists. His remarks stung many scholars within the research community and a great deal of debate arose, involving efforts to make sense of learning in ways that acknowledged both the role of learners and their part in a larger community of learners. Perhaps the greatest debates concerned the implications for constructivism and the nature of science. These debates involved scientists and philosophers, many of whom did not want to relinquish science as an exact way of knowing that was a quest for truth.

As is evident in the chapters of Alfaro and Reyes, the constructivist turn in science education was still in evidence in the mid 1990s, though greater emphasis was being placed on the social aspects of coming to know and researchers like Michael Roth were beginning to employ theories that were well beyond constructivism, advocating a much different form of social learning than had been embraced by social constructivists endeavoring to figure out the relative weight to be given to individual and collective components of learning. Eventually, perhaps by different routes, even science educators who had made their name in conceptions and conceptual change approaches, such as Rosalind Driver, acknowledged the role of communities in individual knowing and adopted social constructivist perspectives. Later, beginning just prior to the turn of the century, dialectical theories that had their origins in the cultural-historical approaches of Alexei N. Leont'ev or cultural sociology of William Sewell came to be used by a number of science educators who wanted to overcome the dichotomous treatment of individual and culture that had marked all other approaches. Dialectical theories sublate the dichotomies, because the polar opposites are recognized as one-sided expressions of larger units— much like the dualistic expression of light as having particle or wave nature when in fact it simultaneously is (a) both particle and wave and (b) neither particle nor wave. The phenomenon of light as a new unit of thought overcomes the particle–wave dualism.

Emergence of Qualitative Genres of Research

Although the Stake and Easley Case studies in Science Education were around in the late 1970s they were not widely available and Tobin was unaware of them in his doctoral studies. However, at the University of Georgia, case study research based on the clinical methods of Jean Piaget and using videotape were well established in the mathematics education program—especially by Les Steffe's research group. In contrast, science educators within the same building were heavily grounded in radical behaviorism and used quantitative data almost exclusively through the late 1970s and early 1980s. The changes that were to occur within the next five years were hard to imagine. However, Hsiung and Espinet were there at the time of change and so was Tobin.

By the time Tobin came to Georgia on a sabbatical leave in 1984 he had already undertaken his ethnographic studies of science classrooms with Jim Gallagher and he was ready to introduce the doctoral students at Georgia to an exciting new way to think about research. Several events occurred to make this more possible than might otherwise have been the case. Russell Yeany became head of the science education department and asked Tobin to teach the doctoral seminar. As Espinet describes in her chapter, one of the studies that was set up at that time focused on one teacher's classroom through quantitative and qualitative lenses. This study involved many of the science educators at Georgia at the time and so too did a qualitative evaluation of the science education program. Finally, as editor of the *Journal of Research in Science Teaching*, Yeany made it clear that he encouraged qualitative studies and research built upon constructivist perspectives. He was inclusive in reaching out to science educators to publish their best work in JRST. Within the field Yeany was a catalyst for changes that allowed numerous researchers in science education to employ qualitative data resources in their research with a sense that they would get it accepted by the leading journals. Although subsequent editors were not quite so welcoming, Yeany's impact in the field had set in motion a movement that would not be stopped by the gatekeepers of science education or federal and state politicians.

Interestingly, on the west coast of the United States, in 1984 Jorde completed a dissertation entitled *An Ethnographic Study of an Urban High School: Science in the School Culture*. The study was in many ways ahead of its time, involving ethnography, studies of an urban school, and the culture of a school. A study with such a title would not be out of place in 2006. However, by the mid 1980s, in science education there was about to be a groundswell of dissertations that employed qualitative data resources. Using the University of Georgia as an example, studies included interpretive research (Linda Cronin, 1986), participant observation (Elisabeth Charron, 1987), and case studies (Mariona Espinet, 1989). As was pointed out by Hsiung, who did a more traditional quantitative study, graduates were equipped to undertake studies that employed a variety of data sources. When Hsiung returned to Florida State University in the mid 1990s she undertook ethnographic research in a high school classroom. Alfaro also undertook interpretive research in his own doctoral research and since then has continued to do collaborative research with teachers in their own classrooms. Espinet also has been involved in collaborative action research with teachers in a European network that is concerned with environmental education.

Michael: In the early 1990s, I was a high school physics teacher doing research in my own classroom and, together with fellow teachers, in their classroom. The teacher as researcher became an important paradigm, though few have contributed to the literature in a substantial way. Jim Minstrel is another one of those who come to my mind. When I switched into the university in the fall of 1992, I continued to do research on teaching and learning in classes I taught, though I no longer was teaching fulltime in the elementary and secondary science classrooms. At that time, it became more popular to do research on teaching and learning in science classrooms—Tom Russell and David Hammer being two other examples of university professors teaching at the high school level and reporting from the perspective of the teacher what they were learning. Then, of course, we began the research and development program on coteaching, which took you, too, into the classroom; and you did it probably to a more extensive degree—over years—than any other science educator that I am familiar with or know about.

CENTERS OF SCIENCE EDUCATION

When did science education begin as a field? Looking at the data from the *Dissertation Abstracts International* historically it is interesting to note that most universities in the United States began to produce science educators from about 1969. However, universities like New York University begin in 1894 and others like Harvard and Cornell show entries in the 1920s. At least in terms of obtaining doctoral degrees it seems as if there was a trend that began in the 1960s for science educators from around the world to come to the United States to obtain a doctorate in science education. Many then returned to their home countries to continue as science educators—involved in curriculum development and university education. Gradually a critical mass of science educators grew in many countries and doctoral programs in science education emerged globally.

Emergence of New Centers

Sjøberg and Jorde clearly were pioneers for science education, not only in Norway, but also within the Nordic countries. As Jorde points out, being one of just a few science educators necessitated that she not have too narrow a focus within the field. She felt a need to carry the torch for science education and with Sjøberg, to assume leadership. Notwithstanding what she regards as a broad focus it is apparent that science education has flourished within Norway and she notes that already there is a third generation of doctoral students. We assume Sjøberg belongs to the first generation and his doctoral students have now supervised their own doctoral students to graduation. Reyes notes a similar situation in Colombia where she was the first person to have a doctorate in science education. Upon her return to Colombia she assumed a leadership role, creating a doctoral program and supervising the first of its graduates. Then, when a review panel closed down the doctoral program she planned with others to create another avenue to create a means for doctoral education to be obtained in science education in Colombia. Here as in other parts of the world graduates from other countries created a doctoral program to allow doctoral degrees to be home grown.

In Taiwan doctoral degrees in science education were initially based in three regions of the country—National Taiwan Normal University in Taipei, National Changhua University of Education, and the National Kaohsiung Normal University. However, with an increasing number of scholars with doctoral degrees in science education the number of universities that now offer a doctorate in science education has expanded to include National Hualien University of Education, the National Chao-Tung University, and the National Taipei University of Education.

In Costa Rica four people obtained doctoral degrees in science education in the United States and since that time they have not sent people out of the country. Since his return to Costa Rica Alfaro has been major professor for approximately 10 students who have completed doctoral degrees in education, not in science education—and not at his own university, but at the University of Costa Rica, in San José.

Doctoral degrees in science education have been available in many European countries, including England, Germany, France and Spain. The developments in these countries, however, have been mediated by different structures and cultural-historical contingencies. In Germany, for example, science education still is located by and large within faculties of science and specific departments. Thus, Manuela Welzel did her doctoral work with a professor whose chair is part of physics rather than education; in most North American universities, however, science educators receive their degrees within schools of education (with the exceptions of a small number of universities, including the University of Southern Mississippi, where Roth obtained his degree in a faculty of science and technology).

One of the influential centers in Europe has been the *Institut für die Pädagogie der Naturwissenschaften* (IPN; Institute for Science Education) at the University of Kiel. In 1966, Karl Hecht founded the IPN within the Institute for Applied Physics; its first two departments being physics and chemistry education. Three years later, the IPN expanded to include two additional departments: biology and pedagogy. In 1971, Karl Frey became the director of the IPN and brought about the interdepartmental collaborations that have characterized IPN to this day. This development was significant in the German context, where much of the work in science education still comes out of discipline-related departments rather than from departments and institutes that focus on science education more broadly. In 2006, the IPN consisted of 143 employees, with twice as many scientific as support staff. Over the years, the IPN expanded its influence, not in the least because of its tremendous research and development output, which, in its 40-year history, amounts to 5,000 publications. As we describe in other parts of this book, key science education centers have emerged in Europe at Kiel, Leeds, and Kings College.

Where Do You Go to Get Your Degree?

Three well known science educators were involved in a debate about which of the science education centers (in the United States) was most influential in science education at the present time and in the past. The debate was good-natured and each mentioned the schools they had graduated from and had taught at. The discussion was far reaching and involved issues of numbers of graduates, what graduates had done with their degree, and to what extent the centers had obtained external funding and published in high-impact journals. Not surprisingly the issue was not resolved.

Ken: I became aware of places where people did doctorates in science education when leading educators in Australia traveled to the United States (mainly) to do a doctorate at either Florida State University, Michigan State University, or Ohio State University. It was early in the 1970s and these people, including Greg Ramsay and Robert Vickery, came back to very senior positions in Australia. In the United States both had experience in nationally funded curriculum projects and their first impacts when they returned were in developing innovative curricula that built upon what they had learned in the US. However, Australia was short of science educators with doctorates and educators from my vintage began to look overseas as well. Many went to Florida State because of its reputation in curriculum development and I felt that I might go there too. However, based on my research on wait time I knew of Mary Budd Rowe and wrote to her and also to Linda DeTure, then a doctoral student at the University of Maryland. In each case I was encouraged to consider a large program and my reading of the literature and advice from the expatriate Americans in Australia was to go to Georgia to study with someone like David Butts who had published in the initial volumes of JRST. Accordingly I made my way to Georgia and it was only when I arrived that I learned of a spirited rivalry with the University of Iowa.

Michael: In my own situation, my apprehensions about returning to the university and my superintendent's advice to consider the *Summer Program for Graduate Studies in Education* at the University of Southern Mississippi (USM) mediated my decision. Furthermore, I had never been aware of differences between universities. In Germany, there did not appear to be differences other than that the Freie Universität (Free University) Berlin was said to have a bad reputation because of its leftist, even communist leanings. So I only realized toward the end of my graduate program that in science education, USM was not "on the map" and that it might have been better for me to go to an institution such as Cornell to do a PhD with Joe Novak. Today I know that the university you attend also plays an important mediating role in career opportunities and trajectories in France. Although some science educators—including my own PhD supervisor Marlene Milkent—claim that it does not matter where you do your degree, I personally think that it does matter. After I had graduated and begun to attend international conferences such as those organized by the National Association for Research in Science Teaching and the American Educational Research Association, I realized the narrowness of my discourse and thinking. The culture at USM was not oriented toward research, and few of my colleagues ever published in a first- or second-tier science education journal being instead more concerned with science teacher education and teaching courses. For example, I was completely unaware of qualitative research or conceptual change theory until after I had finished; and when I encountered constructivism during the 1988–89 stint at Indiana University, it was such a tremendous shock that it contributed to my decision to abandon the idea of a university career and to return to teach science at the secondary level.

An analysis of the data from *Dissertation Abstracts International* suggests that Iowa may have been more productive than the University of Georgia, at least as far as graduating doctoral students is concerned. Since 1970 Iowa has graduated 205 and Georgia has graduated 129 in about the same period of time. However, both pale in comparison to Michigan State, which produced 251 graduates since 1969. Others such as Teachers College (147), Northern Colorado (145), Indiana (141) and Purdue (118) also graduated high numbers of doctorates in science education. The University of Southern Mississippi, from which Roth was a graduate, has produced 128 science educators since 1970.

Tobin graduated in 1980 along with 114 others listed in the *Dissertation Abstracts International* with subject code 0714—the number used to code most science education dissertations. In that year approximately 65 institutions were involved in graduating science educators, including New York University, Temple, Syracuse, Auburn, Michigan State, Southern Mississippi, Illinois at Urbana Champagne, Northern Colorado, Iowa, and Indiana, which each graduated at least three students.

Tobin: When I came to Florida State University the science education had been suspended for academic reasons. I was hired to resuscitate the program. In the period I was there, between 1987 and 1997, 25 science educators graduated with a PhD. This compared to 94 from Michigan State, 77 from Iowa and 55 from Georgia. Hence the pattern is clear, based on the numbers of science educators produced Michigan State appears to have outpaced other universities in the United States. Obviously any argument about the extent to which a science education center is successful needs to take account of many factors. However it is worth noting that centers come and go, in terms of their size (e.g., faculty, graduate students, undergraduate students, external funding, publications in high-impact journals) and quality. Hence, when I left Florida State University the faculty in science education had already discussed the problems of sustaining a viable doctoral program with limited faculty resources to support graduate students through earning of external funds. We accepted the idea that the numbers of doctoral students, especially in state institutions, might rise and fall periodically as a reflection of numerous other factors. The chief concern among those of us who discussed the size of the doctoral program in relation to our faculty was not to become a degree mill. We all knew just how difficult it was to supervise more than just a few doctoral students at the writing stages of their degree, especially if one of them had English as a non-native language.

As Fraser noted in his chapter in part A, the leveraging effects of producing doctoral graduates has a major impact on the field of science education. Since Curtin University was assigned university status in 1987 and therefore could award doctoral degrees, almost 200 students have graduated with a doctorate in science education. The students came not just from all over Australia but also from countries around the world—a sure sign of the commodification of doctoral degrees in science education. It remains to be seen if the numbers of students taking doctoral

231

degrees from institutions such as Curtin decline as the opportunity to study locally increases. In comparison to a doctorate from the United States, a degree from Curtin has several advantages in that students can more easily satisfy residency requirements by interacting with their advisors locally and electronically and the requirements for the degree can be met through a thesis-only approach, hence not requiring attendance in class. Many students are able to maintain full time employment and complete their degree through Curtin. Quality has not been a problem in that Curtin graduates have been recipient of best dissertation awards and several scholars with Curtin doctorates have been awarded early career awards through NARST. Also, graduates from Curtin are currently placed in leading research institutions throughout the world.

Michael: There are advantages but also disadvantages with such an approach. Although I like learning on my own and encourage my students exchange their formal course work for specially arranged courses that allow them to do research apprenticeship, I find that not having access to courses on research methods, for example, hampers students in contexts other than my own. For example, in European countries I have found frequent instances of science educators coming through their science departments who did not have good methods backgrounds and underdeveloped practices. At the same time, I even have written about the limitations of taking courses or reading books on research methods. The students who come through programs such as Curtin are disadvantaged in that they may not be part of an active research group where they can learn how to do research by doing research with peers and professors. Also, I know from experience that graduate students in many universities experience isolation precisely because they are not part of a research group that provides them with the kinds of formal and informal relations that allow them to cope with the stresses and demands of becoming a scholar. I have often heard science educators talk about a pattern of not finishing the doctoral degree when all-but-dissertation graduate students return into their schools and, overwhelmed by the local demands, fail to complete their research and writing it up into a dissertation. Being fulltime at a university until completion certainly is a factor that increases retention.

SCIENCE CURRICULUM DEVELOPMENT

Science educators who received their training in the US and then returned constitute one of the factors for U.S. curriculum developments to make it into the curricula of other though not all nations. Alfaro's early life as a chemistry teacher provides insights into the penetration of *CHEM Study*, one of the high school curriculum projects produced in the United States in the curriculum revolution of the 1960s. Tobin also used *CHEM Study* in Australian schools when he taught chemistry in the late 1960s; and the same curriculum was used in Israel—as shown in the chapter Hanna Arzi contributed (see part E). On the other hand, the researchers at the IPN developed and tested its own curriculum in chemistry ("IPN Curriculum Chemie," to be used in fifth and sixth grade), which, after five years of testing, was published and implemented in 1972.

Jorde's chapter highlights the centrality of curriculum development and design as salient when she did her doctoral degree at the University of California, Berkeley. Robert Karplus, a renowned physicist who turned his hand to the development of elementary science curricula was still a dominant force at Berkeley when Jorde began her doctorate. As was the case in the earlier parts of the book Jean Piaget's psychology was dominant in the *Science Curriculum Improvement Project* (SCIS), one of the most famous of the alphabet soup curricula of the 1960s. SCIS incorporated the learning cycle, an approach to learning that was created by Robert Karplus and his colleagues. Karplus was strongly influenced by Piaget's learning theory and applied it to children's learning of science. With Myron Atkin (Atkin & Karplus, 1962) he developed the idea of guided discovery and then by the late 1960s this had been elaborated into what Karplus and Herb Thier referred to as the learning cycle, which was to become part of the dominant paradigm in science education—research as well as curriculum development. Initially described as exploration, invention and discovery, these labels were clarified to mean exploration, concept invention and concept application. Even today research on the learning cycle is ongoing and the *Biological Sciences Curriculum Study* expanded the learning cycle in its instructional model, often referred to as the "5 Es" (engage, explore, explain, elaborate and evaluate). The middle three terms are similar to the original learning cycle. Berkeley, largely through the leadership of Karplus, Atkin, and Thier remained a powerhouse in science curriculum development with renowned projects such as SCIS, *Great Explorations in Mathematics and Science*, and *Full Option Science System*. Also, due to the leadership of Marcia Linn and colleagues Berkeley has maintained a highly prominent role in research in science education that was initially grounded in Piagetian psychology and gradually has focused on applications of the Internet and technology to the learning of science.

Part D

REDUCTION OF GENDER BARRIERS

GENDER IN SCIENCE EDUCATION

In the early years, the gender differences in science were reproduced also in science education. A quick check allowed us to ascertain that about 10 percent of the articles in *Science Education* during the early to mid-1980s were authored or co-authored by female science educators, where over the past several years, nearly 50 percent of the articles were authored or co-authored by women. This shift reflects a similar shift in the composition of the departments in which science educators find themselves, and where the number of women has increased substantially over the last several decades. In this book, too, there is a higher frequency of male authors in the early parts, but a higher frequency of female science educators in the latter parts. The culture of science education therefore reaches right into the structure of this book. We devote this part D of our book to the reduction of gender barriers in our field.

It is true that there have been female science educators even in the early years of science education of the modern era—that is, after the Sputnik scare. Some of those who distinguished themselves include Jane Kahle, Anne Howe, Mary Budd Rowe, Dorothy Gabel, Marcia Linn, and Frances Lawrenz. But the numbers began to increase as more and more women enrolled in science education graduate programs since the latter part of the 1980s. Science education began to change as the increasing number of females raised the gender-related issues both as object of research and in terms of the structure of and participation rates in the discipline.

Three female science educators, two of them immigrants, studied for their doctoral degrees in the United States, overcame a variety of obstacles in their professional careers, and now enjoy tenured full professorships in major universities. Each of the following chapters has unique parts that address the roles of women in science and science education, publication and writing, obtaining external funds, and catalyzing changes. Two of the scholars are immigrants to the United States, opting to stay in the United States after completing their doctoral degrees. The three auto/biographies raise questions about the identities of science educators because each of these women deviates from what might be regarded as a typical route to becoming a science educator and earning a science education research position. For example, Lee has scant background in science, did not teach in K–12 settings, and does not describe herself as a science educator—even though she publishes primarily in the top ranked science education journals and has obtained millions of dollars of external funding to support research in science education. In contrast, Varelas had a solid background in physics, was certified to teach, and taught science in Greece and the US. Although she studied in a program in curriculum and design without the guidance of a bona fide science educator as an advisor, her advisor had a background in chemistry and strong interests in social learning theory.

Varelas has published in high-impact journals in science education and she, too, has successfully obtained several million dollars in external funding from the U.S. National Science Foundation to support her research in urban schools. Finally, Lynn Dierking, has a strong background in science, taught in K–12 settings, but wanted to study what she called free-choice learning in museum settings that included the participation of families. Such a focus did not fit the template for science education in research universities in the early 1990s and she experienced many obstacles that catalyzed her production of alternative pathways that have led her to a unique appointment as professor of free-choice learning at Oregon State University.

The auto/biographies of these scholars raise questions about the field of science education and what it means to be a successful researcher in the field. As we have in previous parts of the book, we address some of the emergent issues from the chapters in an epilogue that follows directly after the three auto/biographies.

LYNN D. DIERKING

UNDERSTANDING THE NATURE OF SCIENCE LEARNING

Analysis breaks down when we are dealing with complex systems with many interactive loops. In such systems, you cannot just isolate the parts and put them back together, because in isolating the parts you change the system. The system has to be considered as a whole.

~Edward de Bono, 1993

For the past thirteen years I have been associate director of the Institute for Learning Innovation, a not-for-profit learning research and development organization based in Annapolis, MD. Prior to that time I was an assistant professor in early childhood and elementary science education at University of Maryland, College Park and Director of the *Science in American Life* curriculum project at the National Museum of American History, Smithsonian Institution; earlier in my career I was also a classroom science teacher and museum educator. Beginning this fall, I will share a tenured full professor position with John Falk at Oregon State University, in Corvallis, OR, in the free-choice science learning program offered through the science and mathematics education department.

Throughout my career I have been committed to better understanding the nature of science learning and its role in a learning society, by investigating it across the lifetime, in multiple settings, including museums, nature centers, homes, after-school programs and on the Internet, and across varied age groups including among children, adults and intergenerational groups. My work has demonstrated to me the essential role that such learning plays in our everyday lives as we watch television, read newspapers and books, converse with friends and family, visit museums, libraries, parks, and increasingly surf the Internet. In fact, because it is so important, John Falk and I have been advocating the term *free-choice learning*, to replace informal learning, because it focuses on the nature of the learning—nonlinear, personally motivated and involving considerable choice on the part of the learner—rather than where it occurs. Good teachers in all settings, including classrooms, utilize free-choice learning principles in their teaching. Although I am clearly passionate about free-choice learning, I feel that for the most part it is undervalued and poorly understood and my professional life has centered on trying to change that situation.

Three interwoven strands have played important roles in shaping my work and my entry, navigation and trajectory within the science education field: (a) immediate and immersive experiences, (b) supportive mentors, and (c) serendipity and synergy. As I prepared to write this chapter and reflected upon my participation in

K. Tobin, W.-M. Roth (Eds.), The Culture of Science Education, 239–249.

the science education community, I recognize the wisdom of de Bono's quote above, for what is more complex than one's life and the act of introspecting upon it. Therefore, though humbled by the conundrum, and in agreement with de Bono, that it is difficult, perhaps even impossible, to isolate the parts of a system without changing them, for the sake of a narrative and focus, I will discuss each of these strands separately though throughout will also attempt to point out the relationships between and among the strands.

IMMEDIATE AND IMMERSIVE EXPERIENCES

In wisdom gathered over time I have found that every experience is a form of exploration.

~Ansel Adams

On the advice of an enthusiastic and persuasive chemistry professor, who was assigned to be my freshman advisor, I entered the undergraduate program at University of Miami majoring in chemistry. However, after my first year I switched my major to biology, with an emphasis in field biology/ecology, a much better fit for me since I loved the outdoors, was an avid camper, and over my formative years had participated in many extracurricular science activities and programs, many as I grew older, marine biology-focused. Growing up in Florida was also an influence I am sure. Although like many my age who had grown up watching Jacque Cousteau on television, I fancied myself studying marine biology, once at college I quickly became more interested in terrestrial and estuarine systems. The program at University of Miami was excellent, extremely field-based and immersive, with the Florida Keys as one learning resource and the Everglades as another. As I had been in junior high and high school, I was encouraged by my professors and did well. As an honor student I was provided special opportunities, for instance, I spent a summer in the Keys learning about invertebrate zoology and a portion of my junior year conducting original research at a field station run by the university in Ecuador. I earned an undergraduate degree in biology with honors from University of Miami, and then made plans to continue my studies with the intention of earning a PhD in forest ecology at Yale University.

Both as a high school student and undergraduate I had been fortunate to have opportunities to be engaged in research and at some level I had always assumed that I would be a researcher, but in biology rather than science education. Within days of graduating, Wit Ostrenko called me to see whether I would teach a marine biology course for nine- and ten-year-old students in a summer camp program at the museum. I also was a facilitator for the Museum's camp-in program that meant that every other week I spent the night at the museum with about fifty pre-teens, engaging them actively in science and hopefully exciting them about the topic also. It was great fun though exhausting!

The summer job continued into the fall and I was totally captivated and transformed by this early teaching experience so that I decided to change my career path and focus on education. I decided to earn teaching credentials and became a middle and high school science teacher at a parochial school in Miami, teaching introductory physical sciences, biology, physics, and field biology. That decision did not seem at all odd: I was interested in teaching and therefore would become a classroom teacher. At that point in my life I did not realize that the specific field of science education even existed. In fact, as I began to share the news with my former biology professors, I received mostly negative feedback and expressions of disappointment that a "bright young woman" was choosing to be a teacher. I repeatedly heard: "Those who do, do, those who don't teach."

I enjoyed being a classroom teacher, particularly in a parochial school where I had tremendous flexibility and some control over what and how I taught. However at the end of the year, I realized that despite the freedom I had, much of my role was focused on the management and teaching of students, rather than on facilitating the learning of those students. This was a seminal moment—in the process of teaching children in a museum and outdoor settings I had become intrigued by the question of how to facilitate learning and was interested in better understanding how to support learning in these settings though at this point I was still assuming that this would be a sideline interest since my first husband and I were moving to Gainesville to attend the University of Florida.

However, a la de Bono, two serendipitous events occurred that summer which changed my trajectory in ways that later were profound, though at the time I did not realize their full impact. First, right before moving to Gainesville, I was in a serious car accident requiring several weeks of hospitalization. This event consumed my family so much so that when paperwork from University of Florida arrived regarding my acceptance into the forest ecology program, it was disregarded and I never accepted the offer. Second, soon after arriving in Gainesville I wandered into the campus natural history museum to inquire about becoming a volunteer. Once staff there realized that I had experience working in a museum, I was offered an educational programming assistant position with primary responsibilities focused on supervising and training docents and the development of programs for visiting school groups. I held this position for three years, another transformational time for me. Within a year of accepting the position at the museum I was a science education researcher focused on learning outside of schools.

RELATIONSHIPS WITH SUPPORTIVE MENTORS

The greatest good you can do for another is not just to share your riches but to reveal to him his own.

~Benjamin Disraeli

What was the impetus for this shift in my thinking? In addition to the power of immediate and immersive experiences, I have been fortunate to have a number of supportive mentors who have guided, challenged and directed me throughout my career. Enjoying etymology, I discovered that Mentor was the name of the person to whom Ulysses entrusted the care of his son Telemachus when he set out on the "odyssey" that would lead him to participating in the Trojan wars. He was Ulysses' counselor and Telemachus' tutor, thus mentor became a shorthand term for a wise and trusted advisor and tutor.

My first wise and trusted advisor was Wit Ostrenko. Despite the fact that he was concerned about my decision not to pursue a doctorate in biology, in part because I think he felt guilty about having made the call to ask me to teach classes in Miami, he has continued to be a supportive mentor. Fortunately I think over the past five years I have been able to convince him that I have no regrets about that decision and owe him a great debt of gratitude that he made that call.

A second mentor was Edward Munyer, the Director of Interpretation at the Florida State Museum. Ed loved to talk about philosophy of education and in the course of some conversations early in my tenure at the museum he learned of my interests in learning, particularly in settings such as museums and the outdoors. He suggested that I walk across campus and talk to John J. Koran Jr., a professor in science education in the College of Education, who was interested in exploring the use of museums as natural learning laboratories. This was my first inkling that there was a field of science education and it sounded intriguing so I made an appointment with Koran. Throughout the time I was in Gainesville and more recently as Ed moved to the role of director of the Illinois State Museum in Springfield, he has remained a great resource to me particularly suggesting reading materials and forcing me to hone my argumentative skills.

By far the most influential person early in my career was John Koran who became my major professor. I still remember the first meeting I had with him. It was a few weeks before the beginning of the fall semester in 1979. He had recently moved into a new office so boxes were everywhere and there were no seats for visitors. I recall sitting on a table and talking to him for a couple of hours completely fascinated by his ideas. Because I was in limbo regarding the forest ecology program, John convinced me to take an ecology course and an introductory science education course. Although I enjoyed forest ecology and found the science education readings very challenging and far more intimidating than the ecology course, I had changed my mind toward the end of the semester and decided to become a graduate student in science education. I wanted to focus on learning in settings like museums and outdoor settings. Ultimately, I received an MA and a PhD degree in science education under John Koran's guidance.

It is only in the last ten years that there has been an informal science strand within the National Association for Research in Science Teaching, an informal learning environments special interest group within the American Educational Research Association, a policy statement for informal learning published by the National Science Teachers Association and a Board seat focused in that area and a permanent special section on science learning in everyday life in the journal *Science Education*. Thirty years ago one had to be an exceedingly independent thinker to pursue a research agenda in informal settings and there is no doubt in my mind that John Koran was an exceedingly creative and independent thinker. For someone like me he was truly an inspiration. I think it is fair to say that without John's leadership, vision and pioneering spirit, much of the progress that has been made in this area over the last fifteen years would not have been made.

Twenty-seven years ago when I walked into his office, John had the vision and insight to be thinking that museums were potentially wonderful environments in which to investigate learning, an idea in many ways only coming into its own now. Although John is not solely responsible for its acceptance, his work and that of people he inspired, was a strong impetus. It is much easier now to pursue this line of research—people do not necessarily think that you are *that* crazy but thirty years ago you had to be stubborn and you had to be tenacious. John had both of these qualities. He also had a wonderful sense of humor, an ability to always see the positive. When I moved away before completing my dissertation to take a job at the Smithsonian, it was John who continued to have faith in me that I would finish.

It is not that other graduate students and I always agreed with him. John was a strict cognitive psychologist focused on information processing models who had been trained in the behaviorist tradition, and he and I had many a philosophical difference, both about the nature of learning, as well as the most effective methodological approaches with which to investigate learning in such settings. But he was a good listener and excellent sounding board when you were trying to work through a difficult problem. He brought rigor and systematic thinking to a line of research that benefited greatly from his disciplined approach. And he was willing to let me try things even if he did not entirely believe that they were appropriate. At a gut level he understood Disraeli's notion of mentoring being as much about revealing mentees' own ideas as sharing their own.

One sign of John's ability to evolve his thinking and learn from his mentee is evident in the difference between the title and emphasis of my MA and PhD work. Earned in 1981, my MA *The Relative Effects of Recessing as an Attention-Directing Device in a Museum Exhibit* was an experimental study in which controlled groups of students from the local laboratory school were brought to a "treatment area" and given equal amounts of time to observe two different approaches to directing attention in an exhibition. There was also a control group that did not view the exhibit at all. Completed in 1987, my dissertation *Parent-Child Interactions in a Free-choice Learning Setting: An Examination of Attention-Directing Behaviors* was a naturalistic study in which I tracked fifty families from the moment they entered the museum until they left. The focus in both studies was attention, an important research area in cognitive psychology at the time, but the design, methodological approach, and make-up of the studies were quite different.

John was very precise in his thinking and he modeled that precision for me. Even back in the early 1980s he was questioning whether *informal* learning was the best moniker for the kind of learning in which I was interested. In my first publication (under my married name, Shafer) "Learning Science in Informal Settings," John and I discussed the differences between informal and formal learning, focusing a great deal on setting-related differences. As one can see in the title of my dissertation, he and I were using the term free-choice learning setting rather than informal learning setting, a term we discussed and experimented with during the period between my MA and PhD degrees. We were still focused on the characteristics of the setting, probably in great part due to John's behaviorist training. However, this kernel of an idea would evolve to the point that in 1998 John Falk and I suggested that the term free-choice learning be used to describe the nature of a particular type of learning, self-directed and guided by the needs and interests of the learner (Falk & Dierking, 1998). Informal learning instead appeared to refer to a particular kind of setting rather than type of learning

John Koran's independent spirit also resulted in another excellent mentoring activity. John believed in taking his students with him to conferences, workshops and meetings, but not as an adoring backdrop, but upfront, making presentations, co-authoring articles and beginning to establish oneself. Good academic mentors do that now, but thirty years ago such an egalitarian approach was uncommon. I remember the first NARST meeting I attended in 1982 in the Catskills, presenting findings from my MA degree. At that time it was quite uncommon to see a graduate student at NARST, unless they were making the perfunctory "completed dissertation" presentation. After the presentation, I started looking around and realized that there were very few graduate students at the meeting, prominent among them was Jeff Lehman, another one of John's students.

I found two other mentors at the University of Florida, interestingly both in measurement and statistics, my minor focus area. Rodman Webb was a qualitative researcher and because my research interests were evolving in more qualitative directions I took an ethnography class from him in the fall of 1982. The class was transforming in many ways. We had to conduct an ethnographic study and Webb suggested we choose a setting that interested us, sit there, and watch what happens. I chose an area within the Florida State Museum called the *Object Gallery*, which was considered a hands-on area. I sat there on weekends, a time when I was not usually in the museum and through this immediate, immersion experience, I discovered that there was a major group of people visiting this gallery, whom I had not observed when working during the week and at school hour times: families. This group was fascinating to me. Since ethnographers investigated the prevalent patterns observed in an environment, I focused my study on families that ultimately began my career-long focus and work in family learning, both in and from museums, but also in other settings. This study revealed to me that a gallery was only hands-on, if perceived so by the visitors themselves, for despite the opportunity to interact with objects in this gallery many families did not seem to know that they could and so they walked around the perimeter interacting with it as they would in a very traditional exhibition—they looked and pointed but only rarely stopped and talked about what they were looking for.

My second mentor in this department was James Algina, a theoretical statistician who ultimately was a member of my committee and supported my desire to do a naturalistic study. Because of his strong quantitative background and "good housekeeping seal of approval," Koran was willing for me to proceed. He also pushed me to overcome a lifetime of mathematics anxiety by inviting me to be a graduate assistant for a quantitative methods course: I did indeed understand and enjoy applied mathematics and statistics. I also had an "aha" moment, while in his advanced quantitative methods course: theoretical statisticians were in search of the "perfect" data set, rather than dealing with reality. It was at that point I recognized my limitations and interest in the real messy world! Algina was a patient and excellent teacher. For example, because of some theoretical differences between John Koran and me, I ended up having to re-analyze my data late in the process of completing my dissertation. Algina spent hours with me on the phone, tutoring me about generalizability theory that I would ultimately use.

In Gainesville I also met my most important mentor, John Falk, who remains one of my major influences to this day. Falk, associate director of what was called the Smithsonian's *Chesapeake Bay Center for Environmental Studies* (now the Smithsonian Environmental Research Center) was a visiting scholar and faculty member, whom Koran had invited for the spring semester. Fortunately, he arrived at a pivotal point for me; I had just completed my qualifying exams and was beginning to write my dissertation proposal. I was immediately impressed with John's intellect, energy, and scholarship. He helped me to clarify the differences I had with Koran, inspired me to push for those aspects of my work that were essential but different from Koran's, but also wisely counseled me "to focus on winning the war, not the battle."

John was an additional person and a burst of energy, but most importantly he brought additional knowledge and experience in out-of-school learning to a department in which one faculty member was nurturing a fledgling group of students interested in learning in free-choice settings. John Koran was but *one* faculty member interested in the topic and as I learned a bit later in my career, one faculty member, even a full professor, does not make a program. In the majority of my classes I was a lone voice, often finding myself defending museums and other out-of-school settings as places of learning and legitimate environments in which to investigate learning. Although these circumstances honed my argumentation skills and remain useful to this day, I grew weary at times of always being in the defensive mode. John Falk brought validity and credibility to our efforts since he was an established researcher in the area at an internationally recognized institution.

Falk cotaught with Koran a wonderful course on learning in informal settings. Better still were the opportunities to talk with him informally and to observe him at work. I recall vividly how we scribbled down initial ideas for the design of my dissertation on a cocktail napkin. He was generous with his ideas, methodologies, and even with his equipment. He lent me a funky electronic timing device that he had created, which I used to collect my dissertation data. We also shared an office and I was able to learn a great deal from John by listening and watching him at work. He also took me on "field trips," including one to Disney's *Epcot Center*.

Falk also gave me my first job as a researcher at the Smithsonian. It was wonderful to work in an intellectually focused office with others very passionate about out-of-school learning. I worked on the National Science Foundation-funded Community Science Project designed to broker relationships between schools, businesses and informal learning entities. It was an eye-opening experience trying to work with a school system, that although indicating interest in the project was extremely rigid and ultimately set up tremendous roadblocks that made it difficult for us to be successful. The experience reminded me of why I was interested in learning in free-choice settings.

John and I replicated my dissertation research at the National Museum of Natural History, Smithsonian Institution. He led or collaborated on a variety of other research projects and was in the midst of writing a book focused on why people visit museums, how they behave while there, and what they take away from these experiences. Ultimately during this time our professional life together also transitioned from that of mentor–mentee into a more collaborative relationship. I helped to refine and rewrite *The Museum Experience* and to find a publisher for it. It was the first of several book collaborations, the others including *Visitor Experiences and the Making of Meaning* and *Lessons without Limit*.

In *The Museum Experience* we first began to develop a framework for conceptualizing learning in and from free-choice settings and in our next book, *Learning from Museums,* we refined and expanded upon the framework, renaming it the *contextual model of learning*. The framework posits that learning is not an abstract experience that can be isolated in a test tube, or laboratory, but an organic, integrated experience that occurs in the real world and is situated within three contexts. We argue further that learning is a product of millions of years of evolution, an adaptation that permits an on-going dialogue between the individual and the physical and sociocultural world in which she interacts.

The contextual model describes three overlapping contexts: (a) the *personal* context, those characteristics of a person that are brought with her to the experience; (b) the *sociocultural* context that the visitor brings with her and immerses herself in while in the setting; and (c) the *physical* context in which she interacts, both exhibitions but also the space and its amenities also. Factors within each of these contexts are known by learning researchers to be important to the learning process. Learning is the process and product of the interactions between these three contexts and the power of the contextual model is not that it attempts to reduce complexity to one or two simple rules, but rather that it embraces and organizes complexity into a manageable and comprehensible whole. We have also suggested a fourth and very important dimension—time. Looking at free-choice learning as a snapshot in time, even a long snapshot is woefully inadequate. One needs to pan the camera back in time and space so that one can see the learner across a larger period of his life and can view the experience within the larger context of the community and society in which the person lives and interacts.

Beyond formalizing the contextual model of learning, the notion of the importance of longitudinal efforts to investigate free-choice learning, and developing and championing the term and construct of free-choice learning, there is one other product of our collaboration for which I am genuinely proud. In 1986 Falk founded the *Institute for Learning Innovation*, a not-for-profit learning research and development organization based in Annapolis, MD. Although Falk was the founder, throughout its twenty-year history I have played a major role in shaping its identity, products, and reputation. Falk created the Institute after the Smithsonian Institution abolished the Office of Educational Research he had established there. Under our leadership, the *Institute* has become recognized as a pioneer and leader in the burgeoning discipline of free-choice learning and it has enabled us to continue our academic work, while nurturing a research group of seventeen researchers.

SERENDIPITY AND SYNERGY

> Chance favors the prepared mind.
> ~Louis Pasteur

When asked to write this chapter, it made sense to me to develop a narrative and without thinking I chose a chronological approach. Yet I would not be surprised if others were to make a different decision. Life is not lived in such an orderly fashion; I tend to agree with Albert Einstein who suggested that he never came upon any of his discoveries through rational thinking. Our lives take odd twists and turns; we experience life, we meet people and good and bad things happen. By creating a chronologic narrative, the storyteller imposes a purpose, meaning and direction upon the events of one's life. Thus it is important to remember that although it is good to have a goal to journey towards, it is ultimately the journey itself that is of most importance and from which we often draw the most meaning.

My life and my career have been blessed. I am a successful and recognized professional in my field. I can "neatly" describe my work, framed within a sociocultural perspective, which includes the study of learning in free-choice learning settings and the long-term impact of free-choice learning experiences on individuals, families and communities, with a particular interest in the role of gender, race and SES on such learning. A colleague and I at the Franklin Institute Science Museum have NSF funding to investigate a long-held interest of mine: the long-term impact of gender-focused free-choice science learning experiences on girls' interest, engagement, and involvement in science communities and science careers. I am working on a book about family learning, which will summarize much of my (and others') efforts to investigate the nature of such learning over the last twenty years and this fall, I will begin sharing a tenured full professor position at Oregon State University. Described in this manner, my career seems like such a thoughtful, purposeful and straightforward trajectory but of course it neither was nor is. Julie Bishop, the current Western Australian Minister for Education, Science and Training says that synergy and serendipity often play an important role in medical and scientific advances. I believe that they play a tremendous role in life as I hope the following example demonstrates.

In 1989, I became assistant professor in science education at University of Maryland in College Park. I had been at College Park the year before filling a term position as an instructor in early childhood and elementary science education, replacing a faculty member who had left unexpectedly. James Fey, a well-respected secondary mathematics education scholar, was the department chair at the time and he let me know I had been selected and that he hoped I would develop an informal science education program, something I really looked forward to. I loved my work, the teaching and mentoring and I began to develop a research program focused on informal (free-choice) science learning. Unfortunately, unbeknownst to me because of my naiveté about academia and unfortunately the lack of a faculty mentor, two mitigating factors would ultimately make it necessary for me to leave the university in 1992. Though the two factors were different, they were clearly related.

First, the timing was not right, certainly not for me career-wise or unfortunately for the state of the field of free-choice learning, a very emergent discipline twenty years ago. I certainly was not adequately prepared for the task at hand, so unfortunately the opportune chance could not be optimized. If one full professor does not make a program, certainly a program is not going to be established by an assistant professor, without funding and with little departmental support. The feeling I had is best described in a quote from a colleague, unfortunately one who would be critical in building my promotion and tenure case: "I like museums, I go to museums, but this is a college of education, Lynn, and what we do here is train teachers, *school* teachers." I then realized the need to move on or I would be forced to do so.

Second, because the timing was not right for the field or me it was difficult to make a case for the relevance of free-choice learning. Any evidence for its importance or any synergy that I might have leveraged to be successful was minimal to non-existent. Suffice to say, this was a difficult time in my life—I had not really failed at something before, particularly with so little control over the situation and the extenuating circumstances. Fortunately, I landed on my feet, interestingly as director of the *Science in American Life* project at the National Museum of American History (Smithsonian Institution), where I led the development of an integrated middle school science and social studies curriculum! The irony was not lost on me.

There is now both serendipity and synergy working in my favor, and fortunately in the favor of the field that I believe so passionately about and this time I am prepared. Faculty and programs focused on free-choice learning and intersections between such learning and formal learning are beginning to emerge, attempting to take advantage of a quiet educational revolution underway worldwide. However, the centers of this revolution are not the traditional educational establishment of schools and universities, but a vast network of organizations and media supporting an ever-growing demand for free-choice learning. This is an important way that people learn and stay informed about science across their lifespan. In any given week people might read an article in a newspaper about the latest satellite launch, visit a local museum, watch a television special on whales or build a model rocket with their children. At Oregon State John and I will accomplish two of our long-standing professional goals: develop a graduate program for the next generation of free-choice science learning leaders and build a strong collaborative relationship between a university and the Institute for Learning Innovation.

Though serendipity and synergy are in play, in part demonstrated by a great deal of discussion about informal/free-choice learning at the 2006 NARST and AERA meetings, and for once not only in strand sessions, I have some reservations about the conclusions, recommendations and directions being suggested regarding such learning. Although there was high interest in how motivating and engaging these environments and this type of learning can be, there also seemed to be an effort to suggest that given these characteristics, the science education field should recreate such environments and connect to children's daily lives through activities in schools, rather than appreciating that it is the very nature of these environments and the child's *actual* daily life that is so intrinsically motivating and engaging to them as learners; a facsimile of an everyday context is just that, a facsimile.

It was also troubling to see that there was little recognition of the skilled educators, researchers and evaluators, many of them longstanding NARST members, who have spent careers teaching and conducting research and evaluation in these settings. Thus, rather than create a facsimile of an everyday context in a classroom, why not utilize the actual everyday context as a part of the learning experience? The expert group of colleagues in free-choice learning would embrace opportunities to work in respectful and collaborative ways with their counterparts in the formal arena, to create a comprehensive and articulated set of experiences that would engage learners in the exploration of science. This will require major restructuring and rethinking but I believe it is absolutely critical if we are to move forward.

What ultimately troubled me most was that *the* solution again seemed to focus on improving science education in schools, rather than improving science learning across the lifespan. This clearly communicated to me what the next steps need to be for those of us working in this area, along with those science education colleagues willing to truly think out of the box. Three major educational sectors—schooling, workplace and free-choice/informal—significantly contribute to the public's lifelong science learning. In fact, there is growing evidence that demonstrates that the more the three educational sectors of schooling, work and free-choice/informal overlap in people's lives the more successful they will be as lifelong science learners. If this is true, catalyzing interconnections between these three sectors should be our ultimate goal as science educators.

However, as with any autobiography, this is an unfinished and unfolding narrative. And as I suggested at the beginning of this article, through the very act of discussing the isolated parts of the system I have changed them and certainly the act of thinking about my experiences as a science education student, researcher and educator, has sparked my memory and influenced the meaning I have drawn from these memories. The strands of immediate, immersion experiences, relationships with supportive mentors and serendipity/synergy are interwoven and interconnected and will remain so as my narrative continues to unfold. All that I can do is take the next step which for me is to continue to work to communicate these findings outside our insular community and to work collaboratively with educators from the three sectors to develop models of what a comprehensive lifelong science learning program might look like, how it would work and how it would be assessed. A tall order but in my opinion not an impossible task—we understand many of the pieces within the individual sectors, but they need to be more effectively interwoven and inter-connected and most of all there needs to be the will to do so. In closing, I share one of my favorite and in this case apt quote:

> Everything you can imagine is real.
> ~ Pablo Picasso

Lynn D. Dierking
Oregon State University
Corvallis, OR

OKHEE LEE

EMBRACING SERENDIPITY AND CELEBRATING DIVERSITY

My academic career in science education is a combination of a series of serendipitous events and my desire to address diversity and equity. This is intertwined with my identity as a woman, non-native speaker of English, Asian-American immigrant, and non-science major working in the field of science education research. In this article, I describe my entry into science education, progression in my academic career, trajectory in my research interests, and reflections on scholarship and the people involved in my research.

MY ENTRY INTO SCIENCE EDUCATION

My academic career in science education entailed confronting and overcoming not only personal but also cultural and linguistic obstacles. Born in the rural countryside of South Korea, I was sensitive to the inequities Korean society propagated that impacted my day-to-day life, especially for being a girl. I still vividly recall my father refusing to instruct me in Chinese Chess while he encouraged and personally taught this "intellectual" game to my elder brothers. I tried to learn the game over the shoulders of my father and brothers, but could not follow the game when my brothers advanced to higher levels. To this day, I wonder whether I should have persisted and studied on my own despite discouragement, or whether I should have asserted myself and demanded the same treatment as my brothers. Was it my fault, or was the injustice done to me?

As I grew up, I learned to defy, quietly and within limits, the cultural norms related to gender. It is a Korean tradition to visit ancestors' tombs on major holidays to ask for protection from evils and blessings for the good fortune of the family. Only males were expected to carry out this familial duty, whereas females would stay home and clean up holiday banquets. Recognizing this unfairness, I insisted on accompanying the males in my extended family. I was proud of myself when my father granted my wishes.

As I was finishing high school, it was assumed that I would study in a female-oriented field, such as education, psychology, or languages. Not surprisingly, my elder sisters were schoolteachers, whereas my elder brothers were in science-related fields. I completed my undergraduate degree with certification in teaching English as a foreign language in 1981 and my master's degree in educational psychology and instructional design in 1983.

When I completed my master's degree, the expectation for the rest of my life was pretty clear—perhaps teach for a couple of years, get married, have a family,

K. Tobin, W.-M. Roth (Eds.), The Culture of Science Education, 251–261.

and stay home and raise children. However, since I was young, I had always wanted to be a college professor. To pursue my dream, I would have to study abroad where academic careers for women were more accepted. My father granted me one year to prove myself. If I could obtain financial support for my doctoral study, he would yield to my wishes; if I failed, I would follow his wishes. Later in my academic career in the US, people would ask me about my lack of K–12 teaching experience. It was simply not an option at the time because if I had entered teaching, it would have been extremely difficult to pursue an academic career.

I was informed that the odds for a foreign-born doctoral applicant to be offered a scholarship in the field of education at a U.S. university were slim. My elder brother who had recently been accepted for his doctoral study at a major university in the US was instrumental in my preparation to study abroad. After intensely preparing for the TOEFL and GRE tests for months, I obtained high scores on both tests. From the few prominent state universities that offered me scholarships, I chose Michigan State because I was selected as one of five doctoral research interns for the Institute of Research on Teaching (IRT) and because Michigan State guaranteed me a 10.5-month stipend for each of three years. To this day, I am grateful for this honor and opportunity. I also appreciate those who made it happen—Charles Anderson who drew the attention of the selection committee to my application and Jere Brophy and Andrew Porter who were co-directors of the IRT.

When I left my family in my mid-20s, my mother was pleased that I was following my dream and my father was proud of my accomplishments, but everyone was worried about what would lie ahead of me in a foreign country. A major source of comfort for my family was that one of my elder brothers was also a doctoral student in Wisconsin (and to my family, Michigan and Wisconsin appeared relatively close on the U.S. map).

Entering the IRT in fall 1984, I was asked to choose one of two research projects for my doctoral research assistantship, and both projects, to my surprise, were in science education. I hoped that the high quality education I had received in South Korea, my hard work in science throughout my schooling, and the stories that my brothers told me about the wonders of science would serve me well. During my doctoral study at Michigan State, I worked on two research projects in science education. I am grateful for the mentoring and guidance by James Gallagher, Charles Anderson, and Glenn Berkheimer who were the project directors. I also appreciate Andrew Porter for advising my doctoral program of study and Charles Anderson for directing my doctoral dissertation study.

PROGRESSION IN MY ACADEMIC CAREER

It is commonly understood that a doctoral graduate thinks of the university as the birthplace and home of his/her academic career. This sentiment holds true for me as well. Michigan State set me on the professional path I lead today. Not only did prominent researchers across academic disciplines work under the umbrella of the IRT, but also it attracted renowned scholars from all over the world. I was exposed to educational research from a range of theoretical perspectives, academic disciplines, and research methodologies. In addition to building a foundation for my academic career, it was at Michigan State where I met my future husband.

My doctoral degree was in educational psychology with an emphasis on learning and cognition. My doctoral training also included two minors in research methods and measurement/evaluation, and a cognate in qualitative and ethnographic research. My dissertation study examined student motivation and conceptual change learning in middle school science classrooms.

I made two pre-eminently important decisions in my life—coming to the US to do my doctoral study and marrying an American man despite the prospect that my family would disapprove and might even disown me for the inter-racial marriage (which did not happen to my surprise). These two key personal decisions formed my identity and dispositions—these decisions were solely mine and I am responsible for myself (but no one else) in my chosen country, marriage, and career.

After defending my dissertation in fall 1988, I moved to Miami because my husband was a faculty member in the School of Communication at the University of Miami. While finishing my dissertation in spring 1989, I taught a few psychology courses at Miami-Dade Community College (currently Miami-Dade College). During my first year at the University of Miami in 1989–1990, I was an adjunct instructor teaching almost ten courses. Being the only person who had some knowledge of science education research at the University of Miami, I was invited to join a team to write a grant proposal to the U.S. Department of Education for a three-year teacher enhancement program in elementary mathematics and science education. My first foray into grant writing was funded, and I became a research associate the second year. While our grant proposal was being reviewed, the school of education offered me the fulltime position of director of undergraduate advising (a surprise, given that I knew virtually nothing about undergraduate education in U.S. universities). The second year, I worked as both research associate and director of undergraduate advising. During the third and fourth years, I was an adjunct assistant professor working full-time on funded projects. It was not until the fifth year that I was hired as a tenure-track assistant professor.

Although my first four years at the University of Miami were a struggle, I learned the value of research, scholarship, and funding. I realized that publications would be the only way that I would ever become a tenure-track assistant professor. I was successful in getting grants to support my research and myself. At the same time, I was fortunate to continue working with my mentors from Michigan State who helped me turn my graduate research and dissertation study into manuscripts for prominent journals. In 1993, I was selected as a National Academy of Education Spencer Postdoctoral Fellow. The late Samuel Yarger became the Dean of the School of Education in 1992, and I am grateful that he hired me without going through the search process in 1993. Once I was on a tenure track, it took four years to be promoted to the associate professor level with tenure in 1997 and another three years to be promoted to the full professor level in 2000.

In 2000, I was faced with challenges at home and at work. While I was just learning to manage a major research project alone, a family member was diagnosed with a life-threatening illness. It was about this time that my work became more visible, and I had to learn to respond to critiques of my work. Through hardships, I grew stronger personally and professionally. I stayed focused on my family and work, learned to say no to invitations and opportunities, and cut my travel to a minimum. Amid personal and professional challenges, my research became a source of joy and pride. After a few years of slow progress, my research productivity took off, and I expect this increased level of productivity will continue in the foreseeable future.

Throughout my academic life, I have had a series of grants for research and teacher enhancement from various funding sources, primarily the NSF. My focus on funding and research has kept me away from teaching. Although I frequently do not teach "classes," I am a teacher. Through collaboration, the teachers who have participated in my research and teacher enhancement projects and I teach one another. Indirectly I teach the students of those participating teachers. I teach undergraduate and graduate students and post-doctoral associates who work on my projects.

TRAJECTORY IN MY RESEARCH INTERESTS

When a series of serendipitous events have presented themselves, I have embraced such opportunities and expanded my research interests. Through this process, I have pursued broad areas of scholarship integrating cognition, language, and culture in science education in multilingual and multicultural settings. Across such broad areas, I have addressed a range of issues, including curriculum development, teacher professional development, teacher learning and change, student learning and achievement, assessment, urban education, and accountability policies.

My entry into education research in the US started with an ethnographic study of secondary science classrooms while I worked as a doctoral research assistant at Michigan State. James Gallagher, the principal investigator of the NSF project, was instrumental in orienting me toward secondary science education and ethnographic research. This experience enabled me to have a sense of the U.S. education system and provided a foundation for my classroom research.

The second research project, in which I was involved during the latter part of my doctoral study, examined the effect of a middle school science curriculum unit on matter and molecules. The unit focused on students' alternative conceptions and conceptual change learning. Many years later when Project 2061 examined the quality of middle science curriculum materials, they evaluated this unit as one of the best products. From this research, I learned the process of research-based curriculum development and skills for clinical interviews with middle school students.

The two research projects in science education prepared me to conduct my doctoral dissertation study. I examined task engagement and student motivation to learn in the context of conceptual change learning in middle school science classrooms. I was fortunate to conduct the study in conjunction with the research project for which I was working as a doctoral research assistant.

While being involved in two research projects over the period of four years, I made several presentations at conferences, including the American Educational

Research Association (AERA) and National Association for Research in Science Teaching (NARST). However, except for one manuscript that I completed with a fellow doctoral student, I did not engage in writing manuscripts for scholarly journals throughout the period of my doctoral study.

After heavy teaching during my first year at the University of Miami and spending the bulk of my time on undergraduate advising during the second year, I became acutely aware of the importance of scholarship and publications. While managing several teacher enhancement grants that supported me financially, I concentrated on writing manuscripts. With the help of my mentors from Michigan State, I published 9 journal articles in respected journals. I am particularly pleased with the article from my dissertation study published in the *American Educational Research Journal* (Lee & Anderson, 1993). This is the first of my eight articles published in AERA-sponsored journals, to date.

Writing manuscripts taught me several important lessons. I learned how to take the lead for writing manuscripts and to be an independent scholar. I also developed a sense of how to judge the quality of a manuscript for highly competitive journals. I realized that institutions take care of their own interests, not the interests of individuals. In this sense, I learned the meaning of the expression, "Heaven helps those who help themselves." Because I helped myself, my institution was willing to help me. Additionally, I learned that scholarship transcends institutional boundaries and that I would be marketable at any institution as long as I continue to be a productive scholar. Finally, I learned that I enjoy research and writing and that scholarship gives me a great deal of joy and satisfaction.

In 1992, I started working with Sandra Fradd in English to speakers of other languages (ESOL)/bilingual education. I was drawn to this area of research not only because of my undergraduate training in English language and literacy as a foreign language, but also I could identify with this topic as an English language learner myself. We obtained a small NSF grant for exploratory research during 1992–1993, which led to our two major NSF projects from 1995 through 2000. We developed the *instructional congruence* framework (Lee, 2002). This framework expands on the literature on cultural congruence and culturally relevant pedagogy by arguing that effective subject area instruction should consider the nature of academic disciplines (such as science) in relation to students' linguistic and cultural experiences. The framework highlights the importance of developing *congruence*, not only between students' culturally based interactional norms and those of the classroom, but also between academic disciplines and students' linguistic and cultural experiences. It emphasizes the role of *instruction,* as teachers (or educational interventions) explore the relationship among academic disciplines, English language and literacy development, and students' linguistic and cultural knowledge, and devises ways to link these domains. The relative success of instructional congruence depends on both effective instruction within each domain and articulation among the domains.

Grounded in the instructional congruence framework, which is supported by significant achievement gains in science and literacy, we expanded our research with funding from the *Interagency Educational Research Initiative* (IERI) program jointly supported by the NSF, U.S. Department of Education, and National Institute of Health during 2000–2004. Consistent with the IERI's focus on scale-up of educational innovations, we expanded the research in several ways.

First, using a longitudinal design, we involved grades 3 through 5 teachers and their students over the period of three years. We were interested in the impact of the intervention on teacher change and student achievement (and achievement gaps) over the three-year period of science instruction. To meet the instructional needs of the teachers and students from grades 3 through 5, we developed additional units. In the end, we developed two units for each grade level.

Second, we extended our intervention to other research sites with different groups of student diversity. Eugene Garcia from the University of California at Berkeley started the project in San Francisco schools. When he later moved to Arizona State University, he added Phoenix schools. With diverse demographic groups of students from Miami, San Francisco, and Phoenix, we examined patterns of results with teachers and students over time.

Finally, while the research was in progress, high-stakes assessment and accountability policies in reading, writing, and mathematics emerged as major influences on policies and practices at the school and classroom levels. We examined the impact of our intervention within the context of evolving policies on high-stakes assessment and accountability.

From our IERI project, we developed about 20 manuscripts on a wide range of topics. This accomplishment was possible thanks to the research team. I took an active role as mentor and each of the post-doctoral associates and doctoral students published from a few to almost a dozen articles. Of all the publications, the article on scaling-up of educational innovations for nonmainstream students in elementary science education (Lee & Luykx, 2005) is the most significant outcome. This article examined the instructional congruence framework in the context of high-stakes assessment and accountability policies. During the period of our research from 2000 to 2004, we faced the influences of high-stakes assessment and accountability in reading, writing, and mathematics on science instruction in elementary schools. We experienced that scaling up was especially challenging in multilingual, multicultural, and inner-city settings due to fundamental tensions around effective educational policies and practices for diverse student groups. We argue that rigorous attention to such challenges is needed to make scaling up of educational interventions more effective and to answer the question of what constitutes "best policies and practices" for diverse student groups.

On our current NSF project funded by the Teacher Professional Continuum (TPC) program during 2004–2009, I have the privilege of working with Walter Secada in mathematics education and Cory Buxton in science education. The project's goal is to design, implement, and test effective ways for elementary school teachers to teach science to all students, especially English language learners (ELL) students, in light of high-stakes assessment and accountability in science. Basically, the research tests two conventional wisdoms: (1) can English language learners learn academic subjects, such as science, while also developing English proficiency? and (2) can students, who learn to think and reason scientifically, also perform well on standardized tests such as state-wide science assessments? In testing these two conventional wisdoms, we focus on change in teachers' knowledge, beliefs, and practices as they continue their participation in the intervention over the years. This research has several unique aspects that have the potential for wider impact.

First, the core of this research is to design, implement, and test a teacher professional intervention for ELL students in elementary science education within a high-stakes testing policy context. Our intervention consists of curriculum units for grades 3 through 5, teacher workshops spread out over the course of a school year for the period of three years, and research activities that foster teacher reflections on classroom practices and student learning. To fully implement our invention, we have decided to develop a comprehensive curriculum that emphasizes science topics that are assessed annually, along with those topics that are assessed once every three years, according to the state of Florida accountability plan. Three units are developed for each of grades 3, 4, and 5, for a total of 9 units. The units are field-tested based on teacher feedback, classroom observations, and student assessment results. Currently the University of Miami is pursuing a plan to make our curriculum materials commercially available. The evolution of our curriculum development over many years of our research is rewarding, in that the curriculum materials intended as a means to do our research will serve the public.

Second, the evolving policies on high-stakes assessment and accountability provide unique opportunities. In Florida, science has been tested at fifth grade since 2002–2003, but it does not count toward school accountability. Starting in 2006–2007, science counts toward school accountability. Also, starting in 2007–2008, science counts as part of the *No Child Left Behind* Act. In our research, as the third grade students who started their participation in 2004–2006 advance to fifth grade in 2006–2007, they will constitute the first cohort for which science counts toward school accountability. In addition to assessments of all participating students, we also interview a small number of randomly selected students to examine their abilities to engage in scientific inquiry, reasoning, and argumentation. The results will help us test the two conventional wisdoms described above.

Finally, in addition to the longitudinal design, the research also employs a design, in which students are nested in their classrooms, classrooms are nested in their schools, schools are nested in their district, and the district is nested within the state policy context. We are examining the intersections at these multiple levels of the education system that eventually impact student achievement.

While building a programmatic line of research with the NSF funding, I have also engaged in other research and teacher enhancement projects. Two of them are particularly important for my scholarship. One is my fellowship at the National Institute for Science Education (NISE) at the University of Wisconsin-Madison during 1996–1997. Based on an extensive review of literature, I examined equity implications of current conceptions of science achievement and assessment in standards-based and systemic reform (Lee, 1999b). The other is my role as the chair of the *Science and Diversity Synthesis Task Force*, which is a joint project by the Center for Research on Education, Diversity and Excellence (CREDE) at the

University of California-Santa Cruz and the National Center for Improving Student Learning and Achievement in Mathematics and Science (NCISLA) at the University of Wisconsin-Madison. In preparing for a book, I literally read hundreds of articles published since the early 1980s to the present on a wide range of issues on diversity and equity in science education (Lee, 2005).

In short, my research over the years has evolved to encompass a range of issues on diversity and equity in science education. As a secondary area of interest, I have also studied teacher education programs, especially in ESOL/bilingual education. I have been addressing these complex issues using multiple modes of inquiry, including quantitative, ethnographic, case study, and survey research methods.

REFLECTIONS ON SCHOLARSHIP

Like any human endeavor, research and scholarship is a reflection of personal dispositions and guiding principles. I would like to share some thoughts on my research in particular but also on scholarship more broadly.

Personal Dispositions

My background as a female Asian-American immigrant influences my work. First of all, being radical is not in my upbringing, but being persistent is a cultural norm. In addressing diversity and equity issues, educators and scholars may take different approaches. Some may go head-to-head against the hegemony of the mainstream, whereas others may assimilate to the mainstream. I have taken and will continue to take the approach of trying to understand the mainstream and, along the way, change "the course of the river" that has been dominated by the mainstream. One way to make this happen is to build a knowledge base. Research on diversity and equity has a short history in science education (Lee & Luykx, 2006). To make this limited but emerging research more visible in the mainstream research, I have tried to publish my work in prominent journals in education. Having published eight articles in AERA-sponsored journals, I will strive to continue publishing in these and other prominent journals. Ironically, my unconventional academic training has fostered publication in AERA-sponsored journals, as well as others that appeal to broad readership. I do not have academic training in the areas of my research, including science disciplines, anthropology, or linguistics. Also, I do not specialize in a particular research method. However, I have learned enough about these multiple areas of research and research methods, that I can integrate them to meet my research interests covering a range of theoretical/conceptual and methodological approaches.

Second, hard work is another cultural norm in my upbringing, which is further strengthened by my status as an immigrant. When I came to the US, I realized that I would be behind my colleagues born and educated in the US. In order to catch up, I have followed the principle of working one extra hour every day. At this point, I have been in the US and in my academic career over 20 years. What I have accomplished so far is owing to 7,000 extra hours of work. I plan to continue this principle for the remainder of my academic career, which I hope will be additional 7,000 hours for the next 20 years.

Finally, respect for diversity of ideas and opinions is something that I have learned since I came to the US. As I prepared to leave my country, my late mother wisely advised me as follows: "When you go to a new country, you are bound to make mistakes because you do not know their ways of living. If you are nice to people and respect them, they will respond to you in kind." Being an alien in a foreign country is a scary experience. I decided to be a good person and do the right thing, hoping that people would recognize my good intent when I would make mistakes. I have extended this belief to respect diversity of ideas and opinions. For example, when I review a manuscript, I judge it in terms of its quality as a research study, separate from ideological assumptions or theoretical perspectives.

Research Agenda

My first and foremost advice for developing a research agenda is "do what your heart desires to do." Throughout my academic career, I have had a few funded projects simultaneously. I have made conscious efforts to concentrate on my primary line of research separate from secondary or even tertiary interests. Although breadth of knowledge and experience across different areas of research is valuable, one should not lose sight of building a line of research. This requires taking initiative, pursuing funding opportunities, and cultivating working relationships with others.

Building a line of research, eventually, requires funding to support the research. Institutions love money, and one may become attracted to the lure of getting grants. Some institutions or research groups have strong infrastructures to support multiple grants, but it is not easy for an individual or a small group of individuals to manage multiple grants. I keep the principle "funding is a means to do research, but not an end." I try to garner sufficient funding to continue my research, but not to the extent to which management of multiple grants detracts from my research.

Researchers tend to spend a great deal of time designing and implementing interventions, developing instruments, and collecting data (and often too much data). Unfortunately, they do not spend enough time analyzing the data and writing up the results. Once data collection is completed, there is often a false sense of safety that there will be time for data analysis and writing at some point. Unfortunately, when a project is over, people move on to next projects or other urgent tasks. In the meantime, the data are sitting there and gathering dust, and eventually become obsolete. The goal of scholarship is to contribute to the body of knowledge, and insufficient time for writing fails to meet this goal. Particularly, for a multi-year project, it is also important to continue data analysis and writing, so that one can correct mistakes rather than continue to make the same mistakes for the duration of the project. Additionally, one may come up with new insights or research questions, which may lead to additional (and important) studies. If one waits for data analysis and writing until the end of the project, there is no way to correct mistakes or pursue emerging research questions. I remind my research team members and myself that funding comes from taxpayers' money, and that it is our responsibility to do scholarship. I also remind ourselves that educational interventions do good for the participants (often teachers and students), but our hard work does no good for ourselves if we do not follow through on scholarship.

Research results can be written in different forms and for different audiences. I try to write a variety of articles, including empirical papers, literature reviews, essays on theoretical/conceptual issues, and practice-oriented pieces. I also contribute to magazines, newsletters, and newspapers for the general public.

As I manage large grants, the quality of the research depends on the quality of the work by all the research members. I capitalize on the strengths of individuals for specific sets of tasks, while promoting them to learn skills for various aspects of research. I especially try to involve doctoral students in writing manuscripts. Initially, I take the lead for writing, but I encourage everyone to take turns leading. We develop timelines for manuscripts, and I push, cajole, and even harass them to keep the timelines for writing. Personally, mentoring is very important to me, as I think about my mentors who generously supported me throughout my academic career. I would like to do for my own students what my mentors did for me.

Writing

I make writing a priority. In the midst of everyday demands for teaching, committee work, family matters, and other responsibilities, writing can be put aside (i.e., I will get to writing after I finish this and that). Unfortunately, one rarely has the time to write after completing all the other obligations. Instead, I put writing on the top of my daily activities. I am often asked whether there are particular times of a day for writing. My response is that I write whenever I can find time.

Being a non-native speaker of English, writing is a challenge to me. I strive to read and write as well as native speakers of English. A compliment about my writing from a mentor during the first year of my doctoral study taught me an important lesson about writing. He said, "Your writing is good," and I asked him in disbelief, "What do you mean that my writing is good? English is a foreign language to me." He responded, "Of course, your English has errors, but your logic is clear, and I can follow the development of your logic. In contrast, native speakers of English write in English, but sometimes I have no clues about what they are writing about." In my writing, I try to be logical, coherent, and clear about the points I am making. Also, I try to be reader-friendly and avoid the possibility of miscommunication with readers because of the misuse or inappropriate use of a particular word, phrase, or expression. Over the years, I have realized that writing has become easier (although never easy).

I use writing as a way to learn about a new topic. As I write a manuscript, I learn about the topic addressed in the manuscript. Then, I am ready to move on to another manuscript on a different topic. At some point, I am able to integrate multiple (and seemingly unrelated) ideas into a coherent framework. These manuscripts have the greatest impact on my scholarship.

CLOSING REMARKS

I aspire to be a good scholar and to make a contribution to the research in my field. Looking back on how far I have traveled since I started my academic career over 20 years ago, I wonder how far I will travel in the years ahead. I will continue my research on diversity and equity in science education, expand the breadth and depth of my scholarship, continue to work with both old and new colleagues, and hopefully continue to offer insights in this field. I began this journey when research on diversity and equity in science education was in its infancy. To date, the field has advanced, but is still relatively young with many areas as yet unexplored. My hope is that I will see the field continue to grow to maturity and that I will play a role in this process.

I also want to make a difference on a more practical level. I take some pride that our research has benefited the participating teachers and students who otherwise might not have had such opportunities. When challenges sometimes seem insurmountable, I remind my research team members and myself that we do our work "for the good of the children." Also, I take pleasure in seeing junior scholars, postdoc associates, doctoral students, and others grow professionally. They will carry out educational research, hopefully including some of my own research agenda, into the next generation. Furthermore, I hope that through publication of curriculum materials and research results, our work contributes to improving education in the US. In the end, I hope that our work opens up possibilities and offers opportunities for those who have been marginalized and underserved, so that they do not face the injustice of Chinese Chess that I experienced as a young girl growing up in South Korea.

I would like to close this chapter with a personal statement. Based on my cultural upbringing, I am not accustomed to speaking on behalf of myself. I hope that in these pages I do not sound boastful. I often hear comments from junior scholars or doctoral students from minority backgrounds or foreign countries that they would like to learn about senior scholars "who have made it" in academia. I would be delighted if the accounts of my academic career resonate with other senior scholars or offer some guidance for junior scholars and doctoral students from backgrounds similar to mine.

Okhee Lee
Department of Teaching and Learning
University of Miami, Florida

MARIA VARELAS

COLLABORATIONS, MULTIPLE VOICES, TENSIONS, DIALECTICS, AND FERVOR

The Social Construction of a Science Education Career

What story do I want to tell about myself as a science educator, as a physics teacher, as a university professor, but also as a woman, as a Greek American, as a mother, and as a wife of a high-energy physicist? So many identities, maybe so many facets of one identity, a "whole" *me*—identities that have different lifetimes, that have been influenced by different people, different events, have started at different locations, different countries, and different contexts; but somehow they have all come together to "define" me, Maria Varelas, with a nickname by a dear colleague of mine "fervor ball."

SOME OF THE TEACHERS IN MY LIFE

It all started at an all-girls school that I attended in Greece for all my thirteen pre-college years. I came to love physics in high school. Physics gave me a sense of accomplishment. It felt good when things made sense to me. My best friend for 30 years, Vicki Antoniou Douligeris, with whom I shared the same homeroom for the six years in junior high and high school, could not feel the same way. She ended up majoring in archeology, and although over the years we have had a great deal of discussion about contributions of physics to archeology, she was clearly drawn by the "beauty" of relationships of people and places and utensils and building ruins and skeletal remains. As we continued our friendship and our scholarly interests, I came to increasingly appreciate the diversity of various disciplinary practices and their multiple affective, cognitive, and social impacts on people.

What I loved most about physics in high school was the physics class time. I had the same physics teacher for all three high school years—a male, relatively young teacher who knew physics well, but what he was really good at was engaging everybody in class, pushing us forward at our own pace, and creating an environment of respect for those who "loved" and those who "hated" physics. "What do you think about this problem? How would you approach it? Come share with us your thinking. What does this mean?" were some of his favorite sayings. The physics class was all a discussion—a discussion about ways of thinking of ideas, of phenomena, of applications, of issues, of experiments, of observations, of "playing around" with thinking. That is what made me love physics, and not chemistry, and not biology—taught by other teachers, okay teachers, but not like Mr. Kyriakakis.

K. Tobin, W.-M. Roth (Eds.), The Culture of Science Education, 263–274.

I did well in the national examinations that, under the Greek educational system at that time, allowed students to secure a place at the universities around the country—public universities free for all students. I scored well enough to be able to attend the University of Athens, the top Greek university, and, of course, I had chosen the physics department as my first choice. It was the next four years in that university that continued to shape what I wanted to do "for the rest of my life." What I wanted was not just physics—it was not solid state, or nuclear physics, or astrophysics, or high-energy physics, or biophysics—it was teaching physics, physics/science education.

I have a special teacher to thank for this. Kostas Eftaxias is a solid-state physicist who taught several of the introductory and advanced physics courses and teaching of physics courses throughout the four years in the program. He was the one who helped me think about teaching, about students, about physics and students together. And his reputation as a "great" teacher was not only known among the few of us who were leaning towards teaching and education as a career, but also those who were working with other physics groups, in labs, getting inducted in the various other physics "subcultures." Eftaxias's lecture, discussion, and lab sections were packed. I vividly remember one semester, when one of his discussion sections was taking place in a small classroom with rows of tables and a demonstration table along a long blackboard mounted on the wall. When there would not be any more tables to sit at, we would sit on the floor around the room, at times all over the demonstration table and along the blackboard. Eftaxias would need to skip over students to write on the board. But in the midst of so many of us, of the disorder of the room, of the joking and laughing about many topics including physics, he managed to teach, to ask questions, to answer questions, to invite contributions, to excite us, to make us want to be there next time.

As a teacher, he had a particular style. He would start a sentence and he would wait, consciously or unconsciously for us to complete it, to continue it, to get to the next sentence. At times, this would happen as a chorus, a chorus of physics undergraduates speculating about force, motion, torque, electrons, absolute zero, refraction, and friction; or one of us would shout out something and Eftaxias would listen carefully and give feedback. But, he was also the one who kept encouraging us to work together, study together, do physics together. "You can't learn alone" he used to say to all of us, "Use each other's ideas, each other's strengths." And we did. Some of us kept studying together since the first year. The university library, the many cafes in downtown Athens, each other's apartments, the open space around the new university campus were places that hosted some of us in different group configurations at different times solving problems, writing up lab reports, studying for exams, trying to understand lecture notes, and, of course, some "fooling around." I ended up doing my Bachelor's thesis with him on teaching physics to secondary students. My partner, Lena Pavlidou, and I did a thesis together. Eftaxias encouraged us. "You both have similar interests, questions, wonderments—go for it."

Eventually as I taught various-age students, I struggled putting in action those principles that I was coming to think of as important and useful. When I taught undergraduate physics labs I tried to get my students to be critical, to "fool around" with the equipment before they decide how to do something, to work with each other rather than compete against each other for the better grade. I kept telling them "think about this—does it make sense? Do you have a hunch? Different ways of doing the lab are okay." But many of these undergraduate students, like many of my seventh graders—to whom I taught science down the road—resisted the messiness, the working together, the thinking at times. They did so not because they had thought about these and have decided that such an approach or attitude was not "good for them." Rather, they did so because they were used to a different "code," different norms of behaving in school, or relating with the school subjects, of relating to each other when it comes to academic "stuff" and of relating to me, their teacher who was trying to get them to "do science" in a way that made sense to them.

This meaning-making emphasis kept defining me as a teacher and a researcher. For the latter, my doctoral thesis advisor, mentor, and good friend and colleague, cognitive psychologist Joe Becker played a critical role. Although I was a doctoral student in the curriculum design PhD program at the University of Illinois at Chicago (UIC), I "oscillated" between curriculum and educational psychology. I took courses from both departments and I came to explore dialectics, tensions, and multiplicity of voices and approaches from both hugely vast, of course, literatures that I am still immersing myself in. Joe inducted me to Lev Vygotsky and Bill Schubert inducted me to John Dewey. Vygotsky and Dewey, and all other scholars who have written about them became the cornerstone scholars whose thinking would shape my scholarship over the years. But Joe did not only induct me to Vygotsky (and James Wertsch and Alexander Luria) but he also "pushed me" to examine, to question, to debate, to wonder, to articulate, to be dissatisfied with the present state of my thinking and work to further develop it. He is partly to thank for supporting the "fervor" that I have come to live with and nurture over the years—strong and non-concealable passion I feel about issues that I care about and for which I stand up and argue with colleagues, students, friends, collaborators. I have come to think that "fervor" somehow fits my strong commitment to meaning making that comes in different shapes and forms for different people, with different experiences, from different ethno-linguistic backgrounds, from different socio-economical positions, and from different political orientations. Something else I learned from Joe was to listen carefully to others and "hear" what they say or try to say. "Hearing" others helped me realize, accept, and commit to keep exploring and understanding the various ways that various people, young and old, with all sorts of different experiences make meaning, relate, question, and experience science.

A SYNOPSIS OF MY SCHOLARSHIP

My research focuses on teaching and learning science in urban classrooms with linguistically and socio-culturally diverse populations. In much of my research I explore issues that emerge in the implementation of curriculum and professional development projects, multi-year grants funded by the National Science Foundation (NSF). The work that I do is highly collaborative both in the design and implementation of the development projects, and also in the research associated with these projects. Such collaboration allows for negotiation of meaning among practitioners with various forms of expertise; thus, it converts tacit knowledge into shared, articulated knowledge that sustains commitment, growth, and change.

The majority of my empirical work is action research with a deep attention to contextual details. This work involves a theoretical framework concerning the socio-cultural nature of the construction of knowledge, which draws on Vygotsky's theory of the interplay of language and thinking. I have contributed theoretical discussion concerning this framework (Becker & Varelas, 2001), which gives a central position to the dynamic interaction of two dimensions of learning communities: (a) the intellectual-thematic dimension that focuses on the content, ideas, flow, connections among ideas, and concepts; and (b) the socio-organizational dimension that captures the interactions that take place among participants of a learning community. Several of my studies have been concerned with these dimensions, and their interplay, as they emerge and unfold in the ongoing work of the development projects, or in participant teachers' subsequent experiences with their students.

With funding from the Illinois Board of Higher Education, I initiated in 1995 a professional development project (*Inducting Students Into Science [ISIS]*) in which practicing middle school Chicago Public School teachers and I worked together to develop instructional approaches and curricular units that enable students to engage in the interplay of theory and data—an important characteristic of scientific activity. We strove to keep the interplay of theory and data central as teachers and students found ways to negotiate understandings and skills and make science a meaningful practice.

With teacher Barbara Luster, we used the classroom discourse in her African-American eighth-grade class during their study of the circulatory system to examine how various struggles in the classroom led to the development of one dimension (the socio-organizational) of her classroom's learning community while the other (intellectual-thematic) lagged behind. The challenges were related to the lack of differentiation in the language that students and teacher were using, the teacher's emergent subject-matter knowledge, and the teacher's and students' lack of sufficient engagement with the theory-data dialectic (Varelas, Luster, & Wenzel, 1999).

With teacher Edgar Pineda, who taught a predominantly fifth-grade Latino class, we studied mainly the intellectual-thematic dimension, with particular attention to the ebb and flow of ideas as the teacher and students tried to make sense of the concept of friction (Varelas & Pineda, 1999). The study enabled us to identify several challenges that constructivist teachers encounter. One challenge was the teacher's ability to hear students' unformed, ill-formed, partially formed, or even well-formed but not clearly expressed understandings, and to use them to advance the learning of both individual students and the class as a whole. Another challenge was the teacher's dilemma of intervening and using some students' ideas as "hooks" to concepts to be developed versus letting students (especially from ethno-linguistic groups who do not usually speak up in class) express as many ideas as possible without intervention, thus, forgoing potential leads to developing acceptable scientific understandings.

In a third study, again with Barbara Luster, we used the conception of three genres (youth, school, and science) to explore the interaction of the two dimensions of a learning community. As Barbara moved toward more constructivist practices in teaching science, she invited her African-American sixth-grade class to compose creative writing assignments (plays and rap songs) on the topic of sinking and floating. We analyzed this student work (Varelas, Becker, Luster, & Wenzel, 2002) to determine how the three interrelated genres were enacted in a science classroom, and how they shaped what students understood and felt. We showed how students' affective reactions were an integral part of their constructed scientific knowledge. Thus, we argued that "unconventional" (for a science class) forms of student work, such as plays and rap songs, can invite students to bring the very meaningful ways they have of expressing and organizing ideas into contact with the disciplinary knowledge that is coming to them in a more abstract, less experiential form—a form less connected to familiar social and emotional meanings. Furthermore, these forms revealed students' own views on relationships with peers, with the teacher, and with the subject matter.

Another multi-dimensional project—a four-year teacher preparation project funded by NSF in 1999 for over 1.5 million dollars (*UIC-Community College Collaborative for Excellence in Teacher Preparation*)—that I co-led with colleagues in the science and mathematics departments—was also centered on the development of learning communities in and out of classrooms. A particular strand of this project was ten-week summer internships for pre-service teachers in national science laboratories. In a study using data from interviews with these beginning teachers, I explored with my doctoral student Roger House the two dimensions of a learning community as they play out in two distinct identities that these teachers developed—a scientist identity and a science teacher identity (Varelas, House, & Wenzel, 2005). Using two constructs, science as a practice and science as a community of practice, this study exposed ways in which the two identities of these new teachers incorporated conceptions of science that fell on different places along several continua defined by opposing ends, for example, messiness and structure, theory and data, autonomy and authority.

The concept of learning communities has also shaped the on-going five-year, 2-million NSF-funded Graduate fellows in K–12 education project, *Scientists, Kids, and Teachers: A GK–12 partnership with the Chicago Public Schools*. I co-lead this project with colleagues in chemistry (Donald Wink and Marlynne Nishimura), computer science (Tom Moher) and education (Stacy Wenzel). In this project, graduate students in the science and mathematics departments, also known as Fellows, become members of multiple learning communities in the context of reform efforts in Chicago Public Schools. Using their subject-matter expertise, the Fellows work with teachers, and other school personnel, to facilitate change as they themselves learn about pedagogy. These learning communities consist of teachers, students, Chicago Public School support personnel, and science, mathematics, and

267

education faculty who interact with the Fellows as they work together to strengthen teaching and learning of science and mathematics in urban K–12 classrooms. Several interplays are critical in this work—specifically, the ways in which subject matter and pedagogy understandings, and expertise and "noviceness" become intertwined and shape the experiences of all project participants.

Using data from the first round of the GK–12 project—and especially the Fellows' yearlong journaling—former doctoral student Nikoletta Christodoulou and I started a series of studies exploring such issues. The first study focused on the Fellows' curricular conceptions as portrayed in their reflections on events, actions, and issues they noticed in their work in urban secondary classrooms, and how these conceptions relate to both their prior experiences and the projects and activities they pursued in classrooms. Using William Schubert's four curricular orientations (intellectual traditionalist, social behaviorist, experientialist, and social reconstructionist), Joseph Schwab's commonplaces (teacher, learner, subject matter, and milieu), and Dewey's construct of experience, we found that despite their "love" for, and knowledge of, the subject matter, the fellows tended to be experientialists mostly discussing student-subject matter interactions. Furthermore, the study revealed links between the experts' prior experiences and their conceptions, and more importantly links between their conceptions and their work in classrooms. Experientialists engaged in more open-ended projects, relevant to students, with explicit connections to everyday-life experiences. Social behaviorists paid attention to designing "good" labs and activities that teach students appropriate content, lead them through various steps, and model good science and mathematics. Critical reconstructionists hyped up student knowledge and awareness of science issues that affect students' lives, such as asthma, HIV epidemic, other transmitted deceases, and rat infestation.

In 1999, as part of a professional development project, my literacy and language education colleague Christine Pappas and I started designing integrated science-literacy curriculum for primary grades with two Chicago Public School teachers (Anne Barry and Amy O'Neil Rife) that they then implemented. We started this project with internal UIC funding, received a small grant from the National Council of Teachers of English Research Foundation, and finally were awarded a three-year, over one million dollars, NSF research grant in 2004 to work with six teachers and their classrooms. In this project, we work with teachers to develop ways of engaging students with subject matter via integrated science literacy units that combine reading information books, doing hands-on explorations, writing, drawing, and discussing science. The focus is also on the development of a classroom learning community where young children are offered spaces to share and develop their thinking and knowledge. As children become involved in these various activities, we study both their developing scientific understandings and the linguistic registers used to express these ideas.

This project gelled as I kept talking with Chris Pappas in the context of the elementary education programs that we both consistently have taught in, and contributed to their revision. I insisted on the dialectic of theory and data that it is so important, I believe, for all students to experience, including the young ones (despite the contrary flavor the U.S. National Science Education Standards that reserve theorizing for middle school grades) and she kept talking about the role and the importance of good-quality children's information books in their development as readers, writers, and thinkers. We put the two ideas together and soon we had a new passion—something that we both cared so much about to "suffer" through rejections of not one but two grant proposals to NSF before the third one was indeed funded. Of course, every time we revised the proposal we made it stronger trying to address reviewers' and panel's comments. However, I think an important reason for the hardship we faced was the fact that we went "against the grain," so to speak, against the more established paradigms of quantitative, experimental research designs, that NSF has been more familiar with, and against recommended science content standards to "mess around" with interpretive, ethnographic, discourse-based, costly, and time-consuming studies and to attempt "theorizing-richer" science curriculum.

As this work unfolds, we study various constructs. Through the construct of intertextuality—how participants in classroom discourse juxtapose, or refer to, other texts than those they are presently co-constructing—we have been able to see the various connections young children make, as well as their particular inquiry styles in realizing these intertextual links, as they not only observe and describe situations and experiences, but also theorize about them and engage in scientific concept development. We have also shown how the nature of intertextuality changes during an integrated unit as children and the teacher move from narrative-like registers and genres to more scientific-like ones, and how children's funds of knowledge (borrowing Luis Moll's term) allow for complex, hybrid discourses to unfold in primary grade classrooms offering children opportunities to "play with" tentative ideas and words. Paying particular attention to the wordings that young, ethno-linguistically and socio-economically diverse children use to express understandings, and the particular everyday experiences that they bring up, we have exposed the complexity and hybridity of children's language and thinking around complex phenomena, such as evaporation, boiling, and condensation (Varelas, Pappas, Barry, & O'Neil, 2001). Furthermore, we have begun to study how children express meanings in words and in pictures in illustrated informational books that they create as a culminating experience in the integrated units, and what proximal and distal scientist identities children portray in pictures and words before and after such units (a study that doctoral student Eli Raymond-Tucker and myself currently lead). Having a variety of data for a subset of 60 focal children (ten from each of the six classrooms) will allow us to both compose and understand individual profiles of young science thinkers, doers, and talkers in the context of integrated science-literacy instruction, and to explore potential relationships with grade level, ethno-linguistic background, and gender.

Over the years of my life in academia I came to value, work on, and "fight" for collaborative research and development projects that enable me to work together with school teachers, colleagues, and new, "budding" researchers, addressing tensions, dialectics, and emerging knowledge and understandings. But I also consider that there is much to be gained from collaborative work in teaching itself. I find

269

coteaching to be a source of inspiration for engaged scholarship. For example, my coteaching with teacher Jeaneen Benhart in her first grade classroom resulted in developing and sharing with practitioners an integrated unit that used various bodily-kinesthetic activities, hands-on explorations, and information books to offer children an opportunity not only to collect and think about data on various properties of rocks, but also to theorize about the origins of the three rock types and how they relate to their properties, and about the rock cycle (Varelas & Benhart, 2004). My coteaching of a course for seniors in elementary education with curricular theorist, Bill Schubert, helped me think about science relative to other subject matter we teach at elementary school (e.g., the arts, social studies, history), and, thus, ponder how children's "peripheral legitimate participation" in these disciplines may look similar and different from that in science. My coteaching with paleontologist Roy Plotnick of one of the new science content courses for elementary education and non-science majors that a group of us have developed over the last few years continues to allow me to think more deeply about the nature of scientific practice, how it has been shaped over the years, and the who, where, when, and how factors in its evolution.

To recap, my research program has examined practices of teaching and learning science in urban classrooms as teachers and students talk, read, write, do, share, and think science. Using qualitative methods and a range of data-collection techniques, I have collaborated with other educators and researchers to uncover, expose, and celebrate the complexity of teaching and learning, and the cognitive and linguistic resources that ethno-linguistically and socio-economically diverse children bring to and use in science classrooms. In that it involves an integration of teaching, service, and research, this complex interdisciplinary work has been challenging to conceive, execute, and disseminate. However, these challenges derive from the close connection of the research activities with practical questions that emerge in ongoing educational work. The ecological validity makes it worthwhile. Furthermore, the tensions inherent in such work, and in using qualitative approaches in pursuit of understanding of science teaching and learning in urban settings, make it exciting. It is gratifying to be working at a time when the field of science education is becoming progressively more open to such forms of scholarship, as new possibilities are opening up to publish and receive funding for this kind of work. With my research, I strive to contribute to this movement toward diversity of educational research paradigms and possibilities.

MORE ON IDENTITY FORMATION AND EVOLUTION

Growing up in Greece in an era of unrest—the Athens Polytechnic University uprising and the bloody fall of dictatorship, the fighting for civil liberties and constitution changes, the election of the first socialist government—and with parents committed to education despite barely having finished elementary school themselves has helped me become sensitive to, and think about, power, change, social struggle vis-à-vis education. It also has contributed to the fervor that I have kept developing over the years and to the stance I have taken towards teaching and learning, in terms of both practice and research. As Jay Lemke claims, meaning making is "selective contextualization." It is seeing something as a part of some whole and not others; it is differentiating. So, as I have growing up and growing older, I have been selectively contextualizing my experiences, my learning, my aspirations, my expectations, my interactions with others and with texts.

The University of Illinois—a public, urban, land-grant, research intensive institution in the middle of Chicago— also has offered me numerous opportunities to develop my passion for science education, for working with kids and teachers of underserved and underrepresented groups in science, for learning from many colleagues. Although some of these opportunities arose in the context of facing challenges, pressures, agendas, emergent plans, and strategies that needed further elaboration, they, nevertheless, became opportunities for me to take risks, to push myself to articulate my convictions, to work with colleagues, and to construct research opportunities within large, multi-dimensional development projects. In general, UIC supported all my efforts. It was mostly the conversations with thoughtful colleagues in hallways, in offices, or in our classrooms that not only kept me going, but helped me and continue to help me find ways to express ideas that I have been developing. In some ways, I think that some of my colleagues *make the commonplace problematic*—what Fred Erickson calls for teachers to do. It is this problematizing that is a constant source of reflection, renewal, and growth.

It goes without saying, of course, that many scholars have shaped my scholarship, my scholarship of discovery, but also of application, of integration, and of teaching—all four activities that Ernest Boyer specified in his 1990 seminar report as a new way of looking at what is expected of a member of the professoriate. Many of these scholars have not specifically focused on science education, and others have contributed particularly to that field. My voice is heavily populated by the voices of so many others, several of who have also contributed chapters in this book. I will list some who predominately associate themselves with other fields of study and not, or not only, science education: curricular scholar John Dewey and his ideas on the role of experience in education; psychologist Lev Vygotsky and his ideas on the relationship between thinking and language; science philosopher and historian Gerald Holton and his ideas on the private and public practice of science; social anthropologist Jean Lave and her ideas on participation in communities of practice; post-structuralist theorist Mikhail Bakhtin and his ideas on dialogism, on how language is always constituted by and through subjects, and on the unstable meaning of discourse participants' utterances; literacy scholar Kris Gutierrez's ideas on how learning contexts are immanently hybrid, polycontextual, multivoiced, and multiscripted, and the need for attending to "third space"; urban educator and language and literacy researcher Lisa Delpit and her call for providing to students, who do not have them, the additional "codes of power"; sociolinguists James Gee and Jay Lemke and their ideas on discourse and its relationship to identity and affiliation; and I also include science educator Rosalind Driver (whose untimely death stopped short her direct influence in the field) and her articulation of the role of representational systems in learning science.

Beyond scholars though, there have been three other people in my life, my personal life, who have not only made my professional life possible, but who have also shaped important dimensions of and convictions that guide my scholarship. One is my husband Nikos, an integral companion of my life for 24 years now and a high-energy physicist and UIC faculty in the physics department. The other two are my children. Let me start with them. Being a parent helped me see the "other side of the fence" that I could not easily see as a teacher. As a parent I had to put first my kids' interests, strengths, needs, ways of being, ways of making sense, possibilities, and opportunities, putting aside what is "good" for all the kids in the classroom. Before I had kids of my own, I would privilege my classroom "community" over the individuals that comprised that community. Of course, I knew that I needed to attend to my students' individual strengths and needs, but I did not usually become an advocate for a particular child's individuality. Several times as a teacher, I tended to see a particular child's actions, behaviors, attitudes as interfering with the ways I wanted my class to be. As a parent, though, I tended to see my children's actions, behaviors, attitudes as the ones I wanted to attend to, to justify, to nurture, or to change. Thus, my parent identity helped me better address tensions I have had as a teacher, consider necessary dialectics between the individual and the group that a student belongs to, and strengthen my ability as a teacher to juggle multiple voices. But, negotiating my responsibilities as a parent of young school-aged children and my responsibilities as a teacher, a "budding" scholar, an urban educator, a good citizen of UIC and its College of Education was not easy. I missed the annual meetings of National Association for Research in Science Teaching and American Educational Research Association, or shortened my attendance because I could not travel much as I felt I had to be with my children. My identity as a mother was taking precedence at those times.

As for sharing my life with an experimental high-energy particle physicist and academic, conversations with Nikos have helped me get a better feel for concepts, processes, ideas, and practices in a science field that is changing quite fast and has some interesting signatures—experimental collaborations that range from a size of fifty collaborators to a huge number of 2000 collaborators; loads of data; fascinating theorizing; laboratory sites where machines, people, institutions, and nature find ways of living together for years and decades, international collaborations that restrain and enable at times individuals, academic institutions, countries. Nikos also has helped me forming my science educator identity via his own argumentation abilities and skills that have nurtured my attention to meaning making and reasoning. Since our undergraduate years, he and I have been challenging each other to understand, reason, argue, debate, and flesh out arguments, claims, and warrants. This has encouraged me to articulate my scientific thinking and has helped me see critical feedback to my work as invitations to elaborate, to think more deeply about my ideas, to consider and argue for and against alternative positions, and points of view and interpretations.

This has actually become an enabling practice when it comes to my attempts for publishing my work. I remember getting "potentially" condescending, arrogant, comments from reviewers of manuscripts submitted to journals—comments assuming that I was a teacher doing non-rigorous teacher research, comments about deceptive interpretations, comments about who should and should not be part of the authorship team, comments about ignorance regarding the academic writing genre feature of not introducing new ideas in the discussion section. I took such comments as invitations to stand by some of my positions as I also tried to better articulate them in subsequent revisions. Of course, some of the comments hurt at the beginning. So I have been pondering about ways we offer feedback to each other and move our work forward. When I am a reviewer myself, I struggle to balance my own ways of meaning making, argumentation, reasoning, and thinking, with those of the authors, supporting and making room for multiple styles, voices, messages, ways of looking at data, at the same time as I push for articulation, explication, and rich and complex attention to contextualized practices, understandings, dispositions, and thinking. I get upset with mediocrity, but I constantly wonder, and fear, as a researcher and a teacher, whether I am "dismissing" ideas, thoughts, approaches, lines of thinking as naïve, incomplete, underdeveloped, just because I cannot see past my own ways of looking at a phenomenon, a construct, a study. I am reminded that "[l]anguage is not a neutral medium that passes freely and easily into the private property of the speaker's intentions; it is populated—overpopulated—with the intentions of others. Expropriating it, forcing it to submit to one's own intentions and accents, is a difficult and complicated process" (Bakhtin, 1981, p. 294).

WHERE TO?

In science education, now more than ever, we are moving towards questioning what is "good" science education for children, for learners, for people, and we problematize stability, homogeneity, and clear frameworks. We look for uniqueness, for individuality, for change, for variability, for multiplicity of voices, styles, modes of representation, of expression, and of meaning making, for bringing attention to the more "neglected," less "mainstream," more marginalized groups and their participation in science and science learning. We are struggling, though, with the perennial paradox—using existing, familiar, well-defined structures of a sociocultural practice to examine, study, explore, and understand new, unfamiliar, ill-defined structures. But the field of science education is growing, as several of our colleagues work with poor, homeless, or disadvantaged youth in shelters, after-school programs, and classrooms, or English language learners in urban schools, or immerse themselves in the world of videogames, or coteach in "tough" school settings, or challenge hegemony in writing, doing, talking, thinking, gesturing science, or develop multi-faceted, flexible, and evolving constructs of identity. It may not be growing as fast as needed according to some, and maybe faster than necessary to maintain "quality" for others.

I end with echoing some more of Bakhtin's signature constructs. I see the future of science education (with my work blended in it) depending heavily on how much, as a community of practitioners, we espouse "heteroglossia" rather than "monologia," and how well we negotiate between "two forces—a centrifugal force and a centripetal force" as we communicate with each other and the rest of the world. The physicist part of my identity loves to think of these two forces, the former pushing away from a central point and out in various directions, and the latter pushing toward a central point. The educator part of my identity loves to think of the interplay, dialectic, and tension between these two forces. Bakhtin's heteroglossia helps us move toward multiplicity, but not in the form of multiplicity of meanings of individual works, but in the form of different works, different approaches, different constructs where both forces are inherently at work.

Maria Varelas
Professor of Education
University of Illinois at Chicago

STRUCTURING SUCCESS IN SCIENCE EDUCATION

The three female science educators featured in this section all showed significant determination and persistence, and a sense of the game that underpins successful careers. Also, talented mentors who were caring and mindful of what it took to succeed in academe supported them. Being female was salient in each of the cases portrayed in this section—though the salience was different for each person.

ACCESSING DOCTORATES IN SCIENCE EDUCATION

The three auto/biographies in this part D are representative of a contradictory development in science education, where there continued to be barriers to females in the field even though the possibilities for entering it expanded. An increasing number of male supervisors sponsored female graduate students and, through their transactions, mediated the entry and success of female members of the culture.

Like many of the authors featured in this book, Lynn Dierking was involved in science and had no intention of becoming an educator, until she got involved in teaching marine biology to young children at a summer camp. After this experience she changed her career goals, opting to teach science—much to the chagrin of her advisors who felt she was too talented to become a teacher. Maria Varelas also was a scientist, but from the outset she loved to teach and combined her passion with physics and her desire to educate others. In contrast to the others, Okhee Lee was not a scientist and had the goal to be a university professor, avoiding the inevitability of becoming a teacher, which, in her native Korea, was judged to be a profession suited to women. Despite differences in their trajectories and identities in science education, all three scholars are well known researchers—well-established science educators, tenured full professors in research-oriented positions.

Lee and Varelas both came to the US to undertake doctoral studies requiring full support for tuition and a graduate assistantship. Each had social reasons for opting to do a doctorate in the US. Lee's brother was a graduate student in the US and Varelas's spouse wanted to continue his research in physics in the US. There were no viable alternatives in their native countries and the financial packages in countries outside of the United States did not make full time study a viable reality. In this regard they were similar to the international scholars featured in part C, who returned to their native countries after graduation.

Lee studied educational psychology at the Institute for Research on Teaching (IRT) at Michigan State University (MSU). She was offered a choice of working on two projects during her time in the doctoral program—both in science education. Hence, much of her research experience was in science education and when she graduated she showed that she had considerable skill in obtaining external funding in science education and was well placed to publish manuscripts in high-impact journals in science education. Despite a national trend for the gatekeepers of science education to restrict entry to science education doctorates to those with a strong background in science and to hire only assistant professors if they had experience in teaching in K–12 contexts, Lee became known as a leading science educator without meeting either of these criteria.

Varelas first went to the University of Rochester, where she completed a master's degree after two years and applied for and was accepted into a doctoral program in curriculum design at the University of Illinois in Chicago (UIC). Her move to Chicago was catalyzed by her spouse needing to spend time in the Fermi Lab to complete experimental work in his doctorate in physics. Varelas completed her degree after four years, secured a postdoctoral appointment in science education at UIC, and a year later accepted a tenured position at UIC. Although it is known within science education for some universities to hire their own graduates, it is not a common practice, especially as a first appointment. Varelas commenced her journey through tenure and promotion shortly before the birth of her first child and continued it through the birth of her second child and the completion of her husband's doctorate and post-doc in another state. The temporal alignment between family and academic roles was similar to those described in earlier chapters. Clearly Varelas managed her various roles in an accomplished way and was promoted to full professor in a relatively short time.

CREATING FRESH PATHWAYS

Paralleling the gender-related changes of the science education culture there were other developments that may in fact have benefited from the changing gender structure of the field. For example, free-choice science learning as it occurs in museums, outdoors centers, environmentalist groups, and other informal settings appears to appeal to nontraditional science educators. Often women are on the forefront when it comes to militating for the environment, so that it comes as little surprise to find the parallel developments in the increasing interest in informal education and the changing gender characteristics of the field. Similar patterns can be observed in ethnomathematics, for example, where there have been many female scholars (e.g., Jean Lave, Sylvia Scribner, Patricia Greenfield,).

Free-Choice Learning

As a scientist Dierking was accepted into science education but opted to study learning in informal settings, an area that was not mainstream. She identified a university and chose to study with John Koran, who had a strong interest in learning in informal settings. Dierking often found herself at odds with Koran in terms of theory and method. However, she learned from him and others such as John Falk, with whom she still collaborates closely.

Koran and Falk have done a great deal to launch and continuously refine research on the learning of science in informal settings. There is no doubt that these two scholars were productive and got others involved in what became known as informal science education. In North America, museums, zoos, and field centers had been involved in partnerships with schools and communities throughout the twentieth century—however, there was not an active research group examining problems in this area until the late 1970s. The trends elsewhere in the world seem to parallel those of North America. Arthur Lucas, from Kings College in England, also undertook important research in this area in the 1980s, reflecting a scholarly focus on informal science education that continues at Kings College to the present day. Similar trends occurred in other countries. Notable among these is Australia where active groups emerged in several parts of the country—including researchers such as David Symington in Victoria, Keith Lucas in Queensland, and Léonie Rennie on the Australian West Coast. Elsewhere in the world, women science educators such as Revital Tal, Jrene Rahm, and Doris Ash continue this trend by investigating knowing and learning in informal settings (e.g., museums, outdoor centers).

When Dierking graduated and obtained a tenure earning position she found that mainstream science educators did not accept her interest as sufficiently mainstream to justify her spending her time in this way. With her interests in learning science in out-of-field settings Dierking did not fit the mold for university science educators and after three years at the University of Maryland the writing was on the wall—she did not align with the mission of that science education program at that time. This is a little surprising because in 1980 Emmett Wright published an influential study of museum education when he was on the faculty at Maryland. However, by the time Dierking arrived the faculty had changed in composition and its priorities had shifted. Accordingly, she left the mainstream and worked in a variety of institutions in which she could pursue and develop her interests in doing research on free-choice learning. In so doing she was a part of a vibrant group that developed this field internationally. Recently she was appointed to a shared appointment (with Falk) as Professor of Free-Choice Learning at Oregon State University—a first appointment to a position with this title and an indication of the impact that Dierking and Falk have had on the field of science education. Evidently, Dierking's preference for the term free-choice learning in her arguments with Koran was more than just semantics, since the term is now used widely and captures learning across the lifespan, in out-of-school settings, and includes the roles of families.

Creating New Conventions

With the increasing number of women in the field, the old structures broke up, though the network of the "old boys" continued to mark at least some sections of the field. Although all three auto/biographies exemplify the possibility of being and becoming a science educator along non-traditional career trajectories, we exemplify this trend in the person of Okhee Lee.

Lee was born into a culture in which males and females were treated distinctively according to gender. Her descriptions of the ways in which she was raised in relation to her brothers make it clear that analytical thinking was seen as the province of males and many females may have been raised to be wives and mothers. Her independence of mind is clear in her narrative—she was determined to attend university and pursue a career as a scholar and in order to do that she had to leave her country and win a scholarship to support her education. Several factors aligned to structure Lee's doctoral education. First, she went to Michigan State at precisely the same time that James Gallagher was doing his ethnography in high school science classes—affording her an opportunity to be a participant observer in schools for intensive and extensive periods of time. Without well developed English language tools she brought to the studies an external perspective that was valued while providing her with opportunities to get to know science, high schools and issues about teaching and learning. Second, Andy Porter, the co-director of the IRT, was a very successful researcher who obtained large external grants focusing on teaching and learning and associated policy issues. The social bonds with Porter were to serve Lee well throughout her career and also she learned about the importance of securing external funds with well crafted and edited proposals. Third, Jere Brophy, co-director of the IRT, was a leading process-product researcher with solid connections throughout the world—a scholar who knew how to succeed—how to balance his activities to include writing every day and not to spend too much of his time in the office. Finally, as we show in part C, MSU is one of the most productive teacher education programs in the world and its science education group is a "quiet achiever"—producing more doctorates than any other program in the United States. Hence, Lee was in a large program with very productive faculty and graduate students. Although she was not a scientist her research with Charles "Andy" Anderson focused on alternative frameworks and conceptual change, which was rapidly becoming the dominant paradigm in science education throughout the world. Accordingly, the networks in which Lee worked while in graduate school set her up for productive collaborations throughout her academic life.

Lee attributes to her mother the advice to be nice to people so that if you make mistakes they will be forgiving. This advice has been very salient and provides a measure of protection in the peer review activities involved in grant and article review. Lee has learned in her writing and ways of interacting with colleagues to be agreeable and easygoing. Even though she says that from time to time she has been hurt by reviewers' comments, she assures readers that she will not be like those who have stung her with their comments. Lee vows she will provide suggestions for improvement and endeavor to take a perspective that is open to a bricolage of theories and methods. Perhaps the lesson to be learned from Lee is subtler than to be nice to others! She is courageous and willing to take a stand, she is forthright, and she is open. Above all though, she is accepting of others' views and willing to learn from other perspectives. We have both experienced her taking a stand on a variety of issues while being receptive to the perspectives of others.

Lee also did not fit the traditional mold when, after graduating from Michigan State University, she accompanied her husband to Miami where she eventually earned a tenured position, probably because of her grant writing and researching talents. Lee went against the grain for Korean females when she migrated to the

US to undertake a doctoral degree and, prior to that, in her preparation to obtain financial support from the university she attended. Then, as she studied for her degree she addressed the issues of learning a new language and working in science, a field in which she was not well grounded. Finally, when she obtained a tenure earning position in teacher education she applied for external funding of her projects from the National Science Foundation and the U.S. Department of Education, focusing on the learning of science—mainly in elementary schools and in classes in which English was not a native language.

Staying On After Graduation

Becoming a professor in the same institution where one received the doctoral degree may be a mixed blessing, especially for women science educators. Already, there is a tendency in academe to use the vulnerabilities of an untenured professor as a resource in the distribution and frequency of service tasks. Women faculty members more so than most of their male colleagues succumb to perceived expectations and the needs of the collectivity so that they end up accepting administrative and other tasks more readily than males. During graduate studies, the institutional differences between graduate students and professors are in many instances used as resources to view and enact hierarchies. Because of the path dependence of social interactions, these transactions that produce and reproduce hierarchical relations also mediate the interpersonal relations once a person has graduated and now finds himself or herself in the role of a faculty member.

Be careful what you wish for! When Tobin returned to the University of Georgia on a senior Fulbright award he was eagerly anticipating a highly productive year of research. However, things had changed appreciably when he began in a fresh role. He felt different having been away for about four years and yet many of the faculty in the college of education did not regard him as changed. He felt like a graduate student again and he found it necessary to establish his new identity in ways that visiting scholars coming for the first time or new professors may not have to do. He did not regard it as a good idea to return to the university from which the doctorate was obtained—unless there has been a turnover in the faculty. Otherwise, reproduction of the program is a likely outcome. Hence, Varelas broke with this conventional wisdom twice after graduating with her doctorate. First, she returned immediately in a postdoctoral position and then after one year she applied for and obtained a tenure earning position. Perhaps this was a win-win situation because of the unusual situation that had arisen at UIC—where there had been no science educator since John Staver had left for Kansas State University the summer before Varelas came as a doctoral student. Joe Becker, who supervised Varelas' doctorate had a PhD in chemistry and undertook research on cognition, mainly in mathematics education—but also in science education. There were no faculty who identified as mainstream science educators. As a graduate student and then in her postdoctoral roles Varelas was the key science educator at UIC. Hiring her allowed for continuity of production and a continuance of the collaborations she had established with scholars throughout the field of curriculum studies.

PUBLISHING IN HIGH-IMPACT JOURNALS

These days so much is made of publishing in the right places and getting external funding for research. It is a changing façade and it is the case that many of the best-known science educators have not been prolific publishers in high-impact journals and nor have they been in aggressive pursuit of external funding. Yet the height of the bar is changing and there are serious implications of these changes for the field. In each part of the book we explore publishing and funding of research. We do so in this and subsequent parts before examining the trends in a global context. Some of the highly ranked journals based on the citations during the years 2004–2005 in which science educators publish appear in Table D.1. The *Thomson ISI—Web of Science* impact factor is calculated using the equation

$$IF = \frac{\text{total number of citations to journal over 2 years}}{\text{total number of citable articles in journal over 2 years}}.$$

Thus, the higher the impact factor, the higher the rate with which a journal is cited. The impact factor therefore is a measure of the importance of the journal and the article it publishes to the field as a whole. The table shows that in a field of 99 educational journals, *Science Education* and the *Journal of Research in Science Teaching* find themselves near the top 10 percent; in some years in the past, they in fact ranked among the top ten journals. The table also shows some of the other journals in which science educators have published or which science educators cite in their own work.

The trend to make tenure and promotion decisions a function of the number of articles published in high-impact journals is observable around the world. Especially those universities that participate in national and international comparisons or those that want to distinguish themselves as major research institutions expect

Table D.1. High-impact journals in which science educators publish or which they cite[1]

Rank	Name of Journal	Impact factor
1	Journal of the Learning Sciences	2.729
2	Review of Educational Research	1.760
6	American Educational Research Journal	1.383
11	Science Education	1.159
12	Journal of Research in Science Teaching	1.011
38	International Journal of Science Education	0.553
63	Research in Science Education	0.370
83	Journal of Curriculum Studies	0.239

Note 1: The figures in this table are for the years 2004–5 based on a total of $N = 99$ that are listed in the database. The rankings vary over the years

Table D.2. Citation patterns in the four ISI-listed science education journals for 2005

Cited Journal	Citing Journal				
	SciEd	JRST	IJSE	RISE	Total
SciEd	207	142	239	37	939
JRST	231	259	257	61	1331
IJSE	144	72	286	49	772
RISE	14	15	42	26	145
Total	596	488	824	173	3187

their junior faculty members to publish their work in journals that are part of the *Thomson ISI-Web of Science* database. This database has come to prominence initially in the natural sciences, where scholars use it to rank their journals and establish the importance of a scientific work in terms of total citation counts to the work. Sociologists interested in the study of scholarly disciplines use the database to establish the structure of disciplines—the major research centers, groups, and individuals—by means of citations.

In the social sciences generally but in education especially, universities and individuals increasingly are becoming aware of this database and use it as a resource in a variety of decision-making processes—including tenure, promotion, and salaries and salary increments. An increasing number of leaders in the field of science education now advise their graduate students to orient their publications towards the high-impact journals in science education particularly but also in education and educational psychology more generally.

To better understand how science educators are cited within and between journals, we constructed a table that makes available the numbers of citation a science education journal receives from one of the four journals in the field listed in the *Thomson ISI-Web of Science* database (Table D.2). The table shows that most *Science Education* (SciEd) authors cite articles in the *Journal of Research in Science Teaching* (JRST; 39%) and second most cite articles in the same journal (35%). JRST authors, however, cite articles from the same journal much more frequently (53%) than articles in the other journals. Most of the citations of *International Journal of Science Education* authors are nearly equally distributed across JRST (31%), SciEd (29%), and IJSE (35%), whereas authors publishing in *Research in Science Education* (RISE) tend to cite JRST authors. The table supports the contention that JRST authors constitute a more bounded community than the authors constituting the other journals.

The extent to which each of the three scholars published in high-impact journals can in some senses be related to the practices of their chief mentors. Important in Lee's academic life has been her willingness to strive for excellence and learn from her experience. Making sure that she wrote daily and devoted extra time to her studies are just two ways in which she guaranteed she would be successful. Also, targeting high-impact journals as her main research outlets allowed her to stay in touch with the faculty and graduate students from Michigan State University—as they spread throughout the world. The *Thomson ISI-Web of Science* lists twenty eight articles for Okhee Lee, beginning in 1997—twelve in the *Journal of Research in Science Teaching* (JRST), four in *Science Education* and seven in journals sponsored by the *American Educational Research Association*—*American Educational Research Journal* (AERJ), *Teaching and Teacher Education*, and the *Review of Educational Research*. Lee has 222 citations to these 28 articles, the most highly cited dealing with conceptual change in science—one published in JRST (47 citations) and AERJ (32 citations). Lee's key mentors also are well represented in the Web of Science—Anderson (16 articles since 1980), Porter (34 articles since 1967) and Brophy (60 articles since 1970). Varelas has fewer articles in the Web of Science database, nine articles published in high-impact journals since 1991, four in JRST one being her most highly cited work (11 citations), comparable to Joe Becker (14 cited articles since 1989). Dierking has 12 articles included in the Web of Science database, six published in *Science Education* and two in JRST. Dierking's publication profile is similar to her mentors—Koran (11 papers cited since 1969) and Falk (14 papers cited articles since 1976). Dierking's most highly cited work (28 citations), which was coauthored with Falk and published in *Science Education,* examined family behavior and learning in informal settings.

Part E

SCIENCE EDUCATION AROUND THE WORLD

GLOBALIZING SCIENCE EDUCATION

The first auto/biography in this part of the book begins with a statement about the "curriculum wave" that arriving from North America hit Israel in the 1960s. The influence that U.S. science education has had abroad is not isolated to Israel but has extended to other nations as well. U.S. science education has had, one may say, an important mediating role on the development of science education around the world, though, for many (theoretical) reasons, the roles cannot be thought in causal terms.

Michael: I remember my parents talking about the German educational system as picking up revolutionary ideas and developments from U.S. educators that they had already abandoned. I clearly remember how German students and educators talked in a negative way about the introduction of multiple-choice exams on comprehensive tests in medical schools designed to separate the wheat from the chaff, that is, those who would make it into the core program and those who had to pursue other studies. Whereas in all other fields, examinations required the writing of essays or solving problems, the medical education system had taken on this practice—for the ease of administration—that in all other disciplines and contexts was snubbed. To the present day, a number of educators I have met in non-Anglo-Saxon countries critically talk about the influence largely of North American developments on their educational systems and research.

Ken: Welzel's remarks about her teaching of science in East Germany were reminiscent of recent efforts to teacher proof curricula in the United States in the wake of the *No Child Left Behind* legislation. In urban areas like Philadelphia the school system was taken over by the state, which endeavored to teacher proof the curriculum. Efforts were made to have teachers teach by the book and rules were established with testing schedules to ensure that science was taught according to the book—covering prescribed content at specified times and giving tests on pre-assigned dates that assume conformity with the prescribed schedule. As I make sense of this, the folly is to assume that teachers have control over the learning of students and hence prescribing curriculum coverage rates can standardize the science curriculum and thus afford equality for all students.

Although European educational systems differ from the ones in North America, and although educators and researchers of most countries pride themselves of having an approach that stands on its own, the influence of the Anglo-Saxon educational culture on the remainder of the world is undeniable. We already highlighted the fact that individuals from many countries, including other Anglo-Saxon countries like Australia, came to the US to pursue doctoral studies and then returned to their home countries, thereby hybridizing existing practices with elements from the other, their second world (see Parsons, this volume).

The influence of the Anglo-Saxon culture on others is mediated not in the least by the role scholarly publications play and the historical shift to English as the lingua franca of choice. Whereas there are many science education journals published in English, journals published in other languages are rare often delimited to one or two in a language including *Zeitschrift für Didaktik der Naturwissenschaften* (German), *Enseñanza de las Ciencias* (Spanish), or *Didaskalia* (French). Although in some disciplines multilingual journals are more frequent, in science education we are aware of only one, the *Canadian Journal for Science, Mathematics, and Technology Education*. In some countries, like Korea (*Journal of the Korean Association for Research in Science Education*), a national organization produces a journal entirely in the English language to allow its members to publish in this language without having to compete in international journals.

The position of English as the lingua franca in journals is paralleled by its role as lingua franca during many international conferences—e.g., those organized by the National Association for Research in Science Teaching (NARST), *European Science Education Research Association* (ESERA), or the *Australasian Science Education Research Association* (ASERA). There are much smaller conferences that bring together science educators from exclusively German (Germany, Austria, Switzerland), Spanish (Spain, Latin America), French (France, Quebec, some African nations), or Portuguese speaking countries (Portugal, Brazil) so that the language of the conference will be other than English.

The dominance of the Anglo-Saxon science education culture has been supported by other developments, in part designed to transcend the national orientations. For example, the US-based NARST has made tremendous efforts to open up from its initial orientation to national issues and to become an international organization; the 1990s saw substantial drives to recruit scholars from countries other than the US to become members in the organization and its annual conference now is attended by 800 to 1,000 members. That is, there may be an inner contradiction whereby the opening up and internationalization of one national organization actually could lead to a narrowing of possibilities related to the path dependency of a particular community, which now strives to become international. Similar to the narrowing of opportunities and possibilities in the economic markets, where globalization and internationalization of manufacture led to a homogenization of the products, what science education can be—both individually and collectively—changes in different parts of the world.

Despite the existence of trends and structures that allow scholars to communicate in their non-English language, the Anglo-Saxon world remains the main orientation; and in many countries and universities efforts are under way to internationalize education and degrees. In the Scandinavian countries, one can find many doctoral dissertations now written not in the native language but in English. Having to succeed in an international university, scholarship, and labor market—the *Bologna Declaration*, which intends the normalization of university education across Europe, is but one example—favors an increasing trend to make one's work available to an international rather than a national audience. The Bologna Declaration aims at the establishment of a higher education system that facilitates the mobility of people and the transparency and recognition of qualification, quality and attractiveness of European institutions for third-country students. Although students from French-speaking African nations already come to France to do their tertiary studies, the Bologna Declaration will only increase the importance of a lingua franca, which, because of the path dependence of culture, will further underscore the role of English and Anglo-Saxon culture in scholarship generally.

The ways of being of a university-based science educator in Europe have changed over the past decade and continue to change. There is an increasing trend to "Americanize" universities, not in the least by creating new opportunities for realizing career trajectories. For example, the professorial system in Germany was quite rigid in the sense that each university only had a certain very limited number of chairs. An individual could, after doing a PhD and after completing the *Habilitation* (something like a postdoctoral period lasting about five years and resulting in a major published work), apply and submit to the process of receiving the title of professor without nevertheless obtaining a university position, that is, a chair. Some of the changes that have occurred over the past decade include the softening of the requirement of a *Habilitation* for a chairship, and its replacement by the publication for an equivalent number of publications in reputable journals is becoming more frequent. Thus, for example, Manuela Welzel who figures in this part, abandoned her "habil" position in Bremen to take a chair as professor of physics education at the *Pädagogische Hochschule* (University of Education) in Heidelberg, which she obtained based on her publication record rather than on a *Habilitation*.

In some countries, there also exists a tightening of the market for academics. For individuals in such countries it is difficult to find positions and construct careers. For example, many French scientists leave the country because they perceive tremendous opportunities in the US and also in the UK. In education, because of the cultural specifics of the related systems, it is more difficult to obtain jobs in other countries—not in the least because some countries, despite their immigration policies, privilege their citizens. In Canada, for example, all jobs have to be advertised within the country first and international searches may be conducted only subsequently. It is easier for foreigners to complete their degree in a country and then find a job there than it is for someone with a foreign degree and nationality to be considered for a position.

All of these developments, though they might be intelligible and understandable today, could not have been predicted at the beginning of the 20th century, when many natural scientific discoveries and theories were published in the German language. However, academia generally changed in the face of the major cultural-historical events of that century: the two world wars, the role the US played in them, and how its role shaped the economic forces not only at home but of the world as a whole. In addition, the internal structures of academia and the administrative structures of the universities fostered innovation in the US whereas they were more rigid in European countries, continuing to favor a small elite.

The three auto/biographies that constitute this part E of our book provide evidence of the fact that science education does not just exist in and for itself but rather occurs against a background of national and international historical development. Thus, Hanna Arzi writes about teaching while doing her military service, which, of course, evokes the images of the perpetually unstable situation in the Middle East and the struggles between Israel and its Arab neighbors. Manuela Welzel grew up and received her first education in the German Democratic Republic (East Germany) and taught within the communist system. Her PhD years occur against the background of the fall of the wall that had divided Berlin. She studied with one of the famous psychologists in the cultural-historical tradition in East Germany of the postwar era, Joachim Lompscher; but, because his research institute was closed in the wake of the collapse of the communist system, Manuela moved on to do her doctoral studies in what was previously the hostile and foreign West. Justin Dillon's auto/biographical account brings out two important dimension: (a) the increasingly difficult situations of teaching and learning in urban areas, where unemployment and poverty testify to the production and reproduction of societal inequities and (b) the increasing sensitivity to environmental issues on a globe increasingly threatened by its inhabitants.

HANNA J. ARZI

TRAVELS IN AND BETWEEN PRACTICE
AND RESEARCH

I was a high school student in Israel in the 1960s when the great curriculum wave started arriving across the Atlantic Ocean and the Mediterranean Sea, and so the school science to which I was exposed was still shaped by fact-stuffed traditional textbooks. I discovered that a curriculum revolution had occurred when I noticed that my younger sister's chemistry book looked different: It was CHEM Study. In my first school teaching job I was required to use CHEM Study, and from then on, during 13 years as a schoolteacher, change was the norm, and not a single year passed without my teaching a new or a modified curricular unit, usually as a trial teacher. My first non-teaching work in science education, in parallel to teaching, was curriculum development. My first exposure to research-type activities was through formative curriculum evaluation, from which basic and puzzling questions emerged, driving me toward doing research. Personal predilections played central roles in this route as in the diversion I later made away from a normative academic track. Thoughts of becoming a teacher started only during graduate studies in bio-chemistry, but previous experiences as a high-school student had long-term effects on my thinking, hence this is where my present story starts. It is a selective career account in which I give priority to what I regard as formative events, and in which—with the benefits of hindsight—my reflections leap back and forth in time.

MY UNPLANNED (YET FORTUNATE) ROUTE TO SCIENCE TEACHING

Choice of Chemistry Despite Frustrating Teachers

Primary school was fun, but not high school—then the most selective and elitist in Tel Aviv. It appeared that my teachers were knowledgeable, but the feeling was that most did not care much for us. Few exceptions that come to mind are from the junior years: a teacher of Bible and Hebrew literature, an English teacher, and a physics teacher who was particularly clear and enthusiastic—it was good traditional physics teaching and I loved the logic and got great satisfaction from understanding rather than merely memorizing. Thus physics became my first preference when we had to choose advanced-level studies for grades 11 and 12 at the end of which awaited governmental matriculation examinations. Chemistry was very low on my list, but since subject choice came in blocks I could either take advanced-level physics with mathematics and chemistry, or give up physics altogether. Combining my genuine preference for physics with the general pro-science zeitgeist of the 1960s, I chose the advanced-level physical-science class.

K. Tobin, W.-M. Roth (Eds.), The Culture of Science Education, 289–300.

Unfortunately, my hopes for exciting science in the advanced-level courses were not realized, primarily not in chemistry. We were considered lucky to get the senior chemistry teacher who had the reputation of securing her students' highest grades in the matriculation examinations, but I soon found that her "pedagogy" was largely based on oral testing of students in front of the class and on most having private tutors. I therefore had no choice but to give priority to studying chemistry on my own. For the laboratory component of the examination I practiced in the common laundry room on the roof of the apartment building where we lived. Even though this was inorganic qualitative analysis that required little understanding beyond following instructions with a few technical skills, the fun was great and inviting friends to join in made it even better. By being forced to learn chemistry alone, I discovered that it was wonderful, and after considering a few alternatives I decided to continue to study chemistry at the Hebrew University of Jerusalem. While my story supports the educational literature's advocacy for students taking responsibility for their learning, I would wish that this happened with teacher positive facilitation, rather than as a result of a no-choice situation.

Choice of Teaching Despite Initial Antagonism

Disappointing teaching continued in university. Even though most activities in laboratory courses were prescribed, due to my penchant for lab work I could still get satisfaction from merely experiencing things. The fun of laboratory work was enhanced by fresh teaching assistants: Overhearing them discussing their research while I was wiring an experimental system in physics, or waiting for an organic reaction vessel to simmer in a fume cupboard, was a perk of studying in a major research university. The decision to continue to a higher degree with a research thesis was obvious. Interest, rather than employment prospects, was the major consideration, but it was clear to me, as previously in the move from school to university, that a teaching career was not an option.

I discovered that teaching could be a challenge and a joy during graduate studies in biochemistry. Teaching was taken seriously in the department and I applied for a teaching-assistant position for the prestige no less than for the salary. It was in a laboratory course with extended sessions that intertwined theory and practice, and I found myself spending endless hours in pre- and post-session work at the expense of time for my thesis (eventually completed with distinction). I thoroughly enjoyed it and the idea of teaching started floating in my mind, though the realization that I want to be a teacher awaited the next teaching assignment—in the army.

In Israel, although military duty service normally follows high school, prior to university, I continued straight from secondary to tertiary studies as participant of the "Academic Reserve" program in which duty service is done after university, to provide the army with professional officers. Upon completion of BSc and MSc, I served in an applied-chemistry lab of the ordnance corps. Performing tasks like checking stability of TNT stock samples, or analyzing steel in a troubleshooting process, was very different from my academic research on proteins. In hindsight, the almost four years in pure and applied empirical science provided invaluable preparation for science education, giving me a feel for the nature of science and the interplay between research and practice that I could not have developed through studying *about* these issues.

The military service also provided me with thought-provoking teaching experiences, as I was periodically called for lecture series on chemistry and chemistry-related topics in courses for technical officers and sergeant majors. I was instructed to follow syllabi that had been replicated for years, but just starting to teach I realized that neither the chemistry content nor the lecture-type teaching were appropriate. All the students had graduated from vocational schools with no prior chemistry instruction, the required chemistry was vast and understanding could not be achieved in a short stand-alone lecture series. I knew of technologies that had worked before the related science was understood; I knew that my students did their jobs well; I believed they could function even better and have a new kind of satisfaction had they understood chemistry; but what should I teach and how? This nagged at me as I continued with my lab tasks and awaited the next teaching assignment. Reflecting on my different teaching experiences at the university and in the army, I felt a growing urge to know more about teaching and decided to try school teaching when duty service would be over.

To conclude the military story: After duty service I continued teaching for over a decade within voluntary reserve service (even promoted from lieutenant to captain). Thinking in science education has advanced since my first lecture series in the army in 1969; but to this date I do not have good guidelines to manage the dilemmas that intrigued me then.

SCHOOL TEACHING

I started to study education while I was still in the army, taking evening classes in a program for a teaching diploma at Bar Ilan University (chosen for its proximity to my military base). My prime motivation was to learn, not to earn the diploma, hence I gave priority to what interested me and completed the last requirement only after five years, following pressure from my school principal. Two courses made a long-term difference. Educational psychology was my first non-science course since high school and this alone was a thrill. Behaviorism was visible and Bloom's taxonomy emphasized, as could be expected in 1970, and the idea that my teaching could be guided by sets of principles was captivating. Disillusion came along with school teaching, but the introduction to educational psychology led to sustained appreciation of the potential of external input into science education. The significance of the second course—chemistry methods—was due to the lecturer, Yehoshua Sivan. I do not recall Yehoshua presenting highbrow theory, but role modeling enthusiasm for teaching, learning from, and creative adaptation of, multiple sources, including international teacher journals, regardless of hard work and irrespective of the prescribed curriculum (with mild disregard of the establishment). Shortly thereafter we became colleagues, as he recommended to the *Weizmann Institute of Science* to invite me to participate in a curriculum project in which he was involved.

Inside the Classroom

When I first entered a classroom as a schoolteacher in September 1970, my experience as a student was in the back of my mind. I did not want to be like most of my teachers, and I am sure that the familiar observation "teachers teach the way they were taught" did not apply in my case. What I aimed for was a naïve combination of old progressive-type ideals with the then prevailing emphasis on the structure of disciplines. In the absence of guidance (apart from pointing to required textbooks), my previous teaching experiences in non-school settings and my good intentions were not enough. I found my way through reflection on trial and error—many trials and many errors, which bring to mind Myron Atkin's conclusion: "becoming the teacher one wants to be is a lifetime quest" (Atkin & Black, 2003, p. 159). In the first year I felt unsuccessful, though my reputation among students and teachers appeared to be quite good. Over time, self-evaluation improved and eventually I regard the thirteen years of teaching at Herzlia High School in Tel Aviv as a wonderful period.

Comments on curriculum In my first year I was required to teach CHEM Study (using a Hebrew translation) in science-oriented eleventh-grade classes and to use a traditional textbook with watered-down content in grade 10. CHEM Study fascinated me with its logical structure and conceptual clarity, and in teaching it I found myself fine-tuning my own understanding of chemistry. The absence of "applications," let alone links of this American-born curriculum to the local Israeli context, did not bother me because I felt no obligation to stick to the text. I freely intertwined plenty of local materials and can now describe what I then taught as my personal STS version of CHEM Study. Since I took the freedom to make adaptations and since I taught science-oriented students in a prestigious school and initially cared only for what was happening in my classrooms, I had not noticed the "evils" of the curricula of the 1960s that were later critically portrayed in academic evaluations.

Comments on lab work Coming to school straight from a "real" lab, I was bothered by the separation between learning of theory and lab work. With classes of up to 40 students and a single chemistry lab that accommodated about 20, a student spent in the lab two periods every second week; and whatever was done in regular lessons could not be intertwined with lab experiences. Apart from limited time, the laboratory space was not designed to facilitate multiple modes of teaching and learning; thus, for example, bolted-down benches and poor visibility due to tall reagent shelves (used for traditional analysis that vanished upon arrival of CHEM Study) hindered classroom interactions. The school principal was supportive and agreed to remove the shelves and allocate funds for equipment that was not curriculum oriented. But the overall lab design remained inflexible and student lab time remained minuscule, as was the norm across schools—accepted by policy makers and teachers alike.

My different vision of what school lab work should be like was initially based on beliefs and supported by experience in research and applied science. Until I began doctoral studies, I was not conscious of the need for an educational research base, and if I were—there was not enough then, as there is not enough now (NRC, 2006). It is therefore bothering to note the steep decline in the number of lab-related studies since the 1970s, which raises questions on the extent of the match between research trends and needs in science education. As for my penchant for laboratory work: it persisted, and two decades after the first year of teaching I was able to start realizing my dreams of doing it differently as the Director of the Center for Science Education in Tel Aviv.

Crossing the Classroom Door

During the second year of teaching I was invited to join the chemistry group of the Department of Science Teaching at the Weizmann Institute of Science—initially to participate in curriculum development; two years later I became the editor of the journal for chemistry teachers in Israel; after three additional years I decided to do a PhD—all while continuing to teach. The new knowledge and experience thus gradually acquired out of school fed my teaching, but the influence of teaching on the non-teaching activities was much greater: Being a teacher shaped my world-view of education and has underlain all my work.

DEVELOPMENT AND RESEARCH

Experience in Research-Base-Less Curriculum Development

The focus of my curriculum work was a chemistry-oriented physical-science course within a coordinated science program for junior high school. Though done in the mid-1970s, it was part of the grand worldwide wave of the 1960s—a delayed small wave in Israel that started with Hebrew translations of American curricula for senior high school and proceeded to original materials for all ages. Within the Israeli centralized system, the course followed several documents—from policy papers presenting ideology-based grand goals to the specific syllabus that listed content with comments on emphases. Nowhere was research in science education mentioned, which did not bother me, nor seemed to bother my colleagues, as we translated the prescribed syllabus into classroom materials on the basis of our knowledge of science, exemplary curricula, experience, beliefs and value judgment. The burst of research on concept learning occurred after the project had been completed and I then wished it had been published earlier, though I was not sure how it could have guided the development. Over the years, questions have been raised on the extent to which research contributes to practice, and I concur with the observation of Jon Ogborn (2005), based on his forty-year experience in major projects in the UK, that "there are cases where insight, intuition, experience of teaching, and deep knowledge of the subject are at least equally valuable sources of ideas about how to teach" (p. 63). Likewise, Paul Black admits in his retrospective reflections that he had no research evidence for the merits of formative assessment when it became a guiding principle for the UK national curriculum.

293

To affect curriculum, not only research but experience, too, has to build. In Israel, when I was involved with the eighth-grade chemistry-oriented course, there was no experience in chemistry teaching to junior high school students; hence in this respect all those involved in the development were novices. I was also a novice teacher and the youngest on the team and so had overall less experience to offer, though I believe that my enthusiasm and occasional naïve disregard of what was possible according to educational orthodoxy may have had advantages. As development progressed, it so happened that my two senior team colleagues gradually reduced their participation, and eventually I edited the curricular materials and coordinated the formative evaluation, teacher training and nationwide implementation. The course was well received, though not without criticism. For example, some physics teachers did not like our talk of electron flow in electric current that did not conform to conventions in physics, and some science educators accused us for not adhering religiously to "discovery learning." All in all, the course materials (also translated to Arabic) were used widely for two decades, until a new syllabus was introduced in the mid-1990s.

Issue-Driven Longitudinal Research

I have not been involved again in curricular work, but the experience gained was invaluable. A major unexpected outcome was research that flew naturally from the formative evaluation component of the development process. The tangible and practical products of this evaluation were improved curricular materials and in-service teacher training courses, but for me it also was a research generator, raising questions that I could not answer without actually exploring them. For a long time I had resisted doing a PhD because I was happily busy with school teaching and other tasks, but a genuine drive to explore issues that had emerged from my own work changed my mind. In practice, I stayed at the same desk at the Weizmann Institute of Science, just turned from part-time staff member to full-time doctoral student, while at school I shifted from full- to part-time teaching.

Two sets of questions puzzled me: one in regard to students, the other in regard to their teachers. My PhD addressed student learning; the study of teachers waited until I became a research fellow at Monash University (Melbourne, Australia). Both studies were longitudinal: the student study took three years, the teacher study seventeen. Within the time span of the teacher study I worked thirteen years in establishing a center for science education in Tel Aviv, already mentioned above. This work, done as I held the position of executive director, would usually be defined as administration and development, but I count it as my third major long-term study. Following are comments on these studies and the flow between them.

Studying longitudinally student long-term learning My interest in the issues studied within my doctorate originated from being a high school teacher involved in the development of an introductory course for younger students. Prior to the inclusion of chemistry topics in junior high school and prior to becoming acquainted with the notion of preconceptions, chemistry teachers, including myself, treated students entering senior high school as tabula rasa in regard to chemistry. If the earlier teaching of chemistry resulted in meaningful learning, then the tabula-rasa assumption would not hold. This hunch was supported by David Ausubel's learning theory, but the formative evaluation of the eighth-grade course gave evidence only on immediate learning outcomes, not on how much of it remained over time. The research literature did not provide a lead, since memory studies were usually done in psychological labs on small bits of information over short time spans, and I could not find relevant long-term data on school science learning. I therefore embarked on a three-year longitudinal study of learning and retention from grades 8 to 10. It was a large-scale quantitative study, with a battery of instruments providing multiple measures of long-term retention.

Summing up the study briefly: Meaningful learning of introductory science courses indeed makes a difference, yet the longitudinal findings suggest that science learning is cumulative in a more complex way than often is supposed. Curriculum continuity facilitates learning and retention, both proactively and retroactively, and long-term retention has many facets, with gaps between knowledge that is immediately retrieved and functionally available knowledge for relearning and further learning. Our main message for practice: Curriculum developers should strive for related sequences of courses, while teachers should help students grasp course continuity. "Self evident without your study!"—I remember this reaction in a seminar I gave close to the completion of my PhD. Initially offended, with experience in research and practice I have learned that self-evident conventional wisdom in education is one thing, research evidence is another, and applying it is yet something else.

Studying longitudinally change in teachers' content knowledge Through teacher training and feedback in the formative evaluation of the eighth-grade course, it became evident that I had underestimated how difficult it would be for teachers to start teaching a subject (chemistry) in which they had not been initially trained (most trained in biology). Furthermore, some evidence suggested that the conventional wisdom according to which things improve with experience might not always be the case. This was not just disappointing but also puzzling, and so I was driven to a longitudinal study of change in teachers' content knowledge. I did it at Monash University in Australia, and the fact that a research idea that emerged from the Israeli context could be well accepted and its realization facilitated down under shows that despite contextual and cultural differences basic issues are the same across countries. It is also evidence for the openness and hospitality that I enjoyed at Monash University—a stimulating place in the 1980s, with a vibrant group of science educators under the leadership of Peter Fensham.

The study followed secondary school science teachers from preservice training through seventeen years of professional experience. The main data sources were sequences of individual interviews in which knowledge (mainly energy-related) was probed via *concept profiles*—a word-association-type method designed for teachers. Findings show that change is multi-faceted, including details of unused content that fade from memory and reorganization of knowledge structure, with growth beyond simple accretion of facts more likely to occur when teachers teach within their areas of certification. While personal and professional life tracks produce variation between individuals, the required curriculum is the most powerful determinant of teachers' content knowledge, serving as both knowledge organizer and source. Based on the study, a longitudinal perspective on teacher education is necessary, with career-long support for knowledge growth even in teachers' major subjects where expertise is taken for granted.

A note on publications The major parts of the study on student long-term learning were published and the main articles (Arzi, Ben-Zvi, & Ganiel, 1985, 1986) received honors, including NARST awards in 1984 and 1985 for their earlier presented versions. Parts of the 1985–2002 teacher study were presented in conferences and lectures, but I delayed publication, and the more comprehensive article has been submitted only recently (Arzi & White, 2006). The delay started with the belief, supported by evidence, that trend identification before the long-term picture unfolds can be misleading. The extended, probably excessive, delay resulted from my diversion away from the academic track to administrative practice. The research completed by then produced feelings of fulfillment, and I had no pressure to publish for promotion, nor peaceful time for writing. I still have unpublished work in various forms—from conference abstracts to extended drafts.

BACK TO PRACTICE WITH A STUDY OF RESTRUCTURING AND RECULTURING

Unlike my longitudinal studies, the initiation and development of a model center for science education clearly did not subscribe to stringent criteria of scientific research, though features may fit the evolving research approach of "design study." Regardless of research definition—for me it was a significant study. My reflections on this work are situated in my itinerary.

Learning and Research Travels, and Changing Tracks

I was groomed for science education in a science teaching department of a science research institute—an excellent place to grow up in, obviously with a subject-matter focus. In the thesis work I encountered conceptual and methodological questions that were not addressed in the science education literature. I constructed answers from various sources, with crucial contribution from attending the 1982 AERA meeting—my first international research conference. With increasing awareness of the value of external input—from areas outside science and science education, and from contexts other than the Israeli education system, I declined an offer for a tenure-track academic position in Israel (not for the last time) and in 1984 left for almost four exciting years overseas.

Travels started with a British Council scholarship for a two-months visit across science education research centers in the UK. Subsequently I spent a semester at Cornell University, and a Monash University research fellowship brought me to Australia where my major focus was the longitudinal teacher study just described.

When it was time to decide what next, I had not yet fulfilled my desire to travel, learn and do research in multiple contexts, and thus, between firm academic offers in the US and in Israel I opted for the US. But before assuming office, I received an unexpected phone call: It was Haim Harari (later president of the Weizmann Institute of Science) urging me to apply for the position of executive director of a planned innovative center in which he would be chairman of the board. Weighing pros and cons, I felt a commitment to practice, though my notion of practice did not include administration for which I had a pejorative connotation, and I was aware that innovations tend to have short half-lives and some do not survive the planning phase. There appeared to be, however, a big potential and a rare opportunity to put ideas and research into practice at home, rather than suffice with "Implications to Practice" appendages in academic papers (I admit that it was also difficult to withstand the persuasion of Haim Harari). And so, from 1988 until my resignation in 2001, I was the executive director of the center for science education in Tel Aviv (HEMDA, Hebrew acronym), responsible for all administrative and educational activities: three years of planning and construction, followed by ten years of actual operation with continued development accompanied by stabilization.

Establishing a New Center: Features and Underlying Concepts in a Nutshell

HEMDA was established with the grand goal of exemplifying quality science education and providing a model for a different type of regional centers. Underlying the model was the belief in the advantages of the centralization and linkage of material with human resources; hence the center has its own facilities and teaching team. On a day-to-day basis, its original primary task was to serve as the common science campus for schools in Tel Aviv, assuming responsibility for advanced-level courses for their students. Over time, as operation and institutional prestige started to consolidate, activities extended to the further blending of curricular teaching with extra-curricular activities, workshops for teachers and popular science lectures for the general public. The establishment followed a frame recommended by an international commission set by the Rothschild Foundation that initiated the center. As in any transition from policy to practice, modifications occurred; what I wish to highlight are substantial influences of input from the research literature, my personal experience and predilections. This is exemplified through a unique feature of science education—lab work, and a general issue in education—teacher teamwork.

Reconceptualizing school labs It was obvious that the center should have well-equipped labs, but the policy document did not attend to research-based criticism on the prevailing types of school-lab experiences. More bothering was a working paper prepared for budget estimates stating that the center would need X regular classrooms and Y labs—in sharp contrast to my belief in the intertwinement of all school science activities, whereby all lessons take place in laboratory environments of multi-functional classrooms. This had occupied my mind for long, as already

297

mentioned above, later elaborated in a handbook chapter (Arzi, 1998) and recently supported in an NRC report. My attempt to change the concept of school laboratories in the planning of HEMDA was not met by immediate success, not just because it challenged the norm, but primarily because it entailed a huge budget increase. Once approved, the architect had to add a science education lens to his design perspective, also meaning that he had to agree to collaborate. Completion of construction was just the beginning, since dream facilities is one thing, while using them with related change in pedagogy is another.

Introducing teacher teamwork The issue of teamwork exemplifies further the difference between viewing change and innovation as restructuring with tangible products versus a long process of reculturing. The policy document noted that teachers having many colleagues would be advantageous, but did not attend to the literature on norms of isolation and privacy in teacher cultures, meaning that merely placing people together was unlikely to make a difference. Following my then recent research experiences with teachers in the longitudinal study and in the PEEL project in Australia, I was convinced that teamwork was the key to unique qualities at HEMDA. No recipe, however, was available for how to coalesce a group of high-quality individuals: all with graduate degrees in science, yet purposively hired across different ages and career backgrounds—from academe, industry and school teaching. Pulling together pieces from the literature with my research and belief-based hunches, my deliberate approach was to combine formal and informal forms of professional collegiality, including scheduled meetings and imposed team activities, with time and space arrangements for facilitating voluntary collaboration. It was a long and complex route with a range of teacher reactions, gradually leading to multiple modes of teacher interactions (inter alia, coordinated and coteaching of single, inter- and multi-disciplinary subjects, peer coaching and induction of new teachers). Eventually—so I believe—teamwork became a lasting feature of HEMDA.

Comments on challenge and politics I heard people say that academic work is more challenging than practice: As executive director establishing a new institution I had to use all my faculties and to draw on everything I had ever learned and done. The less pleasant part, though not least important, was politics. I discovered politics in education in the 1970s during curriculum development, seeing power play between policy makers, academic scientists and science educators. Subsequently I learned about more interest groups in the educational arena and realized that this is abundant across countries. Since HEMDA aimed for a fundamental change in the practice of science education while breaking away from traditional school structure and culture, politics—macro and micro—was continual and sometimes fierce (too complex to exemplify shortly). In the entrepreneurial and major development phases it was a fascinating challenge, but less so as years went by. In my first encounter with politics, in the context of curriculum, I was an observer; over a decade later, in the initiation of a new institution, I became a participant-observer; in recent years I observe and evaluate, as I do independent scholarly work, teach graduate students about politics (and more) in change processes, and continue to travel.

WRAPPING-UP WITH DEFINITIONS AND SOME ASSOCIATIONS

Writing this chapter made me wonder how to define my work and professional identity. Having been engaged with long-term studies, advocating and discussing longitudinal research in presentations and in writing, and having often referred to Ausubel, I may be labeled "longitudinal-Ausubelian." But I do not see myself a single-method purist, nor a single-theory zealot, and my work never has been method or theory driven—always issue driven. Over time, however, I have realized that long-term projects suit me, as I need long gestation periods and I get satisfaction from looking back at how a complex study evolves and converges. While not attached to a single theoretical perspective, Ausubel's theory has always been a useful frame for my thinking on long-term concept learning. I keep a six-page hand-written letter from David Ausubel, responding warmly to my request for comments on a recently published paper and on a draft entitled "Ausubel Revisited" prepared for the 1986 AERA meeting together with Leo West—then a colleague at Monash University and to this date a close friend. Subsequently, Leo and I met him and his wife for two days at their home in upstate New York—planned as interview that turned into a memorable experience.

Talking about Ausubel connects to Joseph (Joe) Novak, whose name I had encountered as co-author with Ausubel before we met accidentally in 1982 during my first NARST convention. This led to a postdoctoral visit at Cornell University where I evaluated long-term effects of Joe's metalearning course. But his unique contribution was providing affirmation to my work and encouragement to continue doing what I believe is right, even if not in vogue—as he role modeled in his career. I did not lack self-confidence, but Joe's guru-type support made a difference.

Another type of significant contribution came from Andrew (Chick) Ahlgren when he spent a sabbatical at the Weizmann Institute of Science. I was then a doctoral student, and Chick's lectures and personal mentoring strongly influenced the development of insights into what research in science education is, with a focus on educational meanings in statistical analysis that eased a future smooth flow to mixed quantitative and qualitative methods.

In Israel and overseas I have gained from many. Last in my incomplete account in the present chapter is Richard (Dick) White—a colleague at Monash University in the 1980s with whom I am friends and collaborate to this day. Despite the geographical distance, we keep our tradition of extended talks across topics, which started while driving together regularly from Monash to Laverton High School (located at opposite ends of Melbourne) to attend meetings of the first teacher group of the PEEL project.

The issue of how I define my work, from which I wandered to a few professional acquaintances, brings to mind Peter Fensham's (2004) search for the identity of science education, in which he reached an inconclusive answer as to whether it is a distinct discipline. I subscribe to the view that it is not a discipline and have enjoyed working at the junction of multiple fields that inform, or form, science education. I have been sometimes bothered, however, by my lack of expertise across all these fields, aware of the risk of superficiality. A related risk in science education is drifting away from subject-matter issues in the name of need for a broad perspective. The haste in which we sometimes move from one trend to another has resulted in insufficient research base for decision making in important areas. Obviously, new areas deserve attention, so it is a matter of balance— difficult to achieve in science education as in one's personal professional career.

Rethinking my professional identity finally leads me to the challenge of career balance between practice and research. "Career teachers" are commonly categorized in teacher surveys as those teaching for at least five years; "established researchers" were described by Fensham as those publishing beyond their doctoral work. I meet these criteria in regard to my work as practitioner—teacher for thirteen years and administrator for thirteen years (my lucky number), and as researcher. Research work started prior to doctoral studies, almost three decades ago, and has sustained across large overlaps with years of practice. Actual research and research-related activities (like serving on committees within AERA, NARST, NRC, or reviewing for international journals) were done with and without academic affiliation. Since this is not a normative academic research track, traditional identity definitions may not apply. Whatever the definition, my periods of immersion in science and in educational practice, development and research, have been most fulfilling—each separately and interactively. Even though I did not plan my full track ahead, it was my choice to combine travels in and between practice and research.

Hanna Arzi
Tel Aviv, Israel

MANUELA WELZEL

LITTLE WAS PLANNED—ALL SIMPLY HAPPENED

Where are the roots of my interest for science and science education? Sometimes I ask this question and am surprised about the answer and the way of my life. It was and seems to be a way of intuition and used chances. Simply, the things happened to me. First approaches to science and science education I remember lie far behind me in my childhood: My mother loved to show us kids (we have been four, I was the oldest one) "magic things" whenever she saw a possibility. Once, she spun an uncovered can filled with fresh milk, or washing the dishes she produced big bubbles in the kitchen, she turned around a glass full of water covered by a sheet of paper and showed us that no water falls out of it. She produced waves in the bathtub, showed us the rainbow and lightening and thunder above the lake in front of the house. She used a metallic knife or cold water to cool hurts, and my father together with us designed kites in autumn, went with us into adventures at the lakeshore, or let us help repair broken things. Both, mother and father wanted to give us all the possibilities to get a solid scientific basis for our lives, optimism and a cultural background. Therefore, I was also learning three languages: Russian, English, and French. To me all that was producing curiosity in a world of limitations. These limitations are grounded in the fact that I was growing up in the eastern part of Germany, behind the wall between east and west, between unlimited possibilities at the one side and social and political borders on the other. But, in our world we learned to use and find opportunities for individual development and an enjoyable life. It was always necessary to find solutions. During my time at school, my curiosity always was fed. My first physics teacher was a woman who loved doing experiments in the classroom, and she was able to give us the chance to conduct scientific experiments by ourselves. To discuss phenomena and to understand qualitatively what happens have been my favorite interests. Thus, when I was fifteen years old I had positively experienced the world of science and it became obvious that I should study physics to become a teacher. To study languages was the second alternative, but under the given political conditions this did not make sense at the time.

But, there was a second field of activities, which influenced my possibilities. When I was a young girl, I liked sports and practiced regularly and as part of professionally trained teams: eight years in gymnastics and three years of judo. Doing sports, I experienced and realized the relationships between theoretically described principles of motions and my own body—I am still able to feel details of motions I did decades before in my childhood—and at the same time the importance of discipline and hard and steady work for success. Looking back I can resume regional and national championships in both disciplines. Thus, I can say that I was a fairly successful sports girl.

K. Tobin, W.-M. Roth (Eds.), The Culture of Science Education, 301–309.
© *2007 Sense Publishers. All rights reserved.*

BECOMING A TEACHER

At the age of eighteen I started to study physics and astronomy at the University of Jena, 300 kilometers away from my family. The combination of these disciplines was brand new at the time; we have been the first group of students studying this combination of subjects. Astronomy was a regular one-hour subject in tenth grade for all students and was seen to be the most prestigious of the natural sciences because it combines knowledge from all the natural sciences and mathematics. The studies included a balanced mixture of theory in physics, astronomy, and mathematics, laboratory work, sky observations, school practice, and active student life. We experienced intense communications and social life during this period. The conditions have been perfect to become self-sufficient and socially responsible. Nearly all students were living in dorms and sharing rooms and time. I did not have problems studying my subjects and enjoyed genuine teamwork and friendship. Here, at the age of twenty-one, I found my first love—a physicist—and married him. After four years I did my diploma and started to experience first practice as a young teacher. Yes, although I studied actively and successfully, I felt a fear of not being perfectly prepared to teach, to answer all the questions about physics and to stand the demands of serious everyday life in a school.

Where to go after having studied was not alone the choice of the candidates: the government sent young teachers to the schools it selected. We could apply only for some regions in the country and hoped that our wishes would come true. I was privileged because I was married: I got an appointment in a small rural school in the southeast of East Germany. One year before, my husband got a job there as a physicist in a local plant.

According to the national school system, I not only taught physics in grades 6 through 10 and astronomy in grade 10, but also mathematics and Russian language. For teaching the different subjects detailed compulsory curricula existed for all schools in the country. "Detailed" meant that all contents were prescribed week by week; it meant proposed teaching methods, the same material in all well equipped schools, for each school year one official text book, central exams for all students at the end of the tenth school year, obligatory in-practice teacher training and governmental visits to control the teaching practice. I learned to translate the official idea of teaching objectives and teaching practice into my individual school practice, and to find my private ways of following ideas—I did not lose my enthusiasm for scientific phenomena. So, together with the students of my classes I experienced the magic and the fun scientific phenomena offer inside and outside the school building. I liked the large amount of time we had in the curriculum for teaching sciences. I missed a bit more freedom. But the situation we had gave us safety within the political system. Fortunately, the natural sciences were not political and it was easier for me to live with than with social sciences. The students were "my children," and I was happy having good friends in my colleagues. With these friends, generally more experienced teachers than I who loved their profession, I could discuss what I did and what I wanted to do behind the classroom doors. I received valuable hints and was encouraged to try new things, to vary my actions, and to observe the results in the learning of my students.

During this time—I was still working one year—I had my own child, my daughter Maria. After five month at home I brought her to the Kindergarten (as it was usual—non-working women were called lazy and not shining examples) where she spent the days and I continued to work as a teacher. Despite this condition I loved to teach physics and astronomy in the small school. As it was in my own childhood, I could see that the students liked the experiments very much, the scientific phenomena always around, excursions outside school, but they hated to learn theory and to calculate. Soon, I found myself within the conflict of offering attractive teaching-learning contexts and fulfilling the official curriculum. Daily I struggled with expectations, official plans, with students in the stage of adolescence and only less time for my own family. The things had to be organized effectively. Optimistic as I always was and still am I learned it in a relatively short period of time. After three years I was settled and after six years I felt strong enough to ask for more. I was interested in learning more about educational research and teaching and learning. In the meantime, I divorced and raised my daughter as a single mother.

WANTING MORE—THE WAY INTO RESEARCH

In 1988 I was offered the opportunity to give to my life a new direction: It happened to me that I was asked to qualify myself doing postgraduate studies and a doctoral dissertation in the field of educational psychology. Taking that chance, I moved to Berlin to start further studies in different educational disciplines, philosophy, and psychology and to do my own research at the East German *Academy of Educational Sciences*. Here I received important input for my future professional development from Joachim Lompscher. At this time he was one of the leading educational psychologists in the country and, to me, a trustworthy supervisor and teacher. He had the competence of being an attentive listener and adviser. He was full of knowledge and humanity. To me he became the model of a real scientist. Around him interesting and impressive young researchers were learning and working. From this time learning processes and activity theory have been the foci of my interest. I received a good introduction into the theoretical work of important scientists of this field and started to read: Piotr Galperin, Alexei N. Leont'ev, and Lev Vygotsky. In my own studies I started to empirically investigate students' preconceptions of certain concepts of mechanics. I designed and discussed research design, conducted structured interviews with fourth grade students, and learned to analyze empirical data. Within the research institute of Joachim Lompscher I experienced for the first time scientific discussions and tried to participate actively.

Unfortunately the Academy was closed after two years. The reason was the break down of the political system in East Germany. The wall between the political systems of East and West had come down and many institutions were closed because of political reasons. Although I had the possibility to continue my research at another university in Berlin, I tried to find an alternative to accomplish my scientific profile. Good fortune was with me when I found an advertisement in a newspaper in 1991: At the University of Bremen they looked for a scientific staff member for working in a research project on individual learning processes in physics

303

and for completing a doctoral degree in the field of physics education. I applied, got the job, and moved to Bremen—at first without my daughter who stayed with my parents.

MOVING FROM EASTERN TO WESTERN GERMANY

New World and New People

From then, in Bremen, the world—my exciting new western world—expanded tremendously: unbelievable feelings and experiences of real freedom in thinking and acting, always and every day new perspectives, differently thinking people, and breathtaking advantages crossed my way. First I felt like a stranger but after some months I was really happy with all these challenges around me. I wanted to learn and have experience in the world of scientists of different disciplines in a dynamic Institute of Physics Education. Theoretically, too, I entered a new world: I came to know constructivism, and followed the radical ones based on Humberto Maturana and Ernst von Glasersfeld. Both scientists I learned to know personally in 1992 and 1994; both of them impressed me lastingly. They were thinking about thinking and behavior and the theoretical and empirical discourses about it. That opened to me new perspectives and a new view on the processes of learning and teaching. Positively, I also remember the intensive contacts to neuroscientists and psychologists, the international researchers who worked in Bremen, and those who have been contributing to an interdisciplinary research network. Intensive and never-ending discussions on research and education practice theory of learning and meaning construction have been the standard. The main initiator of the progress we achieved within the institute was my supervisor Stefan von Aufschnaiter. He was the one, who always encouraged me to take any learning possibility I had, and to whom I still am grateful. He gave me the push and the energy to go forward.

Learning Process Studies

I started my research in Bremen in the field of individual learning process studies. This was a new step into detailed observation and description of human actions, what was at this time in Bremen the main focus of physics education research. We wanted to understand and describe the processes of individual learning in authentic school settings. My idea was to describe individual processes of thinking and learning at short time scales and their interdependence during interactions while learning physics. The research questions have been: How are new meanings generated during the learning processes of students in experimental classroom settings? How can we describe these individual processes? How can we describe changes within those processes qualitatively and quantitatively? and Do individual students' learning processes mutually interfere while learning together?

I learned to use new research methods. We videotaped small groups of students within the physics classroom and invented possibilities to analyze the behavior and actions of the students in order to understand the students' individual thinking and meaning building processes. We tried to find acceptable, that means rather objective and reproducible, ways for video data interpretation. We intensively discussed heuristic systems of categories. During this intensive work within the institute we together could accomplish a rather coherent constructivist theory of learning. We found and described learning dynamics within different time scales, and levels of cognitive complexity that have to be approached and reached to get certain learning results (von Aufschnaiter & Welzel 1999; Welzel, 1997). Within the national and international science community, we discussed our findings with others—and often felt misunderstood. I asked myself frequently, how to get people from outside of our small and informed community into our thinking, how to get them to understand our ideas? This was a question that was and is hardly connected to constructivist theory. Learning was and still will be an individual process, based on individual experiences in the world. What can we do, to communicate our ideas and findings, to become understood? We have to find appropriate contexts that the "others" are familiar with, and we have to communicate more than once. We have to become familiar with the world of experiences other people have constructed during their lives. So, I never finished to search for the right language and effective communication contexts. I defended my theses in 1994. It was a case study about learning physics and the role of interactions.

Teaching at University

I also had to teach at the university, and I liked to do so. Here, I could apply what I have learned through research. That was the next and very important factor for my further professional development. To understand teaching as an appropriate arrangement of contexts to situate students' learning processes, and not as the transfer of pre-formulated knowledge, was an always interesting task and motivated me to do further research. Besides, the students themselves knew and experienced "knowledge-transfer-instruction" only during their own school career. To get actively involved and responsible for their own learning was rather new to most. But it worked. I "taught" the physics laboratory for engineering students, tutorials in physics to them, but also a project for preservice physics teachers. The future teachers had to prepare a course on electricity, to learn about students' learning, to observe students' behavior, and to reflect on the things that happened within the physics classroom. The important link between research and practice was realized. Until this day, this intensive linkage—understandable as a feedback system—has been extremely important in my approach.

International Affairs

Joining the scientific community through the membership in national and international scientific societies, I experienced different cultures of scientific discourse. That was truly interesting and enriched my own thinking and living basically. I met Wolff-Michael Roth in 1995 at a national science education conference and afterwards during a presentation to which he was invited in Bremen. He, from the first moment, was to me an extraordinary scientific personality: he crossed the borders of the mainstream of thinking within a single discipline. He expressively linked his own experiences of knowing and learning to research results of different social sciences and philosophy. He seemed to me to be driven by the idea to get to know the mysteries of human thinking. He was working always and everywhere productively, he was discussing intensively, and he was extremely involved in pure research. He was interested in introducing young researchers into the international science community. After meeting in Germany we decided to work together for a while in Canada. Thus, in February 1997 together with my (new) family I stepped on an airplane and traveled to Victoria into my next adventure. Again, an exciting opportunity crossed my way. For two intensive months I was a colleague of Michael, I tried to speak English only and I was working on interview video data he got after teaching mechanics within Canadian physics classrooms. The discussions of the data interpretation we had, sharpened, I think, my and our view on students' behavior, the interpretation of the students' actions and the reconstruction of their thinking. We found that interview situations designed for assessment also are meaning-construction processes (Welzel & Roth, 1998).

During the same year I attended the NARST and the AERA conferences that took place in Chicago. Participation in these large conferences made a strong impression on me. To an international (worldwide) audience I could present my research results. I was grateful for getting this opportunity, and I got the feeling of growing.

From this time my own feeling about the interplay of my competency, self-sufficiency and social involvement reached a high level and brought me into the realization of new ideas. Back in Germany, and together with a psychologist, and on the basis of her and my work, we designed a training program on teaching and learning for young teaching staff members at the university, mainly young scientists. Into this program we put all things we both had learned about learning and teaching. We were so successful that another university (Hamburg) hired us to deliver the courses there.

Once, I believe it was in 1996, a request for partners for an European project on learning within laboratory situations reached me. This request came from Marie-Genevieve Séré in Paris. Because having the impression of an expertise within this area, I asked my supervisor, whether we should take the opportunity for our institute to get involved. His positive reaction took me to Bordeaux for a preparatory meeting; as a result our institute participated for two years in an interesting international project on "Labwork in Science Education." Scientists from France, Denmark, Italy, Great Britain, Greece, and Germany worked together. Never before did I think about the cultural differences between people of European countries, who nevertheless think they work in the same research contexts. Through intensive discussions we had to acquire a common scientific language and to respect and understand different cultures. That was difficult at times, because all members of the project have been recognized as experts within their countries and institutions. Now, all of the members had to open their minds much more than they expected. Together and step-by-step we learned to understand each other, to get insight into the work and the research of the others, and to use the concepts in a way everybody of the project can live with. We published research about objectives of labwork (Welzel et al., 1998), the image of science teachers in the participating countries, labwork activities, and a number of case studies on learning and teaching in experimental labwork situations. To experience this learning process within this community was extremely important to me. It showed me a way of acting in an international network.

My time at the University of Bremen lasted almost nine years. These have been years of intensive learning, of enjoyable work, of friends, enthusiasm and a time with a new beloved family. Elmar Breuer, also doing his dissertation in physics education, being at the same institute and involved in the same vivid contexts, became the next important person in my life. He accompanied the highs and lows that are common when writing dissertations. We lived, discussed, and worked together. Together, we worked on physics textbooks for schools; and we struggled with the difficulties existing in so called communities of "experts" and with different ideas about how to write textbooks. Our slogan always has been: "It is a game. Let's try to find and define the rules." Often in connection to our profession, we traveled through the world, practiced our language abilities, discussed about teaching and learning, helped each other and tried to be good and helpful parents to Maria. Thus, professional and private things have been linked together during this important period of my life. Elmar, after three years within the institute, became a teacher at a *Gymnasium* (grammar school and highest of the three-tiered German school system), but he always supported my professional activities and until today we share joint physics education projects.

PROFESSORSHIP IN HEIDELBERG

Physics and Physics Education at the University of Education

In 1999 an interesting job announcement landed on my desk. A full professorship for physics and physics education at the University of Education in Heidelberg tempted me. At this time, I was well settled and happy being a recognized member of the Institute of Physics Education in Bremen. But, the announcement hit me— again something happened to me. I dreamed of supervising once and somewhere my own group of young scientists, of building my own research field. I was ready to try an application. I tried it—and finally I got it!

So, my family again moved—to Heidelberg. Elmar obtained a new appointment in a private *Gymnasium*; Maria changed school. She was now a student in a gymnasium three years before the *Abitur* (graduation).

In Heidelberg, a new life was awaiting us. I came into a very small group: one professor, one staff member, and one technician. There was no research profile, no tradition in supervising young researchers, the colleagues did not really participate in the science education community and had not even been building that vision. I remember my first talk to the rector of the university; I was introducing myself—it was in his office—and within the conversation he surprisingly said that he would expect me to make the physics at this University of Education so attractive that the students would come here because of me. Obviously, the university had high expectations for me. What a challenge that was for me!

First time on my own, I was responsible for inservice teacher training. I had to give regularly lectures and seminars and had to accompany student teachers in their school practice; and I wanted to begin a research group. The focus for the research should have been the transfer of the results of learning process studies I brought with me from Bremen into the design of learning contexts especially in teacher training. Step by step, together with my new colleagues and always in my mind what I have learned in my life before, I became familiar with my new situation and developed new ideas. Step by step interesting possibilities appeared to turn these ideas into reality. More and more I dedicated my time to the profession. The whole process went faster than expected. Now, I am a part of a network of education and science education. I am linked to people, projects, and institutions. I am as I always was: active and curious.

I have been at this university for more than six years now. Systematically I could establish research and developmental projects for science learning in and out of schools, at different age levels, and within international co-operations. I am working in the field of teacher education, and to implement video techniques into this practice field (Welzel & Stadler, 2005). Surrounding me, a number of young scientists work and learn. Two doctoral students now have completed their theses under my direction—one on educational reconstruction of physics labwork for physics students, and one on contextual conditions for physics education in the German *Hauptschule* (the lowest of the three-tiered German school system). At the moment my team consists of eleven researchers: five doctoral students are working on their theses. They are involved in different projects and communities. Always, the main focus for the work in Heidelberg was and is the scientifically accompanied transfer of research results into all-day learning contexts in and outside of schools and the use of video for research and teaching practice. I am also actively involved in the administrative work of the university.

Selected Project Highlights in Heidelberg

I am involved in many different but connected projects, Here, I only want to give a glimpse into these activities. As described, I like very much science phenomena. Thus, I am always trying to infect others with this fervor. The first have been my colleagues Klaus Scheler and Ludger Fast. In 2001 together with preservice student teachers we organized an interactive science day for families with their children. During one semester using a seminar on interdisciplinary teaching and learning we prepared such a day, the students advertised it in a newspaper and designed and printed flyers, and prepared interactive experiments. Finally, for one Saturday we had hundreds of people in our university doing experiments and being excited about science phenomena. That was really exciting and for me the initial project that was directed by others, which led us bring our ideas of fascinating people for science into the region. In-between, three science days with a growing number of participating colleagues and students took place. Even school classes from Heidelberg, some German science centers, and colleagues of other than physics disciplines offered interactive science stations.

These science days provoked further reactions. We were asked to install a small science center in Heidelberg. So, we have been actively involved in the planning and organization of the *ExploHeidelberg*, a very small but thematic science center; it is based on science education research results. It has only 250 m², a multimedia lab, a wet lab, and an interactive exhibition. There are about 1000 visitors every month. The visitors can experience more than fifty phenomena on "perception with eyes and ears, and motion." The science center is actively involved in preservice and inservice teacher education.

The science days and the *ExploHeidelberg* brought us in contact with other colleagues. For several years now, we are involved in a project on the education of Colombian street children. This project now is well established within the international projects of our University. After a first visit in 2001, Elmar Breuer and I are traveling every year for several weeks to Colombia and work with student teachers and street children. It is an international collaboration of a German University of Education and an *Escuéla Normal Superior* in Colombia in the field of teacher education. Within this project, *Patio 13—School for Street Children*, Colombian and German institutions collaborate intensively. One important objective of this project is to develop and evaluate together with Colombian preservice teachers teaching sequences for street children. These teaching sequences are be taught by Colombian student teachers to street children in Colombia. To get more information about learning characteristics of street children compared to those of the student teachers of our course, we documented the learning processes of both groups through video ethnographies and through observation protocols. Differences between the learning processes of preservice student teachers and street kids became visible. The characteristics found are the basis to develop a specific pedagogy for street children. I could add some more very interesting projects, with Austrian, Swiss, French, and German partners. But I want to stop it here. The things are in motion and development.

I have the peculiarity of being able to successfully and rapidly suppress negative experiences to go forward constructively. Indeed, the past has not been easy all the time. But my personality allows me to invent, discover, and enjoy the world around me with never ending curiosity. I like to be that way and I want to stay that way as long as I can. I want to enjoy the little and big mysteries and surprises, I want to work with people and enthuse them about science and science teaching and learning.

Who knows what the life will bring next?

Manuela Welzel
University of Education
Heidelberg, Germany

JUSTIN DILLON

AN ORGANIC INTELLECTUAL?

On Science, Education, and the Environment

The contested and diverse relationships between science, education, and the environment have provided the context for my own personal and professional development throughout my career as science teacher and academic. With that in mind, I examine the nature of scholarship and quality in educational research through the lens of the day-to-day work of a "brain worker."

ON THE INCREASING INDUSTRIALISATION OF ACADEMIC LIFE

Education is the silver bullet. Education is everything. We don't need little changes. We need gigantic revolutionary changes. Schools should be palaces. Competition for the best teachers should be fierce. They should be getting six-figure salaries. Schools should be incredibly expensive for government and absolutely free of charge for its citizens, just like national defense. That is my position. I just haven't figured out how to do it yet. ("Sam" in "Six Meetings Before Lunch," *The West Wing*)

I should not be doing this. My priorities should be obtaining grants from research councils and writing papers for peer-reviewed journals. In the UK Research Assessment Exercise (RAE), which determines the "visions" and practices of all universities and most of their staff, book chapters in edited volumes usually do not count for much. During my eighteen years at King's, the RAE has steered academic life, reduced collegiality, and seriously affected research practices. The British scholar Anthony Clifford Grayling, writing in 1997, lamented the "increasing industrialization of academic life," noting that "Most scholars are academics now, and not all academics are intellectuals":

In the new climate of research ratings, the cultivation of intellectual virtues, and the organic rather than forced pace of enquiry, is discounted. So the intellectual scholar, a person occupying a place apart, is a rarer creature now, even though there are many more universities. (1997, n.p.)

I don't feel in "a place apart" so I guess that I haven't reached the category of "intellectual scholar" yet but it seems like a reasonable goal to aim for. Some years ago, Stephen Ball described me in my appraisal as "an organic intellectual." Not being familiar with Antonio Gramsci's writings, I took it to be a kind reference to my scholarship in the field of environmental education, whereas it was actually meant to signal my ways of working as being collegial rather than isolationist.

K. Tobin, W.-M. Roth (Eds.), The Culture of Science Education, 311–322.

Notions of what counts as an intellectual differ depending on who is doing the defining, Gramsci thought everyone was an intellectual, or at least had the potential to be so. In line with the thinking of Edward Said, Henry Giroux argues that educators are "public intellectuals" challenging the *status quo*. Giroux argues that such public intellectuals engage in intellectual practices that refuse:

> both the instrumentality and privileged isolation of the academy, while affirming a broader vision of learning that links knowledge to the power of self-definition and the capacities of administrators, educators, and students to expand the scope of democratic freedoms, particularly as they address the crisis of the social as part and parcel of the crisis of both youth and democracy itself. (Giroux, 2003, n.p.)

Such public intellectuals, Giroux argues, "interpret and question power rather than merely consolidate it." One reason that I became a teacher and, subsequently, an academic is to promote social and environmental change. So, while I *should* be writing another paper for a peer-reviewed journal or filling in a grant application form, I would argue that I *ought* to be writing for a wider audience, questioning and interpreting power.

Typically, the invitation to write this chapter came while I was attending a conference. I was in a wet and windy San Francisco, attending the annual meeting of the National Association for Research in Science Teaching (NARST). I say "typically" because conferences seem to be the place where one can engage with people and ideas for hours on end without the distractions that usually get in the way of being an academic. It's at conferences where you meet like minds and decide that you really should edit a book, write a paper or put in a research proposal. It's also at conferences where you're on public display and one can offend and discourage both deliberately and accidentally. Joy Carp, who used to work for Kluwer, once told me how she could recognize academics' nationalities by the way that they asked questions at conferences: U.S. scholars, she said, tended to ask gentle, congratulatory questions; Germans asked technical questions; and the British would begin with, "Well, that was a very interesting presentation" before asking a question that would completely expose the lack of theory, rigor, or whatever needed to be exposed.

In typical U.S. style, NARST hands out awards each year for a range of "bests." This year the award for "Outstanding Contribution to Science Education" went to David Treagust from Curtin in Australia, someone I have known for many years and one of my referees in my recent, successful bid for promotion to Senior Lecturer (what every country in the world apart from the UK would call Associate Professor). David, a former president of NARST, made a gracious acceptance speech acknowledging the value of collegiality throughout his career. I bumped into him a few days later at the annual meeting of the American Educational Research Association (AERA) and congratulated him on his award. The first thing he said, in that unique Australian/Yorkshire accent, was "I meant it, you know?" reiterating the value of colleagues around the world. Both AERA and NARST continue to provide an opportunity to stimulate and catalyze my professional interests in science and environmental education. More importantly they provide an opportunity to be collegial—to meet, eat, and drink with friends and colleagues from around the world. They provide an opportunity to listen, talk and think—three aspects of the job that never get their fair share.

ON NAMES

I was actually born Robert Boyd. I was adopted a few months after being born in June 1957. More adventurous parents might have chosen "Sputnik" as my middle name not Simon. But Justin Simon Dillon I became, sometime around about October 1957. My parents were both educators: my father trained as a teacher of arts and crafts during the World War II and my mother taught what was termed at the time "remedial reading." They both had had other jobs earlier in their lives, my mother had worked in the pottery industry (I was born in Stoke-on-Trent, in "the Potteries") and, at one time they ran a chemist shop (a pharmacy). As science teachers were in short supply in the 1960s, my father retrained as a biology teacher and spent most of his days in "secondary modern" schools, teaching students who had failed the 11-plus examination and who were, generally, given a second-class education. Without a science degree, my father's chances of teaching in a grammar school, working with those students who had passed the 11-plus, were slim if not non-existent. I passed the 11-plus—actually, being born in June, I took it when I was ten. Unlike most of my primary school friends, I went to Newcastle-under-Lyme High School, about thirty minutes away from home by public transport. It was my parents who decided that the other, closer grammar school was definitely not where I was going to spend my formative years. School choice has been an issue in England for much longer than people would have you believe.

ON LEARNING ABOUT SCIENCE AND THE ENVIRONMENT

Until now I had never given much thought about when my interest in the environment began. Although there is a literature on "significant life experiences," I have never found the idea that particular events in one's early years result in life-changing decisions particularly convincing—a case of *post hoc ergo propter hoc* if ever there was one. I do, however, remember having a nature table at home—full of pinecones, leaves, and other souvenirs and specimens. I had stick insects, the odd rabbit, hamsters, guinea pigs, and the like, but I was not a dutiful owner.

Teachers in the 1960s and 1970s were not well paid. For a few years my father bred Mongolian gerbils as a sideline. They make ideal pets—awake during the day, clean, bright and they do not bark, bray, hoot, or otherwise disturb the peace. Local pet shops bought them in substantial numbers, and at one time, I think we had more than forty cages in the back bedroom. Unfortunately, my father's attempt to boost the family income came to a fairly sudden demise. Mongolian gerbils breed with alacrity: demand, and thus price, diminished to the point at which it was no longer a financially viable concern.

313

The post-Sputnik angst and the ambitions of the Labour Party to modernize Britain (the so-called "white heat of the technological revolution") led to physical science and technology becoming more prominent in the public consciousness. The rise of science has not been free from critique and criticism and this partly stemmed from concerns, often justified, about environmental impacts of new products and processes. Rachel Carson's *Silent Spring* was published in 1962, when I was five, and during the 1960s and 1970s, growing numbers of the public were becoming anxious about where science and society were headed. Environmental issues were increasingly reported in the media and the UK created the first Ministry of the Environment in the world in 1970 (which had been designated European Conservation Year).

Environmental education, which was to become my intellectual home some years later, was in its infancy in the 1970s with international gatherings in Belgrade (1975) and Tblisi (1977) setting the scene for future developments in the field. I did not therefore have any environmental education myself in any formal sense although the informal education provided by the media was probably highly influential in shaping my views.

My formal science education, post-11, involved a couple of years of "science" followed by a year of separate biology, chemistry, and physics. At the end of the third year in secondary education (at the age of 14) I had to select what options I was going to choose to go along with compulsory English, mathematics and religious knowledge. At this point, biology and I parted company for several years: I opted to take physics and chemistry instead. In a boys' school in the 1960s and 1970s, biology was perceived as a "girls'" subject and one which few adolescent boys would risk choosing (even if their father was a biology teacher). Chemistry and physics seemed to leave more doors open—and medical schools did not require school-level biology, not that I ever really considered being a doctor.

Two years later I emerged with a good crop of "Ordinary level" exam passes. As with the vast majority of my peers, I stayed at school and went onto take four "Advanced levels," chemistry, physics, mathematics, and general studies for two years. Again, being a boys' school, science subjects were the norm. I was actually better at English than I was at science in those days but the job prospects for English graduates at the time were not as good as those of science graduates.

Mathematics teaching at the school varied substantially and any love of the subject that I had faded fast. Chemistry was taught well and, in the post-Sputnik era, was supported by new teaching materials from the Nuffield Foundation. Nuffield Advanced Chemistry involved a substantial amount of practical work—the *sine qua non* of English science education. The cost of teaching Nuffield Chemistry must have been quite high—lots of ground-glass-jointed boro-silicate glassware and other state of the art apparatus.

Physics was taught by several staff at least one of whom was sufficiently acerbic and insular to make the decision of which subject to study at university relatively straightforward. I went to the University of Birmingham, sufficiently far away from my parents to allow independence but close enough in case of emergencies. In those days, the vast majority of students lived away from home, paid no fees and received a grant from their local authority to cover rent and food. Things have changed substantially since then.

From what I can gather, Birmingham University's chemistry department was typical of UK chemistry departments. The teaching varied from reasonably effective to appalling. One academic was famed for never making eye contact with the students during fifty-five-minute lectures. In general, the lectures and the laboratory work bore little relationship to each other. The contrast between chemistry at school and chemistry at university was marked: At school, your work was marked regularly and teachers had effective teaching styles and knew who you were. At university the opposite was true. I am afraid that, in my experience, the quality of university pedagogy is substantially lower than that provided in schools.

Growing concerns about the environment in the 1970s led to university courses focusing on the impacts of chemistry. One of the two essays that I wrote in four years at university focused on the environmental impacts of lead in petrol, an issue that was making the headlines at the time. The opportunity to research a topic, drawing on scientific papers and relating the chemistry to public health and government policy, was a highlight of my undergraduate studies. The other essay that I wrote, in 1976, was on gender and science, part of the Inter-faculty Studies course that undergraduates had to take. 1970 had seen the publication of *The Female Eunuch*, by Germaine Greer turning her into a "public intellectual," an academic who was able to shape and influence society through her writing and speaking.

ON BECOMING A TEACHER

During my final year at Birmingham I decided to apply for teacher training. I could not envisage working for a chemical company and it was increasingly unlikely that I would do a master's or a PhD in chemistry. I was turned down by my first choice of institution, the London Institute of Education ("better candidates available") but was accepted at Chelsea College, also part of the University of London.

The preservice year (1979–80) was probably the turning point in my life. The Centre for Science and Mathematics Education was pre-eminent in the UK and many of the staff had been heavily involved in development of Nuffield Science. Two eminent figures in science education, Paul Black and Jon Ogborn, were joined by Arthur Lucas, one of the "founding fathers" of environmental education, at the end of my one-year course in June 1980. Jan Harding and John Head proved instrumental in developing my thinking about gender in science education and the personal response to science. John Barker, an inspirational biology educator and oenophile turned into a life-long friend and Bob Fairbrother initiated me into the complex world of assessment. Such a concentration of talent (in such an unprepossessing building—a converted factory) would have been hard to find anywhere in the world. However, my father, a trained and committed science teacher, had probably never heard of them or their contributions science education, working as he did in the second tier of science teaching. The distance between science teachers and science education researchers was greater then than it is now.

315

The one-year preservice course involved teaching at two schools, in classroom groups (three to four student teachers with a small number of students) and solo. About half the course was spent in school. My first teaching post, which I took up in September 1980, was as an assistant teacher of chemistry at the John Roan School in the southeast of London. The head of science at the school was Brian Matthews, who later went on to become the main science educator at Goldsmiths College in London. Brian led an innovative department and was committed to a socially critical form of science education. The first-year science course, for pupils aged eleven, included such activities as trying to identify the contents of tins without ever being able to see inside or open them—an attempt to focus students on thinking about the nature of science.

During my early years teaching science in London I attended some meetings of the British Society for Social Responsibility in Science, led by Joan Solomon. The radical science strand within British science has always been on the fringes and was never in a position to have much influence. Nevertheless, the experience shaped my view of science, the environment and scientists. I realized that my understanding of chemistry improved radically as a result of teaching the subject. Both Nuffield O-level and A-level courses offered a range of options that students could choose. I was able to teach historical topics at O-level - making sodium by electrolysis—and mineral process chemistry at A-level, extracting metals from mineral ores. Together with friends from Chelsea College days, I began to organize field trips to some of the UK's most beautiful and geologically interesting places, ironically, not far from where I was born.

During my time at the John Roan School, I carried out my first piece of rudimentary research in science education, looking at the attitudes of pupils towards science and scientists, a topic of concern even then. After four years at the school, I was recruited to the Inner London Education Authority's Science Support Team. The team contained several innovative science educators (including Jonathan Osborne) and engaged in curriculum development as well as providing cover for examination classes in schools. I gained experience of teaching in six schools across London and was involved in making a video about the iron and steel industry in southern Wales as part of a major review of the secondary science curriculum in the days before the implementation of the national curriculum.

One of the schools that I was supporting, Eltham Hill, offered me the head of chemistry position, which I was happy to accept. I led a group of committed chemistry teachers, all of whom were older than me, and, together, we made chemistry a popular subject, not an easy task in an all-girls school in the 1980s. I also argued successfully for the implementation of technology as a separate subject in the school—one of the first all-girl schools in London to teach the subject.

During my time at Eltham Hill I took a two-year, part-time master's degree in science education at King's College. Doing the MA was one of the wisest choices that I have ever made. It opened up my eyes and brain to a broad range of literature and enabled me to meet teachers from around London and beyond. As well as teaching courses in research methods, recent developments in science education and curriculum studies, I wrote a 15,000-word dissertation on "Technology, Culture and Gender," thus pulling together several interests into one piece of work.

After four years at Eltham Hill I was encouraged to apply for the vacant head of science position at Kingsdale School, labeled by the *London Evening Standard* as the Inner London Education Authority's "School of shame" as a result of its poor record for behavior and attainment. Situated in a leafy suburb, many of Kingsdale's pupils came from poorer areas several miles away. This was, by far, the hardest job of my life and despite the skill and collegiality of the staff—it was one of only two London schools to have a bar—the school struggled to maintain any semblance of a decent education for all its students. The Authority's response to the challenging situation that we were in was to employ "Inspectors Based In Schools" (the IBIS team), an interesting innovation that did not last beyond the break up of the ILEA in 1988.

ON BECOMING AN ACADEMIC

After a year at Kingsdale, Rod Watson, who had succeeded Jan Harding as the main chemistry educator at King's, contacted me. Rod had already encouraged me to get involved in the PGCE chemistry course as a mentor of student teachers and through assisting with geological field trips. When he received a grant to set up a National Environmental Database (NED) project, he asked me to join the team at King's, sharing the teaching of the preservice course and working on the NED project. However, it was quite a risk—I gave up a permanent post for a contract that lasted three months with only the possibility of an extension.

The difference between working as a schoolteacher and working at a university was quite remarkable. Sunday afternoon was free of the stress revolving around the difficulties of teaching in inner-city schools and the regimented life that they force on staff and students alike. Going to work became a pleasure. As Dylan Wiliam, a colleague at the time, put it: "When I was a teacher, all the reading and writing I did for an MA was a hobby—now I get paid to do it all the time." Another colleague complained to our then-head-of-department Arthur Lucas that she was so busy "doing" things that she did not have the time to think. Arthur replied that that was the one thing he paid her to do; and that is primarily how I see myself—as someone who is paid to think.

A recent review on the death of the English intellectual notes that the term "intellectual" has several meanings including the sociological one—"brain worker." "Lecturer," my current job title, is almost completely inappropriate as I rarely lecture—virtually all the teaching I do is leading workshops, seminars, and discussions or conducting supervision.

Arthur Lucas was highly influential in encouraging me to devise and teach a master's course on environmental education. That in turn led me to read more literature and to develop a line of work that has dominated my career so far. Arthur was a natural leader and would spend time wandering the corridors late at night to see who was working and how life was going. His influence on my academic life and my confidence was profound.

317

ON BEGINNING RESEARCH

The first funded research that I was involved in came out of a very small project initiated by Rod Watson. Rod had taught in Spain earlier in his career and had developed a range of contacts in Spanish universities. With the aid of a small grant from the British Council and a Spanish funding agency, we were able to set up a link with the University of Malaga. Together with Teresa Prieto, Rod and I carried out questionnaire surveys of students looking at their understanding of everyday processes such as burning. The project lasted for three years and resulted in publications in *Research in Science Education*, the *Journal of Research in Science Teaching* and *Science Education*. The studies also provided data for use in preservice teaching and in research methods workshops. I have never engaged in such a cost-effective study since!

A disadvantage of being on temporary contracts is that you feel obliged to say "yes" to any teaching or other work that comes along. As a result, Martin Monk—with whom I shared an office—and I taught far more than we should have and did far more consultancy work than was the norm. Martin played a key role in my early development as an academic. He had been an outstanding teacher when I did my MA and continued to provide support and guidance whenever we worked together. His knowledge of the history and philosophy of science seemed limitless and he had a better understanding of psychology and sociology than the rest of us in the science education group.

A significant amount of my time was spent on preservice teacher education—teaching and visiting students in school. By then, it was a government requirement that preservice teachers spent twenty-four weeks of their thirty-six-week course in school. Although in some ways laudable, the net result was a major inequity between the quality of mentorship received by our students. As a science group, we decided to produce a book of theory-based practical activities that was published by Falmer Press as *Learning to Teach Science*. Written entirely by King's science staff, the book was edited by Martin and me. Both Martin and I were clear about what we wanted and edited other peoples' and each other's contributions—quite heavily in retrospect. This initial experience of editing gave me the confidence to develop critical lines of thinking about science education and science teacher training.

In response to comments from inspectors that our preservice teachers did not appear to do much reading during their one-year course, Meg Maguire and I set about putting together an in-house collection of papers on educational issues. The collection of papers was enough to convince the Open University Press of the potential for a book, written entirely by King's staff, aimed at the preservice teacher education market. The first edition of *Becoming a Teacher*, published in 1995, contained twenty-three chapters. The tone of the book was set out in the introduction:

> In putting together this book we have tried to emphasize the three Rs: reading, reflection and research. Good teachers are able to learn from their experiences, reflecting on both positive and negative feedback. The best teachers are often those who not only learn from their experience but also learn from

the experiences of others. Reading offers access to the wisdom of others as well as providing tools to interpret your own experiences. We have encouraged the authors contributing to this book to provide evidence from research to justify the points that they make. We encourage you to reflect on that evidence and on the related issues during the process of becoming a teacher.

The book sold well and OUP commissioned a second edition, published in 2001 and a third edition is due out in 2007. I have sworn never to do another edition of the book after each edition has been published—editing the writing of one's colleagues can be difficult, embarrassing, and time-consuming.

INCREASING INFORMALITY

Rosalind Driver's appointment, in 1995, to the Chair of Science Education at King's strengthened the science unit that already contained well-known figures such as Philip Adey, Paul Black, Arthur Lucas, Jonathan Osborne, and Michael Shayer. Ros was a "person" person, interested in people as well as ideas, and able to engage with anyone she met. Though her time at King's was relatively short, she brought a sense of clarity and determination and played a key role in the department as well as in the science unit. She joined John Head as one of my PhD supervisors and offered firm and wise guidance.

Ros' illness and untimely death in 1997 diminished the world of science education immeasurably. Earlier in the year she had been the recipient of NARST's Outstanding Contribution to Science Education award. She received a standing ovation from the audience in Chicago. I was privileged to be asked to contribute to an edited collection highlighting the impact of her work on the science education community *Improving Science Teaching Through Research,* which was written in an innovative and collegial manner: the authors produced first drafts of their chapters in advance of a writing workshop held in Leeds. The rest of the authors shred each chapter, sometimes quite finely: the end product is certainly much better than the first drafts that we bought to the table. I have used this technique of producing a book subsequently together with Dick Gunstone and Debbie Corrigan from Monash (King's and Monash are twinned). This time the writing workshop preceded the 2005 *European Science Education Research Association* (ESERA) conference in Barcelona.

Ros had played a key role in the creation of ESERA, which was founded at the European Conference on Research in Science Education held at Leeds in April 1995. The final conference session involved vigorous discussions about "Europe," "science education," and "research," which have different meanings in different European countries. John Gilbert ably chaired the session, managing to cajole and coax agreement among a very diverse group of science educators. Rick Duschl—the only U.S. science education academic at the meeting—and I counted the votes for the first ESERA board.

Jonathan Osborne and I spent a good deal of time at the 1998 NARST conference trying to identify potential candidates for the vacant chair of science education. One of several names that kept coming up was Rick's, then at Vanderbilt University. He was appointed to the Chair of Science Education in 1999.

A year after Rick's appointment, he was approached by colleagues in the US who wanted King's to join in a consortium with the University of California Santa Cruz and the San Francisco Exploratorium to bid for funding from the U.S. National Science Foundation. A year later, the Centre for Informal Learning and Schools (CILS) was created, with a grant of almost $11 million. The creation of CILS radically affected the nature of the science unit at King's. Eight PhD students and three postdoctoral researchers all looking at aspects of learning in schools, museums, and science centers have broadened the focus of the science group's interests.

As a result of CILS, our relationships with institutions such as the Science Museum and the London Natural History Museum (NHM) have become much stronger and mutually beneficial. CILS has also facilitated collaborations with other departments within the university. Together with a colleague from our mechanical engineering department, Mark Miodownik, I have just been awarded £82,000 by the UK Engineering and Physical Sciences Research Council to support an innovative project called "What can the matter be?" The aim of this project is to investigate ways in which engagement with contemporary culture enhances the public's appreciation of science and engineering. We intend to develop tours of the Tate Modern, which will cover three themes: chemistry/images, materials/form, and engineering/installations. An MP3 tour—for iPods and the like—will be downloadable from the Internet.

Another outcome of CILS was the opportunity to evaluate the *Permanent EuropeaN resource Centre for Informal Learning* (PENCIL). PENCIL consists of fourteen institutions from twelve countries working together developing and sharing good practice. Our task is to evaluate the progress of six of the pilot projects and to look at the development of the network as a whole.

RESEARCHING EDUCATION AND THE ENVIRONMENT

I was lucky enough to be able to attend NARST and AERA virtually every year. I attended the early meetings of AERA's Ecological and Environmental Education Special Interest Group in the 1990s, which drew together some of the most well known environmental educators from Australia, Canada, the UK and the US. Over time I developed a strong collaboration with colleagues at Bath University and the *National Foundation for Educational Research* (NFER) in the UK that also have benefited some of our doctoral students.

In recent years I have worked with colleagues at both institutions on a range of research projects funded by government agencies and non-governmental organizations. Two of the projects involved literature reviews which provided me with the need and the time to read widely in the field of outdoor education. If I had to choose one publication that I think represents my best piece of collaborative work it would be the *Review of Research into Outdoor Learning* published by the Field Studies Council (Rickinson et al., 2004). Literature reviews, which again don't seem favored by the RAE, actually provide opportunities for scholarship of the highest order: reading, interpreting, and synthesizing hundreds of studies is an inordinately challenging task.

Subsequently, Bath, King's, and the NFER were commissioned to carry out a study into *Learning in the Outdoor Classroom*. This action research project coincided with me developing cataracts in both eyes resulting in a severe deterioration in my eyesight and a slowing in my work. The wonders of modern science and medicine coupled with a free public health service have transformed my life.

EDITING, SCHOLARSHIP AND QUALITY

In 2005, a year after I became Secretary of ESERA, itself a great honor, I was invited to become one of the editors of the *International Journal of Science Education*. The journal has five editors working with the Editor-in-Chief, John Gilbert. With fifteen editions of 225 pages each year, the workload on the editors and the reviewers is substantial. The growth in the number of science education journals and the pressure on academics to publish or perish has its critics. A. C. Grayling describes the situation:

Coteries of dons write in impenetrably specialist codes for internal consumption only, in hundreds of journals and monographs. (*Peccavimus omnes*) (1997, n.p.)

Editors and reviewers are seen as the gatekeepers of scholarship in the community of practice that is science education, trying to improve or, at least, maintain, the quality of science education research. This might not be the ideal system for encouraging change and diversity among the field. Editors are not elected by their peers and reviewers may hide behind the cloak of anonymity. It is by no means a level playing field. Jim Shymansky, NARST's outgoing President at the 2006 meeting where I was invited to write this chapter, raised the issue of journal article length during his awards-luncheon speech. His point, tellingly made with data from the *Journal of Research in Science Teaching*, was that papers have got longer and that the total number of papers being published in *JRST* has not increased concomitantly. Shymansky's beef was that papers are too long; but the real issue is whether the days of paper-based journals are numbered. The answer is almost certainly "yes," the issue of article size thus becomes almost redundant.

More recently, I have been involved as a coach in the ESERA Summer Schools, in which sixty or so doctoral students from around Europe get together for a few days of workshops, lectures and discussions about science education. It is activities such as these that help to develop the community of practice that is science education. The challenge of understanding and critiquing the work of colleagues from varied cultural backgrounds provides intellectually stimulating opportunities from which everyone seems to benefit. ESERA has grown now to be a strong organization able to organize biennial conferences and summer schools and with the financial reserves to support colleagues from the central and eastern parts of Europe, where science education is fledgling and poorly funded.

I have been lucky in working with excellent doctoral students of my own while, ironically, struggling to find time to complete my own doctorate which looks at the role of middle managers in science teachers' professional development. Not having a PhD is a significant barrier to promotion in the UK system. As one of my colleagues once said, having it is not the issue; it's not having it that is the problem.

So, after 26 years as a science and environmental educator, I look back at what has been a challenging, enjoyable, and eventful career. There is, I hope, much more to come. The day before I concluded this chapter, I was attending a meeting of a group of science educators from across Europe who had been brought together by the Nuffield Foundation. We were looking at the challenges facing science teachers and science education generally as we move in to the twenty-first century. So, what have I learned? Well, "Sam" was right, "Education is the silver bullet. Education is everything. We don't need little changes. We need gigantic revolutionary changes," still. And they will only come about when we as a community have enough evidence, commitment and influence to make them happen.

Justin Dillon
Kings College
London, England

TRENDS

Traveling abroad and even just looking at cultural-historical events and development in other countries changes one's gaze. An important lesson one can learn traveling and seeing science educators around the world is that of the contingency of selfhood and community; rather than being the result of internal (psychological) and external (sociological) determinations, science educators and the science education community continuously develop at the very moment that they concretize themselves. This contingency also extends to the collaborations science educators engage in, both within their nation and with their foreign colleagues. Finally, by traveling and looking abroad, we can learn about some of the major forces that are at work in shaping (though not in a deterministic way) the field of science education and the typical auto/biographical possibilities the field offers to its members.

CONTINGENCY OF SELFHOOD

Anything from the sound of a word through the color of a leaf to the feel of a piece of skin can, as Freud showed us, serve to dramatize and crystallize a human being's sense of self-identity. . . . Any seemingly random constellation of such things can set the tone of a life. Any such constellation can set up an unconditional commandment no less unconditional because it may be intelligible to, at most, only one person. (Rorty, 1989, p. 37)

Manuela Welzel entitles her auto/biography "Little was Planned—All Simply Happened," thereby bringing to the foreground the contingency of her professional career, an aspect of her Self. The theme of a professional career as a contingent trajectory leading to a contingent Self also is apparent in the titles of the chapters that Michael Roth and Okhee Lee contribute. Thinking about individual lives used to be in terms of ultimate causes; in religious thought, individual lives still are thought in terms of a *telos*; and in popular culture, among those individuals who consult horoscopes, the determination of an individual life and therefore of Self is a form of marking sense of one's life. However, serious philosophers and scholars have given up the idea that an individual life is *determined*; rather, contingency makes us take this or that direction whenever there is a branch in the road, to take up on Robert Frost's poem "The Road not Taken"; and our lives are full of roads that we have not taken, the ones that are more or less traveled, depending on the contingent choices we have made. Doris Jorde meets a Norwegian at Berkeley, falls in love with him, and then moves together with him to his native Norway; Jorde explicitly articulates the event as an unpredictable one that has given shape to her life as a science educator.

The contingency of individual science education careers and therefore—because individuals are constitutive of the community—the contingency of the science education community should be apparent. There is hardly an auto/biography that does not tell the importance of this or that science teacher, this or that mentor, or this or that fortunate event. All of these encounters and experiences could have been otherwise. There always are, to paraphrase Robert Frost, two roads that diverge; but we must not be sorry that we can travel only one. *This* one, ours, always is the one less traveled by, and it always makes all the difference. Each science educator, in his or her way, a Hanna Arzi, who decides to straddle what sometimes is a boundary between research and teaching—as we point out in part A, Michael Roth for a number of years was researcher all the while teaching full time in a college preparatory school. Each science educator, in his or her way, is a Manuela Welzel, who is full of enthusiasm for the work she does, and in so doing, not only changes the world around her but also changes herself. Each science educator, in his or her way, is a Justin Dillon, an organic intellectual, weaving the discourses of the ivory tower with those other discourses surrounding the tower and making it possible. (It is not only in James Joyce's *Ulysses* that intellectuals live in towers, from which they have to descend if they want to be part of life.) In each life of a science educator, the possibilities of the science education culture realize themselves—and this is so when life is easy and enjoyable and when there are hardships. Both the good life and the moments of hardship realize possibilities that exceed but are concretized by the individual.

Science education, as other communities, confers awards to some of its members who thereby come to stand out, if only for the moment of the award ceremony. Many contributors to this volume have received awards, for example, from the National Association for Research in Science Teaching, American Educational Research Association, European Association for Research on Learning and Instruction, Association for Science Teacher Education, or other professional organizations. Some of these awards are for individual papers or articles; others honor the lifetime contributions a person has made. Although these awards are conferred to individuals, they have to be understood as the result of a constitutive relation between individual and collective. This is so because

> [t]he difference between genius and fantasy is not the difference between impresses which lock onto something universal, some antecedent reality out there in the world or deep within the self, and those which do not. Rather, it is the difference between idiosyncrasies which just happen to catch on with other people—happen because of the contingencies of some historical situation, some particular need which a given community happens to have at a given time. (Rorty, 1989, p. 37)

The narrative forms available to the individual are an important aspect of Self. Thus, all the auto/biographies are expressions of narrative forms and narrative contents available in the community. Narrative forms and contents already exist as collective possibilities, for otherwise one person's auto/biography might not be intelligible to another. The very notion of language as a medium and tool embodies its collective nature—there are no private languages. Our way of writing the term *auto/biography* is intended to highlight the fact that all autobiography is biography, is a form of narrating the life of a person, a possibility that is available to others, too. Each element of an auto/biography, however idiosyncratic it appears, concretizes a general possibility that is intelligible and comprehensible as soon as we take the person's frame of reference onto the community and the world.

This world looks different whether we are rooted in North America or in Europe, Asia, or Africa or elsewhere in the world. Thus, Justin Dillon refers us to Joy Carp's ability to discern the origin of a science educator by the way he or she contributes to a public event. This comment shows that there are notable differences that go beyond the language and accents characterizing a person but that mediate the form of participation, the style of asking questions, the particular types of comments a science educator will make. These differences allow us to understand that there are characteristic ways within each nation that differ from the way science education is talked about in other nations. Welzel refers to such differences when she accounts for a project that brought together science educators from several European countries, all thinking that they are working on the same topic: laboratory work. But a science educator doing research on laboratory work in Germany brings different goals and perspectives than science educators in France, Spain, or the UK, even though all have an expressed interest in laboratory work. Because the objects of productive activity stand in a dialectical relation with the subjects of this activity, the differences in the objects also reflect differences in the (human) subjects, and therefore, the Selves and identities that come with them.

Justin Dillon questions whether he is an *organic intellectual*, thereby raising the question whether being an intellectual generally and an organic intellectual particularly is a possibility within science education. Here, too, there are apparent differences in the way an individual can be across nations and continents. In North America, one can note an anti-intellectualist stance, whereas in European and Latin American countries, science education professors may act and describe themselves as intellectuals. The anti-intellectual stance is quite noticeable in the peer review process, where authors are held to write simple texts rather than texts that make use of the full spectrum of possibilities that the English language offers to authors. Who science educators can be, in and through the texts that they publish, therefore is constrained by structures and people in the community. Science educators are not totally free to be who they want to be, but they are who they are always within a community.

CONTINGENCY OF COMMUNITY

The language (form of discourse) that develops within a scientific community is as contingent as the individual biographies and as contingent as the community itself. This is so because the three—individuals, language, and community—stand in a mutually constitutive relationship. Who anyone can be is mediated by the possibilities a community offers to the collective as a whole; but the possibilities themselves are the outcomes of the actions of individuals. The language science educators use to constitute their objects (research, teaching, theories, concepts) and the life narratives of scientists and science educators constitutes a set of possibilities; new forms of language are continuously created in and through the work of individual science educators. Yet even these new forms of language are acceptable only because other science educators also recognize them as possible ways of articulating science and science education. When this recognition is not the case, then science educators may find it difficult if not impossible to publish their work; and work that is not published has little if any impact on the field.

Welzel presents the case of the work done in the institute where she obtained her PhD. Strongly informed by the radical constructivism that Ernst von Glasersfeld and Humberto Maturana advocated, the researchers around Stefan von Aufschnaiter developed a different way to think about learning and development. In a series of MA and PhD theses, the group articulated how from the engagement (agency, schema) with material objects (resources), new, higher-order schema come to be constructed. Despite considerable work within the group, the reception in the science education community has been marginal. It may be that the re-orientation of the science education community toward social constructivism, sociocultural theories, and cultural-historical approaches in the mid- to late 1990s mediated the lack of reception. If radical constructivism had gained a stronger foothold in the community, the studies Welzel and her colleagues conducted in Bremen might have shaped science education more than they ultimately did. It is not surprising, perhaps, that this type of work has been all but discontinued as the leader of the group (von Aufschnaiter) has retired and the students and postdoctoral fellows have gone on to focus on different aspects of science education.

Radical constructivism certainly made an impact on science education in the 1980s and von Glasersfeld, in particular, gave influential keynote addresses to NARST and AAAS and engaged in debates with science educators and philosophers. From 1993–2004 only fifteen articles specifically used radical constructivism as a keyword in the *Thomson ISI-Web of Science* database. However, more than two hundred papers used the terms constructivist or constructivism in high-impact science education journals from 1989 to the present time. In 1991 Grayson Wheatley's widely cited article on constructivist perspectives on science and mathematics learning laid out the basic premises of radical constructivism and was published when interest in constructivism in science education was at its peak. Two sets of forces may have taken attention away from radical constructivism in science education. Conservative groups with backgrounds in the philosophy of science strenuously attacked radical constructivism in journals and books, taking strongly polarized positions that tended to support the powerful political forces advocating objectivist science and the teaching and assessment of it in traditional ways. Tobin was involved in some of the early debates, but saw them as a tiresome distraction from his research in classrooms. He was much more interested in pressing von Glasersfeld on the roles of social interaction and society in learning and figuring out how social action connected to what teachers knew. From his standpoint the power of radical constructivism was epistemological, drawing attention to the salience of what the learner knows and can do in making sense of experiences and creating understandings of them. The ontological implications were important too because, from a radical constructivist perspective, sense making and viability of knowledge would reflect the social and cultural histories of learners. Accordingly, making sense of science in classrooms would be polysemic. Within science education the conceptual change group was focused on the pathways from what was known toward canonical understandings of science. Gradually, with the exception of the conceptual change group, science educators moved toward the use of sociocultural frameworks, with Roth forging new pathways, especially in his book *Design Communities*.

A field mediates the possible identities its practitioners may evolve, and the evolving identities mediate what the field can become. Thus, after Doris Jorde meets and marries a Norwegian fellow graduate student, she moves to Norway and, by working with Svein Sjøberg, makes a mark on science education in that country. Being part of a first generation of science educators, and given the (historical) path dependence of any culture, science education in Norway is tied to both their names and work in essential ways. With the coming of Welzel to Heidelberg, substantial changes have occurred to physics education at the university, in the city, and therefore to science education in Germany more generally. More so, as the collaboration with a Columbian teachers' college shows, Welzel's presence in Heidelberg led to an international cooperation that piggybacked on an existing relationship between two universities. Again, the contingent development of the individual and the contingent development in the field are intertwined and unpredictable, though fortuitous.

INTERNATIONAL COLLABORATIONS

Previously we have had adequate testimony to the fact that Leeds and King's College were leading centers in the UK. When Ros Driver moved to King's College, as noted by Dillon, she joined a very strong group of science educators that were making their mark nationally and internationally. Dillon notes somewhat wryly that King's was twinned with Monash University in Australia—presumably because of shared interests in conceptual change research and science teacher education.

The three scholars we included in this section have been educated in their home countries and have for the most part made their major contributions at home while participating in the international community. Arzi, like so many of her colleagues from Israel, has traveled a lot and in so doing has made connections with some leading science educators. These connections were facilitated by her participation in NARST and the Special Interest Group on Conceptual Change in AERA. For example, in the US she connected with Joe Novak, perhaps the leading conceptual change (alternative frameworks) researcher in North America. She did a postdoctoral study with Novak and effectively became part of his extensive international network that included David Ausubel and the group at Monash University. As Arzi notes, her affiliation with the Weizmann Institute also served her well in that she met and was influenced by notable scholars such as Chick Ahlgren, who made early contributions in the field of learning environment research and in his roles at

327

AAAS, is well known for his contributions to *Science for All Americans*, regarded by many as a forerunner to the United States' *National Science Education Standards*. Ahlgren assisted Arzi to appreciate the value of mixed methods in research in science education.

Although he was not mentioned in Arzi's chapter, Pinchas Tamir, from Israel was a perennial globetrotter and made connections with science educators all over the world—at an individual and institutional level. He created social bonds that served many contemporary science educators well for more than forty years. Not only did he visit many science education centers around the world, but also he brought many notable scholars to Israel, to benefit emerging scholars such as Arzi, Hofstein, and Lazarowitz and the generations of scholars that followed. An analysis of Tamir's publications in the Web of Science database show that he is one of science education's most prolific scholars who continues to publish in spite of poor health for at least a decade. Tamir's research focused on science inquiry, laboratory work, and science assessment—supporting strong connections with science educators at the University of Iowa, Michigan State University, Stanford, and Berkeley.

Like numerous others included in the book, Arzi used *CHEM Study* when she taught science in Israel. Her remarks about her involvement in curriculum are reminiscent of the science educators in Part A, who were extensively involved in curriculum development and adaptation. Arzi has a commitment to lab work and its inclusion in the curriculum and like Sjøberg, she was oriented toward connecting science to technology and society. Dillon's experiences with curriculum reflect him being situated in the United Kingdom, which had its own curriculum development era associated with large projects developed by the Nuffield Foundation and the Schools Council. His strong interests in environmental science were included in his work with curriculum and his innovations in science teacher education.

FORCES SHAPING SCIENCE EDUCATION

Science education, as any field and culture, is a dynamic entity and process, continuously undergoing change all the while being reproduced in the concrete lives of its members. As in any other field, one can note forces that appear to shape the ways science education is enacted and the outcomes of the productive activity of its members. Among these we include publication in high-impact journals, the appearance and existence of journals, the renewal through new members entering the field, and access to the resources funding agencies set aside for particular fields to support research and development.

Publishing in High-impact Journals

Dillon seems cynical in his use of phrases such as publish or perish and noting that he should not be writing a chapter for a book or reviewing the literature because these are not valued contributions to the field of academe. What is valued and who values it? To get a crude measure of the articles published in the high-impact science education journals we searched the *Thomson ISI-Web of Science* to see which articles were published in science education's high-impact journals: *European Journal of Science Education, International Journal of Science Education, Journal of Research in Science Teaching, Science Education,* and *Research in Science Education.* The yield was 4698 articles. Not surprisingly the number of publications emanating from the United States is very high in comparison to other countries (46 percent of the total). After the United States are England, Australia, Canada, Israel, South Africa, the Netherlands, and Spain. Based on the articles included in the citation database (ISI, Web of Science) some patterns can be found to support the dissertation data we discuss in the previous part to this book. Based on publications in high-impact science education journals a cluster of institutions stands above the others—consisting of Curtin University (Australia), University of Georgia, Purdue University, Hebrew University (Israel), Arizona State University, University of Iowa, and Kings College (England). In a cluster beneath the top tier, each with about the same number of listings as the other are: University of Wisconsin, University of Victoria (Canada), Florida State University, University of Michigan, Michigan State University, University of Texas, University of California at Berkeley, and Leeds University (England).

Among the most frequently published authors are Roth (66), Tobin (63), Lawson (60), Treagust (48) and Yager (41). Each of these scholars is part of a community that can be thought to comprise coauthors, cited scholars, and citing scholars. Lawson, the first recipient of NARST's *Distinguished Contributions to Science Education Through Research Award,* explored the role of cognitive development in science learning from a Piagetian perspective. His articles, especially through the 1970s and 1980s, shaped the field of science education in the US and beyond. Among those who have coauthored with Lawson and/or cited his work are John Renner (US), Mansoor Niaz (Venezuela), John Staver (US), Ed Marek (US), Wolff-Michael Roth (Canada), Larry Scharmann (US), Michael Shayer (UK), Peter Preece (UK), Uri Zoller (Israel), and Michael Abraham (US). The group of researchers that explored science education has been dominant in science education for at least three decades. However, the group contained people who branched out. For example, David Treagust, the most recent recipient of the *Distinguished Contributions to Science Education Through Research Award,* also undertook research from a Piagetian perspective. Over time Treagust became affiliated with the conceptual change group, being influenced by the research of Peter Fensham, Rosalind Driver, and Peter Hewson.

To obtain insights into the articles on conceptual change published in the high-impact science education journals we searched the *Thomson ISI-Web of Science* obtaining a yield of 591 articles—almost 13 percent of the 4698 articles in the database. The most frequent authors using one or more of the keywords (or their derivatives) conceptual change, misconceptions, and alternative frameworks were Treagust (Australia), Roth (Canada), Hewson (US), Lawson (US), Keith Taber (UK), David Brown (US), Joe Novak (US), and Chin-Chung Tsai (Taiwan). The interconnections with the Piagetian group are evident with Treagust and Roth being represented in both groups and with Novak, Hewson, and Brown showing the foundations and emergent qualities of research in this area. Tsai's inclusion in the group reflects his long list of publications and, in this case, self-citation.

Some of Tobin's early research also involved Piagetian perspectives and the scholarship of American scholars such as Lawson and Renner and a less well-known Australian from Monash, Richard Tisher. Since he was involved in classroom research he was interested in the ways in which participation and learning were mediated by formal reasoning ability of students. Hence, in his quantitative research he wanted an easy-to-administer pencil and paper instrument that could produce measures with sound claims to validity and reliability. With William Capie, Tobin pioneered the use of two-tier multiple-choice items that cut down on guessing as a factor in measuring formal reasoning ability. While Tobin and Treagust were colleagues at Curtin, Treagust began to use two-tired tests in his research on conceptual change and with his graduate students continues to use them to this day. Tobin was to move away from Piagetian based research and the citations of his work and coauthors show evidence of his trajectories over time.

Tobin has coauthored multiple articles in high-impact journals with six key science educators: William Capie (US), Barry Fraser (Australia), James Gallagher (US), Deborah Tippins (US), Campbell McRobbie (Australia), and Wolff-Michael Roth (Canada). These scholars were central figures in Tobin's growth as a science educator and reflect a twin research agenda in research on teaching and learning and learning to teach. Unlike Lawson and Treagust, who maintained a strong focus over about four decades of research, Tobin has changed the theoretical frameworks he has used and the research methods. These changes are reflected in citations to his work. In addition to the coauthors mentioned above, those who have cited Tobin's work on multiple occasions include Stephen Ritchie, David Treagust, Okhee Lee, and Michael Bowen.

Michael Roth also has changed the foundations for his research in the two decades since he completed his doctoral dissertation. Included in Roth's coauthors and citing scholars are Michael Bowen (Canada), Campbell McRobbie (Australia), Michelle McGinn (Canada), Tobin (US), David Hammer (US), Keith Lucas (Australia), Chin-Chung Tsai (Taiwan), Angela Calabrese Barton (US), Joseph Krajcik (US), and Sasha Barab (US). Although Roth began his research within a neo-Piagetian paradigm, his move to become a full-time teacher in a secondary institution brought about his interest in the social and linguistic foundations of knowing and learning, with an associated shift in his reading and citing preferences. These changing preferences also are reflected in the changing research interests, which entailed collaborations with science educators who normally would be considered as doing very different work. Thus, with Barton he shares an interest for critical ethnography and the rethinking of scientific literacy in terms of social justice, issues that the two worked out in *Rethinking Scientific Literacy* for which they received the AERA Div-K research award. Bowen and McGinn were graduate students interested in everyday cognition in scientific laboratories and engineering firms. Hammer and Roth share that they have done research out of their own physics classrooms and that their work has been published in journals associated with applied cognitive science, learning science, and educational psychology, including the *Journal of the Learning Sciences* and *Cognition and Instruction*. With McRobbie and Lucas in Australia and Tobin in the US, he has had common interests in classroom research. With Tsai, Krajcik, and Barab he belongs to a group of science educators also involved the field of the *learning sciences*, with its own journals, institutional structures, and research interests (which run more along the lines of applied cognitive science than traditional science teaching concerns). That is, the high rates of co-citations with science educators normally doing quite different work are the result of Roth's multiple divergent interests and participation in different research communities.

Robert Yager, from the University of Iowa has been president of the major science education organizations—NARST, ASTE and NSTA. Also he has been recipient of numerous career awards, including the *Distinguished Contributions to Science Education Through Research Award.* He is well known for his work that connected science curricula to technology and society and for applying constructivism to science education. In the 1980s, following the report *A Nation at Risk,* Yager was involved with colleagues at the University of Iowa, notably John Penick, in a project called *The Search for Excellence.* On the basis of this very prominent project Barry Fraser and Ken Tobin launched an interpretive study of exemplary teaching and learning of science and mathematics in Western Australian schools, encompassing grades K–12. Yager has been active internationally and has doctoral students from around the world—obviously a key factor in establishing the high status of the University of Iowa's doctoral program in science education.

In many parts of the world universities are placed in a dichotomy—teaching institutions or research institutions. In the US faculty in research intensive universities (often referred to as tier 1 or 2) are expected to publish in high-impact journals and receive regular support from external sources for their research. Similar requirements have spread throughout the world and there is also a tendency to hold all faculty members to the same requirements, even if they are in teaching universities or universities not designated as research intensive. This trend works in opposition to a trend to create boutique journals to cater for different roles, interests, and talents of science educators in universities. Accordingly, scholars from all universities submit their best work to the journals listed in the *Thomson ISI-Web of Science.* Scholars work out a pecking order and send their work to the top journal and if it is rejected to the journal that ranks second—until finally the piece is accepted. This practice tends to create a problem of the journals in science education being much the same as one another. It was precisely this problem that led us to create a new journal, *Cultural Studies of Science Education.* As Dillon notes in his chapter the editors of the high-impact journals were becoming gatekeepers in the field—deciding what would and would not be published, what theoretical frames could and would be used, who would be cited, and how long articles could be. The level of editorial control over articles varies from reviewer to reviewer and from one journal to the next—however, our experiences were that micro-management was common, editors often scolded authors for submitting work written in a particular way, and the editorial suggestions for acceptance were often tantamount to re-authoring of manuscripts. We found many of these problems transferred from one editor to the next and the practices were quickly disseminated through the editorial board and the reviewers selected to participate in the peer review process. From our perspectives it was time for changes that would provide for an alternative genre that supported sociocultural perspectives and encouraged authors to author their texts and for publications to be launching pads for broader and deeper conversations.

Creating a New Journal

We had a sense of psychological theories dominating the published literature in science education. Even though there were landmark papers that included strong sociocultural emphases these were diluted with the mass of papers grounded in cognitive "between-the-ears-and-under-the-skull" models for learning. As readers we wanted an alternative where thoughtful sociology frameworks would be used to explore science education in each of the articles. Also, we wanted a review system that would have peers who knew the field and could credibly review the papers submitted. As authors we had experienced the ad hoc process of having reviewers request that their favorite perspectives be included as a condition for publication—this on top of editors requesting that their preferred frameworks also be included in a revised version of the paper. These were just two of many compelling reasons to create yet another journal in science education.

We were adamant that *Cultural Studies of Science Education* would include a different approach to the editing process. As the foundation editors we decided to read all papers submitted and decide whether or not they were potentially publishable in the journal. As the editors of the journal we would make the publish-or-not-publish decision and would not leave that decision to the editorial review board using some vote counting procedure. Our process is to read a manuscript and then give the author feedback on the extent to which a manuscript fits the mission of the journal. If a sociocultural framework is used to examine critical issues in science education then we provide suggestions to the author on how to revise the paper to make it suitable for publication. For the most part we annotate a PDF file containing the paper so that specific edits and comments are referenced to the specifics of the submitted paper. In the process of providing feedback to the author we are especially mindful that our role is to assist the author to author the paper in ways that produce an acceptable paper that conforms closely to the mission of the journal. We do not want to assume authoring roles that truncate the agency of the authors. Instead we offer suggestions through successive revisions of papers until we reach a stage where we are willing to publish the paper.

Once we reach the stage at which a paper is accepted for publication we send the paper to the forum, which is an extension of the review process in which scholars interact to produce a new paper that begins with the original paper and extends theory, method and key ideas beyond those addressed in the paper. The paper is a starting place for a larger dialogue about the issues arising from the paper. We select participants for the forum based on our sense of who would be best taking into account the authors' suggestions. In most instances the forum consists of two to three authors and an equal number of outside experts. In selecting outsiders we sometimes go outside science education, selecting scholars from sociology, anthropology, and linguistics, for example. Through our outside selections we use the forum for leading scholars to bring to bear their latest thinking of critical issues in science education. Hence, in the first volume of the forum we have had scholars such as Allan Luke (Australia), Joe Kincheloe (Canada), and Margaret Eisenhart (US) from outside of science education and some of the leading science educators participate in the forum. Also, we recognize the opportunity to bring into the forum leading junior scholars as well and we are conscious of our responsibility as editors to nurture growth within the field.

Unlike most science education journals, we do not have a page or word limit on articles that can be published in the journal. We encourage papers to be fully developed and argued without being redundant. Accordingly, much of our feedback to authors after the initial submission is to flesh out parts of the paper that are not fully argued. Because the forum is an established part of the journal, it is also possible to leave certain issues to be addressed more fully in the forum. Hence it is not necessary too include everything about a study in the paper because side issues can be introduced in the forum and issues that could go deeper can go deeper.

Already we have had some junior scholars who have written fine pieces that incorporate sociocultural frames decide to submit their work to one of the high-impact journals because the promotion and tenure process in their universities does not give sufficient weight to a paper published in a journal that is not included in the *Thomson ISI-Web of Science*. Hence, papers that would be ideal in CSSE are submitted to *Science Education* and *JRST*, where they are or will be published. To be clear, even though it was the preference of authors to submit to CSSE the forces to publish in a high-impact journal with a high rejection rate was overpowering. They could not risk negative votes from colleagues for submitting work to journals not included in the *Thomson ISI-Web of Science*. Of course we anticipated this as an issue and as editors of the journal we have plans for CSSE to be a high-impact journal because the articles we publish will be highly cited and salient to science education and the social sciences more broadly.

Getting off to a Good Start

As Dillon emphasizes, getting off to a good start in the field of science education necessitates getting a good job and then earning tenure. The pressures might then be characterized as publish or perish, or in terms of where, when and how much to publish? Except for positions like Dierking's, which was outside of a university environment, most new hires have to teach, do service, and publish. No matter whether the university is research intensive or teaching intensive, the requirement to publish seems universal. Hence, learning to write for journals is a priority in graduate school—and if the first step in the process is securing a job then having publications in relevant journals prior to applying for the first job is just good common sense.

We are struck by the fact that so many graduate programs in North America are out of synchrony with the necessity for students to graduate with publications in order to secure a good position in a research-intensive university from the outset. Doctoral degrees have been laden with coursework and students are required to write to pass courses with the papers they write having little resemblance to what might be needed for publication. This needs to change. Instead of writing to pass a course, students should learn to write as scholars to contribute to the field. Perhaps graduate school could have as its mantra to shorten the time taken to make a transition from student to colleague!

Consistent with scholars like Okhee Lee, who works an extra hour every day, we advise graduate students to write every day—setting aside time every day to write, including creating new text, editing and polishing. We regard it as essential for graduate students and all scholars for that matter to write consistently and thereby get better at writing for peer review. Whether or not a graduate program of studies affords such writing it is important while in graduate school to get experience in writing research papers, literature review syntheses, and critical reviews of manuscripts submitted for publication. With the plethora of journals now available and online blogs, it seems likely that most graduate students would be able to apply for their first position as an assistant professor with evidence that they can publish their work in peer reviewed journals and can participate effectively as peer reviewers.

When Tobin left the University of Georgia, after spending an extra semester as a research associate after completing his degree, Bill Capie wished him well with the comment that it was time to sever the umbilical cord. That never really seemed like a very sensible comment because building and maintaining social networks may be the most important part of becoming a scholar. Also, since Tobin and Capie were to coauthor twelve articles with Capie being first author on only one, the umbilical cord metaphor seemed highly inappropriate. What had happened of course is that Capie had treated Tobin as a colleague from the very beginning and all the writing they did together was focused on publishing in journals, monographs and technical reports. Tobin's dissertation then consisted of a manuscript style dissertation in which a collection of papers, published or submitted, from a dissertation study were collected together in the dissertation between introductory and concluding chapters. Rather than separate from Capie it might have been more desirable to have laid out plans for active collaboration, to establish active research sites in Australia and the United States, and to continue the classroom research they initiated in Georgia.

Recently, we have heard from junior scholars that they are being encouraged to show their independence as researchers—to publish work that is different from what they did in their dissertations and to show separation from their doctoral or postdoctoral advisor. We regard such advice as superficial and potentially deleterious not only to the scholars concerned but also to the field of science education. The field desperately needs programs of research that are ongoing and intensive and we need scholars to be supported by networks of science educators—much as we rely on one another even though we both have been full professors for many years. On both counts the advice is not well targeted.

Most effort in social science research, as Lee noted, is spent on accessing and analyzing data sources. Very little of it gets written up into peer reviewed papers in journals—high-impact or otherwise. Hence, one outcome of a good doctoral program would be a set of publications and a dataset that can be used for ongoing analyses. Especially in the first few years when it is imperative for junior scholars to write and submit the work it seems a no-brainer to continue to work from an existing database. For example, consider the database from our NSF and Spencer sponsored research in urban high schools. Many doctoral students obtained their doctoral degrees from this research, including Gale Seiler, Rowhea Elmesky, Melissa Sterba, Beth Wassell, Sarah-Kate LaVan, Sonya Martin, Linda Loman, and Stacy Olitsky. Since these scholars graduated and went into university positions several of them have been discouraged from continuing to analyze these data—the priority for others being that they establish fresh research and publish from new datasets with different research agendas. Even so, we have continued to network as scholars and many papers and books have been written from the dataset, including the doctoral graduates mentioned earlier, and others such as Roth and Tobin, Stephen Ritchie, Cristobal Carambo, Jennifer Beers, Kathryn Scantlebury, Catherine Milne, Tracey Otieno, and Regina Smardon. Furthermore, three more doctoral dissertations will be based on the dataset, which is used for dissertations for Carambo, Anita Abraham, and Clare Tracy-Stickney. Even with all of this intensive research there are still many more years of research that could be supported by this dataset, even if more resources were not added to it. To us it makes more sense to encourage junior researchers to continue where they left off and to focus on doing good scholarship—field work is just one component and asking junior scholars to start afresh is a recipe for missing tenure of struggling to make it.

There may well be instances where it is highly desirable for a junior scholar to separate from an advisor—reasons pertaining to lack of productivity or perhaps exploitation of the junior scholar. Similarly, some datasets may not support ongoing research. What makes sense in offering advice (and receiving it) is to consider each case on its merits and to discuss alternatives before allowing the junior scholar to make the decisions. We wonder if much of the well-intended advice given to junior scholars is not well researched and may not be grounded in accurate data and associated interpretations. Any one-size-fits-all assertions would qualify as good examples of what to avoid—such as, the only way to get tenure here is to publish a book. To the advice giver—get the data and show junior faculty the trends in science education within this institution and in comparable institutions in the nation and worldwide. To the advice receiver—ask questions about the productivity of peers locally, regionally, nationally and internationally. Also, be sure to interrogate the advice to ensure that your own interests are being protected and that in accepting the advice you are not steering your career down a pathway that is not right for you.

External Funding Shapes Interests

As we mention in the previous section, the external funds available to support research often come at a price of the researcher having to adjust her interests to fit the specifications of the request for proposals. It does not seem fair to the individuals concerned to name names here, so we will not do it. Instead we ask you to take a peek at the top grant getters in science education and ask what they have done with the money to improve the field of science education. We worry that the priority often seems to come down to spending the money and supporting graduate students and post docs. To what extent has the field of science education been advanced by the scholarship of the last forty years? We leave this question as a rhetorical challenge. But in doing so we note the imperative many feel to align their priorities with institutional priorities to bring in large grants. Being successful can have many other alignment costs. For example, there have been trends in North America and elsewhere for priority to be given to projects that include professional development and curriculum development.

Also, in the US there has been a tendency for federal funds to be used to commission reports from experts, often including scientists, social scientists, and science educators. The nature of the report reflects the initial selection of personnel to serve on the committee and then the ways in which differences are dealt with within the committee. The cycle continues when the reports become essential frameworks for requests for proposals for federal funding of projects. In this way, reports produced with federal funding have become gatekeepers on federal funding by ensuring that proposals are consistent with them. Examples include reports produced by the National Research Council—recommending National Science Education Standards, a need to employ evidence-based scientific research (especially quasi-experiments), and publications that embody a conceptual change approach and the latest advances in neurosciences.

Dillon raises questions about what it means to be an intellectual worker in this day and age. One thing is for sure: it is essential that science educators learn to navigate the waterways from which external funding to support scholarship might be obtained. Getting a balance seems imperative—rather than jump at the biggest and most available pot of money, perhaps the best payoff on a career sense it to take the time to develop that rationale and methods you subscribe to and present a case that reflects your own ways of being as a science educator. The forces that shape science education can be daunting if you find that you do not easily fit the mold that appears to be there for all to squeeze into. We always encourage our graduate students to pursue intellectual rigor that relies on diversity in frameworks, methods, and ways of characterizing the field of science education—replete with its attendant problems and potentials.

Part F

OPENING UP THE CONVERSATION ACROSS CULTURES

SCIENCE EDUCATION AND COLONIALISM

Historically, initially philosophy, then natural philosophy, and finally science was the domain of men, who, for a variety of reasons, limited access and excluded women from actively participating in their shared endeavors. Male philosophers and scientists have developed specific forms of discoursing and thinking that express a distancing and cutting off from emotion not typically found in the way women relate to the objects of their knowing. It therefore comes as little surprise that some of those scholars who critically interrogate science, conducting careful cultural-historical analyses of it, have come to use adjectives such as *phallogocentric* to characterize science and its mode of operation. This adjective, a new creation, points to the male (phallus) and linguistic (Gr., logos) orientation of the science disciplines and their epistemic products.

Historically, too, the roots of the sciences as we know them today and their roots in a particular form of reasoning, lie in the Greco-Judean culture and tradition, although this culture found its spreading particularly enabled by the empires that adhered to Christian and Roman value systems. When the great empires began to subjugate and colonize the lands of indigenous peoples, they also attempted to export their knowledge and belief systems by eradicating as much as they could of the indigenous forms of religious practice and ways of knowing. The order of the day was not to integrate these peoples into new state forms but "make them like us as much as possible." In Canada, for example, the children of the First Nations were forced to go to residential schools often run by nuns and priests, where they were not allowed to speak their languages; and they were beaten and abused when they did speak their mother tongue.

Whereas assimilation to the Western culture in the Hispanic world began with the arrival of the Spanish, the active policies and practices continued into the 20th century. Canada, though nowadays a country that is very open to diversity, forms of knowing and religious practice, and different forms of culture (e.g., Sikh members of the Royal Canadian Mounted Police may port the traditional turban rather than the Stetson typical of the police), is a prime example of the abuses aboriginal peoples have undergone around the world. Policies regarding aboriginals in Australia, apartheid in South Africa, and the inequitable treatment of people of American Indian origins in Central and South America serve as examples of internationally dispersed endeavors to eradicate indigenous (religious, knowledge, cultural) practices. We use Canada as but an example to highlight the state-operated machineries that subjugated and assimilated indigenous peoples rather than allowing them to thrive in multicultural societies. In this subjugation, Western forms of knowing including the sciences, were imposed upon aboriginals, who, in the process, were forced to abandon their own ways of knowing especially about their natural environment that had developed through thousands of years of lived experience in often-hostile circumstances (e.g., Inuit, Dena, and Innu in northern Canada).

In Canada, the *Gradual Civilization Act* was passed in 1857 to assimilate Indians to the Western, largely British culture of the state. During the period from 1870 to 1910, the clear objective of both missionaries and government was the gradual assimilation of Aboriginal children into the lower levels of a hierarchical society. From the 1920s on, attendance for Aboriginal children aged 7 to 15 was compulsory, and priests, Indian agents, and police offers often took them from their homes by force. By 1931, there were 80 residential schools and in 1948 there were 72 with a total student population of nearly 10,000 students. The last federally run residential school in Canada did not close until 1996—Gordon Residential School in Saskatchewan. Students were not just forced to attend schools, but also to abandon their language and creed. In addition to these what might be considered symbolic forms of violence, it became evident by the 1980s that the Aboriginal students were subject to sexual and physical violence as well. Today, Canada as a nation is engaged in working with the Assembly of First Nations (negotiations, reconciliations) to work through this sad and tragic aspect of its history. This begins with an active recognition of the events:

> Sadly, our history with respect to the treatment of Aboriginal people is not something in which we can take pride. Attitudes of racial and cultural superiority led to a suppression of Aboriginal culture and values. As a country, we are burdened by past actions that resulted in weakening the identity of Aboriginal peoples, suppressing their languages and cultures and outlawing spiritual practices. We must recognize the impact of these actions on the once self-sustaining nations that were disaggregated, disrupted, limited or even destroyed by the dispossession of traditional territory, by the relocation of Aboriginal people, and by some provisions of the Indian Act. We must acknowledge that the result of these actions was the erosion of the political, economic, and social systems of Aboriginal people and nations. (Department of Justice, Canada: http://www.justice.gc.ca/en/dept/pub/dig/healing.htm)

Clearly, the colonization of Aboriginals, which has lasted until the end of the 20th century—and, as some of our Aboriginal graduate students emphasize, continues to the present day—has had its effect on the traditional forms of knowing, in particular the traditional ecological knowledge. It also had its effect on the recruitment of teachers generally and teachers of science particularly: as Ngati Kahungunu/Ngai Tahu (Elizabeth McKinley) describes, she is one of but a small number of Maori science teachers, and Eileen Parsons makes it apparent that African American science educators are still underrepresented among science teachers in the United States. In Canada, the proportion of Aboriginal teachers is only between 20 and 30 percent of the proportion of Aboriginal children aged 0–14 in the population. Although we do not have precise numbers, this under-representation also is the case in for First Nations scholars in Canada or Aboriginals in Australia; in Canada, as Statistics Canada reports in its "ABCs of Educator Demographics" shows, the proportion of Aboriginal teachers in the education sector is lowest for secondary and university professors.

It is not quite clear whether the tide is turning, but there are clear attempts to rethink the relationship between traditional ecological knowledge and scientific knowledge. In Canada, for example, an internationally known, outspoken advocate of the Aboriginal cause in science and science education is Glen Aikenhead. Key to his approach is a more symmetrical treatment of ways of knowing, achieved by treating Western science as a subculture of Euro-American culture (Aikenhead, 1996). Glen explicitly works to make, as he states on one of his websites, Western science and engineering accessible to Aboriginal students in ways that nurture their own cultural identities; that is, so students are not expected to set aside their culture's view of the material world when they study science at school." The extent to which such approaches lead to a greater retention of Aboriginal individuals in science, science teaching, and science education has to be seen over the coming decades. But Glen's initiatives show that there is a change in the way different forms of knowing and the culture of Aboriginal peoples are acknowledged in states that heretofore had active colonial and colonizing school policies.

We are far from claiming that colonialism has disappeared and that there is no cultural bias in modern nation states. But it is clear that disciplines such as science education now provide possibilities for individuals of Aboriginal origins, including African Americans in the US, to contribute to the field and to bring about changes in the school curriculum that go at least a little way of rectifying past inequities. Pauline Chinn, Ngati Kahungunu/Ngai Tahu (Elizabeth McKinley), and Eileen Carlton Parsons are but some of the names of science educators with traditionally *othered* cultural origins that have made their names in the field.

Evidently, it is not just individuals from Aboriginal cultures who bring about the change. Even Western culture changes, and, as seen in the efforts by the former geneticist David Suzuki, scientists, too, contribute to change our understanding that the role of traditional knowledge can play in the human society–natural world transactions. Suzuki has been an advocate for the inclusion of traditional ecological knowledge to understand better the problems humanity faces—global warming, sustainable pisciculture and forestry, genetically engineered foods. In appreciation of his efforts Canada's First Nations people have bestowed upon Suzuki five native names.

Other signs of change can be found, for example, in a fisheries research institute located at the University of British Columbia. There, the seventh ranking Chief of the Hesquiat First Nation on the west coast of Vancouver Island is an adjunct professor. He has been an instrumental force bringing about the inclusion of the integration of First Nations' knowledge and values with modern ecosystem science. Some of the scientists, in a project called *Back-to-the-Future*, actively work with First Nations people to evaluate and institute ways of including traditional knowledge to environmental research and resource management.

The two auto/biographical narratives in this sixth and last part of the book are full of details that articulate and exhibit the experiences of non-whites growing up in nation states and societies modeled in the European tradition. The hardships and injustices they faced in the course of their lives parallel those of the still active colonialist cultures as we evoke them here. That they have made it into the ranks of university professors, however, should not be taken as a sign that the chances are equal for their younger cultural peers in today's societies—despite the changes that have occurred. They have made it, if we can say so, despite and in the face of inequities and forms of social injustice. That is, although their auto/biographies are testimony that the possibilities for realizing careers in science education and science education identities are changing, opening up, a lot of work remains to be done—precisely because of the path-dependent, contingent nature of any community of practice or culture.

In postmodern literatures, speaking *for* or *on behalf of* another, especially a member of one culture or group doing so for or on behalf of another group, has been subjected to (sometimes-heavy) critique. Thus, many feminists suggested that only women could talk for and on behalf of women, because of their particular experiences. From such a perspective, the actions of white scholars and scientists such as David Suzuki and Glen Aikenhead might be considered inappropriate. But one can see and frame their contributions in ways other than as attempts to speak for and on behalf of Aboriginal peoples. One can see them, for example, as working within Western science and culture to prepare the ground for change and for the possibility of Aboriginal scientists and science educators to be heard and comprehended. Without the preparation of the ground, claims of traditional ecological knowledge to constitute a valuable form of knowledge might find it much harder to be accepted—see, for example, the somewhat skeptical perspective Svein Sjøberg expresses in this regard. There are considerable numbers of science educators who, more frequently in camera than publicly, denigrate the contributions traditional ecological perspectives can make to our collective understanding of the natural world.

342

ELIZABETH MCKINLEY (NGATI KAHUNGUNU/NGAI TAHU)

BODIES OF KNOWLEDGE

Narratives of Colonialism, Science, and Education

E nga mana, e nga reo, haere mai. E nga kaiako putaiao, tena koutou. E nga akonga, tena koutou. E nga kairangahau marautanga putaiao, tena koutou. E aku hoa, tena koutou. Tena koutou, tena koutou, tena koutou.

He Maori toku papa.

Ki te taha o tona hakoro. Ko Aoraki tona maunga; Ko Rapaki me Tuahiwi ona marae; Ko Tuahuriri tona hapu; Ko Ngaitahu tona iwi.

Ki te taha o tona hakui. Ko Nga Waka-a-Kupe tona pae maunga; Ko Wairarapa tona moana; Ko Moiki me Tuhirangi ona marae; Ko Ngati Hinewaka me Ngati Hikawera ona hapu; Ko Ngati Kahungunu ki Wairarapa tona iwi.

Ko Liz McKinley taku ingoa.

Tena koutou, tena koutou, tena koutou.

I begin this chapter with a *mihi* (greeting) that pays respects to all of us as people and to our languages. I especially acknowledge those people who work with us: teachers, students, colleagues and friends in the science education community. I follow this short greeting by citing a shortened form of my *whakapapa* (Maori ancestry), both mythical and historical, which signals to people my ancestors, where I come from and where I belong in this Maori world. This form of introduction is commonplace in Maori society where we do not ask, "Who are you?" but instead ask, "Where are you from?" Maori who hear my introduction recognize the representations of the self from the geographical locations and people mentioned, and can locate me in an historical narrative that is often connected to them and their people. This form of introduction includes tradition, history, geography, language and ancestry. It is simultaneously a national narrative and a personal identity— about Maori as a collective and me as an individual. This introduction may seem strange to readers who think knowing me is knowing my name, my professional position and having some personal information, such as how many children and whether I like red wine or not. So why do I feel compelled to tell you who I am (again) in a language the vast majority of readers have never seen let alone understand?

Within my culture, to be known *as* Maori is connected to your *whakapapa* or ancestry, and it speaks about your existence in two ways. First *whakapapa* encompasses the idea of kinship or whom you are related to by blood. Today we conceptualize it as including genetic inheritance. Maori use *whakapapa* to show connec-

K. Tobin, W.-M. Roth (Eds.), The Culture of Science Education, 343–353.

tions both within (generational) and between (intermarriage) tribal groupings. Secondly, *whakapapa* is about a person's cultural and spiritual ancestry. These can include origins and explanations for diverse things such as flora and fauna, parts of the human body, words and speaking, and the cosmos. Cultural values and practices such as history, *waiata* (songs), and mythology are also included. Spiritual ancestry is about the *mana* (prestige) of your *tupuna* (ancestors)—which is carried forward through the descendants. *Whakapapa* books, found in families, contain genealogies as well as family and tribal history, alongside the cultural and spiritual ancestry. All this suggests that *whakapapa* is inclusive of, but wider than, a discourse on corporeality. The work of Frantz Fanon recognizes this, pointing out that an analysis of the corporeal schema without situating it within a historico-racial schema is inadequate. For Fanon the historico-racial schema is the colonial encounter and how the relations between the colonized and colonizers is shaped in historically specific ways:

> Below the corporeal schema I had sketched a historico-racial schema. The elements that I used had been provided for me not by "residual sensations and perceptions primarily of a tactile, vestibular, kinaesthetic, and visual character," but by the other, the white man, who had woven me out of a thousand details, anecdotes, stories. (Fanon, 1967, p. 111)

The inadequacy of an analysis of the corporeal lay in de-personalization of the *experience* of the body. Fanon argues here that the body is about its history as well as its corporeal nature. *Whakapapa* is both noun and verb. It is a genetic map and a condition of being. Knowing your ancestry or *whakapapa* in such a way is important to many Maori and forms the basis of much Maori researching and writing. And so it has been with me.

People who have read my work know that issues of identity, and the questions associated with it, have driven my teaching, research and scholarship. For example, I have been concerned with how the identity of "Maori woman" has come to be constituted in contemporary times and what connection, if any, that it has with our colonial past. The questions are placed in a wider, and often deeply felt, politics of the relationship between the indigenous peoples of Aotearoa (Maori) and the descendants of the British colonial settlers (Pakeha). Academic work based on identity has been supported by wide-ranging political and social change that Aotearoa New Zealand has undergone in the last thirty years—change focused on finding ways forward as a bicultural country that recognizes and respects both signatories of the Treaty of Waitangi—the Maori chiefs of New Zealand and the British Crown—that was signed in 1840 and allowed for British settlement. For Maori educationalists, these politics have largely been conceptualized as language (and knowledge) revitalization through establishing an education system from preschool to tertiary based on Maori philosophy and values, taught through the medium of *te reo Maori* (Maori language) and the development of all the supporting policy, curriculum, pedagogy and assessment and evaluation that goes with this form of change.

There are two main strands to my work that I see as related but unfortunately have me publishing in two different areas, such are the structures of academic disciplines. The first area concerns questions of subjectivity and relate to teaching and publishing in the area of cultural and policy studies. Much of this work arises from questions surrounding identity, both self-identity and identification with and by others. This line of inquiry has allowed me to expand my interest on the relationship between (Enlightenment) science and Maori (woman), in historical and contemporary times. Secondly I have focused on indigenous science education. As a result of being only one of few Maori science teachers in the country I have been fortunate in being at the forefront of Maori science education initiatives, such as establishing bilingual (Maori/English) classes and the writing and implementation of our first national curricula in *te reo Maori* including science. But first I will give a brief account of how I have come to this point.

SCHOOLING THE SUBJECT OF SCIENCE

Narratives about schooling can be seen as a means by which we retrospectively make sense of our own construction within educational contexts, discourses, norms and practices. Our bodies have been schooled into particular subjectivities and our stories about schooling embody some of the complex disciplinary techniques deployed in schools and other educational institutions. My focus here is not so much at the process by which I became academically successful but the conditions that have led to my emergence as an academic and the effects those conditions have produced. My strong sense of identity, my experiences of life, and my educational success have helped to create an ambivalent relationship with the academic community generally. While I had to believe that schooling would make a difference for my life, on the other hand I was being schooled in difference in contradictory ways. Schooling is a site for much that counts as identity-formation, which is done through repeated acts of norming. Through repeated processes of interpellation we come to be schooled into particular identities even as they may be resisted and transformed.

Academic success is an accomplishment one has to achieve in order to become an academic. For many minority students' educational journeys can begin with expectations of mythical proportions, which people believe will create upward social and economic mobility. High achieving students often have behind them a family that has a strong belief in the dominant discourse of the education system. Furthermore, there is a belief in equality of opportunity for all regardless of who you are—a form of invisibleness for many minority students—and that meritocracy is the critical ingredient to get a good job. This was certainly strong in my own background. As products of the depression years in New Zealand both my parents found their schooling falling short of what they desired. My Maori father never made it to high school being absent more than he was present. I always got the impression that it was about the confines of the classroom as well as the punishment for speaking Maori at school. On the other hand, my Pakeha (white) mother had one year of high school but wanted so much more. Her desire was to go nursing but coming from a large, working class family found her aspirations unattainable. She left high school to work in a fabric shop, married young and devoted her life to bringing up six children and working part-time at any available job to make ends meet. My mother, in particular, was very disappointed with her education and she was very determined to make sure her children had a good one.

I attended high school in the early 1970s and as a bright Maori girl I was placed in the top academic class through the schooling practice of streaming (tracking), where students were separated into different classes based on intelligence, which was common practice in schools at the time. Between this practice and my success at and interest in hard science subjects—mathematics, physics, and chemistry—I spent most of my years at secondary school as the only Maori (woman) student in many of my subjects. The fact that this may have been unusual never really occurred to me. While there were a significant number of Maori students in the school, most of them were in the commercial or technical classes where they were never expected to stay at school long enough to sit the national qualifications. For our family it was different. There was an unspoken rule that everyone did the study and no one asked to leave school before completing Year 13. Whereas my parents knew the value of a good education they never questioned what it contained, only that we had the opportunity (and they ensured we did the work) to do academic courses. While my father taught us some of the values of Maori, he refused to teach us *te reo Maori* saying that we needed a good Pakeha education to succeed. He was right in one way. In the end they were proud of their achievement—every one of us continue to have careers in teaching, nursing, and the computer industry.

At the same time as strongly believing in the equality of the educational system, the educational journey constantly reminded me that I was different. I recall on my last day of high school, when I had collected an academic award at the school prize-giving ceremony that afternoon, a Pakeha woman coming up to me afterwards to tell me I was a credit to my race. It was intended as a compliment. However, when you have been brought up in the house of a mixed marriage and with stories of my (white) mother being treated differently to my Maori father, you become aware of the realities of what it is to be different. For example, my father recalls, as do other Maori of his generation, being punished for speaking Maori language at school and would tell his children about it. When they were first married my mother would look for rental accommodation, which suddenly became unavailable when they saw my father. These, and other racist events, form part of your everyday life because it is part of (normal) family life. Many teachers forget that minority students can be politically astute because of this, sometimes more so than their teachers and it can lead to behavioral difficulties if students detect forms of racism that teachers do not recognize. The idea of Maori and achieving academically was, in the woman's mind, an oxymoron. I never believed (and nor did my family) that living in a brown body meant I could not think.

Such constructions were not just part of life in the 1960s and early 1970s. In the late 1980s, after I had been teaching at a girls' high school for three years, the principal approached me one morning break and asked if I considered myself to be Maori. She said she needed the information to fill in statistical information for the department of education. However, gathering statistical information on staff and students in New Zealand schools is standard yearly practice. So what had she written in the previous two years? The only explanation I could understand was that my principal saw my success at being the head of the science department in the school and being Maori was contradictory. Questioning my Maori woman identity had never really occurred to me as I carry the visible markings of a Maori, and I had never considered I had the choice. I had never really thought how my identity as a Maori woman could be a question for someone else. The question made me think that I am positioned in people's minds identified by visible, racial markings, and that these markings are constantly drawn upon to distinguish me from other science teachers or academically successful girls. I have constantly been reminded in my lifetime that there is knowledge about me that somehow relates to the visible characteristics on my body. How was it that I might be known to people by how I look? What was the contradiction of Maori woman and science or science teacher? I felt I was often alienated from being Maori (by both Maori and Pakeha) through being interpellated into the discourse of being bright or being associated with science. The notion of identification (by others) and self-identifying is an interesting distinction and in my experience, constantly overlap and intertwine. This disjuncture has proved to be a productive space for my academic work.

QUESTIONING THE STATUS QUO

As I attended university and the level of specialization in science increased so did my sense of isolation from other Maori (women) students. I completed a degree in chemistry and in 1979 entered teachers college to complete a graduate diploma as a high school teacher. As a beginning teacher in 1980 I was introduced to the *Learning in Science Project* (LISP), begun by Roger Osborne and Peter Freyberg at the University of Waikato. While I found these resources useful in preparing some of my classroom activities I found the research did little to assure me regarding questions I was forming in my mind regarding Maori and girls in science education. For example, in taking up a leadership position in a large urban state girls' school in 1986, I had been welcomed as a role model for the many Maori girls attending the school. However, in my twelve years of teaching, fewer than five Maori girls made it to my Year 13 Chemistry classes no matter how hard I tried in encouraging, helping, or engaging in out of school activities with them. I came to the conclusion quite quickly that "role models" could never really work without attending to larger social, economic and cultural educational changes.

While at the school, I was fortunate enough to be involved with the establishing of one of the first high school bilingual (Maori/English) units, which was set up in response to an emerging Maori language immersion early childhood and primary (elementary) schooling. The Maori language immersion schools were a significant development and questioned the educational system at its very core. The establishment of a Maori language educational sector had potential for transformative change and left explanations of role models and barriers in its wake. At the same time Beverly Bell, who had been involved with the LISP work, had been appointed

347

as the science curriculum officer for secondary schools in the *Curriculum Development Unit* (CDU) of the Department of Education in the mid 1980s. She began promoting the development of bilingual science education and actively sought out, and sponsored, the few Maori science teachers into national curriculum development work under her direction. (I still keep an archive of those early attempts, which I look back upon fondly as naïve and unsophisticated, in the hope that a student may one day be interested in analyzing this journey through the texts produced.) It was at one of the CDU meetings that I met Pauline Waiti, another Maori science teacher, with whom I would later work. There is little doubt that this association and continued work with Beverly Bell over the years initiated and supported my career in science education research.

Being appointed to a position at the Hamilton Teachers College in 1991, now the School of Education at The University of Waikato, I pursued postgraduate studies while working fulltime. In 1989 Beverly Bell had moved to Waikato University to the *Science and Mathematics Education Research Centre*. During that first year, and only having just started my studies, Beverly invited Pauline Waiti and I to write with her an article for the *International Journal of Science Education* about what was happening in Maori education in New Zealand (McKinley, Waiti, & Bell, 1992). This article was an opportunity to review the culture and science education field and to provide a rationale and document some of the changes occurring in New Zealand with respect to Maori science education. I did not fully realize at that time that what was happening in New Zealand, and the ambitious target that we had set ourselves, was so far removed from what was happening elsewhere. Correspondence from people in the field over recent years has made me think that the article was in advance of its time. There was very little work being done on indigenous education as the field was mostly concerned with multiculturalism. While multiculturalism was around in New Zealand, it failed to find a lot of traction due to the bilingual and bicultural developments. It was widely argued and accepted that if we could achieve a bicultural nation between Maori and Pakeha, multicultural status would be relatively easy to attain.

In the 1980s and 1990s in New Zealand, Maori education was being developed in directions that were radical and ambitious for indigenous peoples in a colonized state. Aotearoa New Zealand has a very unique context in which to work with our colonizing partners, unlike Australia, Canada, and the USA. The nation was established by treaty and Maori were guaranteed, among other things, equality and possession of their *taonga* (treasures) and the British settlers were given the right to settle here and call it home. While the treaty was breached, in the last thirty years we have been aiming to rectify many things, including the near death of the Maori language. These circumstances have had a huge effect on my work and the way that many international indigenous scholars see New Zealand as a pioneer in the indigenous education movement.

The initiation of a Maori medium schooling sector in the early 1980s was targeted at placing Maori philosophy, culture, and language at the centre of education instead of being the pieces added to curriculum. My involvement in curriculum work during this time, and the increased awareness of schools and the curriculum not delivering for Maori students, meant I was invited in 1992 to be involved with the writing of the national Science curriculum. Furthermore, when the first Maori language science curriculum (Putaiao i roto i te Marautanga o Aotearoa) was initiated in 1993, Pauline Waiti and I co-directed the development. I subsequently wrote up the process and analysis as my master's of education thesis and published an article in *Research in Science Education* (McKinley, 1996). The paper explored the issues involved in writing a curriculum document against all odds—no technical language and little cohesion around curriculum direction for Maori medium schooling. Where did one begin?

While I was involved in some very exciting and far-reaching changes in New Zealand, my interests and work made it difficult for me to fit into the science education research field as it was being developed at the time. Our aim as Maori in New Zealand at the time was to make significant changes at a societal level. For example, our aim was on the use of Maori language and the inclusion of Maori cultural knowledge in science classrooms in order to challenge the basis of taken for granted knowledge in our classrooms. It is not about rejecting Western scientific knowledge in our classrooms, but to develop a schooling experience for our students that encourages and is supportive of them to live as Maori (McKinley, 2005a, 2005b). Our experience of 165 years of engaging with the colonizer has shown that by adding Maori bits to a Western curriculum does little for the health of Maori language and knowledge, and less for Maori students. Under current circumstances where Pakeha still tend to control educational decisions the dominant group gets to decide what parts of another culture are included into the mainstream by "policing the boundaries of cultural intelligibility" (Fuss, 1995, p. 143). The additive approach to curriculum sees Maori culture as fixed in artifacts and customs, and untainted by its historical locations, safe in some romantic, mythical past.

In the 1990s, there was much interest in promoting science as a cultural phenomenon and to build on children's existing knowledge and experience. However, there were some basic assumptions that underpinned this goal. For example, were our science educators educated in knowledge of the Other to be able to recognize or value children's existing knowledge? For me, the issue of Maori children (or ethnicity) not forming a category of analysis in the LISP work in New Zealand is not surprising. For many Pakeha New Zealand researchers it would be a difficult task to recognize an answer that may indicate the influence of a child's cultural background. For example, some Maori students bring to their science lessons the belief that inanimate (from the perspective of Western modern science) objects, such as carvings, stones, and rivers, have a *mauri* (living spirit). Western modern science considers this view as unscientific and can be dismissed as children's incomplete understandings, infantile knowledge (complete nonsense), or even prescientific knowledge. It certainly would not fit what would be expected as a child's explanation of scientific phenomena. It is most likely to be dismissed as an anomaly. The effects are that this Maori knowledge would be seen as a nonsense answer as the teacher could not respond. It is unsurprising that the contents of science curricula were in the past, and are largely still, Eurocentric or Western in orientation.

349

While there is much in this field that could be attributed to ignorance, some science education researchers have gone much further and have argued for a cultural exclusion theory of education. For example:

Why people feel driven to assert equality of achievement between cultures is itself interesting. It seems more sensible to say that some cultures do some things well and other cultures do other things well. European Jewry has had (but only since the mid 1800s) terrific success in fostering scientific talent, it clearly has had no success in fostering sporting talent. The Hmong people of South-East Asia have wonderful handicraft traditions but little achievement in technical areas. The medievals built gracious cathedrals, but they did not master science. Some cultures have outstanding musical traditions, while other cultures barely rise above noise production. (Matthews, 1995, p. 217)

Matthews implies that some students are naturally good at particular curricula while presumably being naturally bad in other areas. In other words, people are born with deficiencies and capabilities according to our racial backgrounds, and some result from culture, creating a hierarchy of peoples, which will affect the education we receive, and hence the careers to which one can aspire. In addition, Matthews' choice of example reflects a further hierarchy of knowledge as the white races are naturally good at science and building gracious cathedrals—that which requires intellect (and an implication of attaining high culture as well)—and the Asian Hmong people are naturally good at handcrafts or manual work. In other words, Matthews is marking out subjects where white peoples are shown to be naturally dominant over others (read non-white groups) because they are more intelligent. The effect of this statement is to construct knowledge of peoples in order to justify the absence of different groups of people in science and to maintain the status quo. Such views, and the fact that publishers allow them public space, suggest there are still very deeply held and racist views on non-white peoples in science and science education.

Views such as this just reinforced my view that there is more to getting Maori (or minority) students engaged in science education than those issues identified in the science education literature. One of the weaknesses in the field during the 1980s and 1990s was that it treated girls and minorities as a single, unified and homogenized group of others. There was an emphasis on deficit theories, located in the person (girl or minority student) or culture, and rarely questioned other aspects such as the nature of science knowledge being taught. Maori girls' science education seemed to be excluded from science education literature instead of being doubly included through culture and girls. No matter where I looked in the science education literature I found I had to divide myself to fit. I existed under the rubric of Maori and girls—the excluded bodies of science education—but at the same time, science education saw itself as inclusive because the field is about learning (universal) science. In theory I occupied more than one place but as a Maori girl/woman I was not visible in any of them.

A QUESTION OF KNOWLEDGE

In an attempt to answer some of these questions I embarked on a doctoral study towards the end of 1995 (at the age of 40) while I continued to work full-time, not uncommon among education doctoral candidates in New Zealand. After completing my master's thesis one of my supervisors, Sue Middleton, gave me Edward Said's (1978) *Orientalism*. In reading *Orientalism* I began to find ways to historicize colonialism through the body and the unconscious, and to connect them to the present. Said showed that "the Orient" was fabricated from an interrelated web of writing from literary, historical, scholarly, political, military, and imperial administrative accounts, which demonstrated the complicity of academic forms of knowledge with institutions of power. What I read seemed to resonate with Aotearoa New Zealand's experiences where the coinciding of colonialism and the Enlightenment science project is evident in our own history. Internationally, the age of enlightenment coincided with imperial projects of discovering new lands for settlement, exploiting resources, expanding the boundaries of the empire (both land and human empire) and the conquest of nature.

I was interested in how knowledge produced in the past could continue to influence how people know others today. My experience of twelve years as a Maori science teacher suggested that the literature that surrounds the issue of minority (or Maori) lack of participation and achievement in science subjects was too simplistic in its approach. Said had argued that there were two levels to the knowledge of the Other that operated simultaneously.

> The distinction I am making is really between an almost unconscious (and certainly an untouchable) positivity, which I shall call *latent* Orientalism, and the various stated views about Oriental society, languages, literatures, history, sociology, and so forth, which I shall call *manifest* Orientalism. Whatever change occurs in knowledge of the Orient is found almost exclusively in manifest Orientalism; the unanimity, stability, and durability of latent Orientalism are more or less constant. (Said, 1978, p. 206, original emphasis)

This framework gave me an insight into how as a bright Maori girl people could see my achievements as unexpected beyond the blind alley of an ideology of racism. It is not that racism does not exist, however, racism can often lead to entrenched positions that are diametrically opposed and suppress agency. I found I could begin to connect the ideas in Said's work to my personal questions on how I can be known today and, in particular, the use of Said's latent and manifest Orientalism that enabled (at least) two bodies of knowledge to circulate at once. Alongside the insights of Said I looked at the work of poststructural, postcolonial, and feminist theorists and combined them with Maori cultural theory to extend and localize the methodological framework. Michel Foucault, Jacques Derrida, and Gayatri Chakravorty Spivak enabled me to explore ideas of discourse, power and deconstruction, while Homi Bhabha and Frantz Fanon allowed me to develop an agential position in discourse for the colonized in *returning the gaze* by invoking the unconscious to explore the possible workings of latent knowledge.

351

My doctoral thesis, completed seven years after beginning, explored how 'Maori woman' is situated in a network of writing that captures and fixes the marked body as a sign of negative difference in both colonial (historical) and postcolonial (contemporary) discourse (McKinley, 2003). The thesis, entitled *Brown Bodies, White Coats*, firstly examined how Maori women were brought under intellectual control of the emerging scientific academy. Drawing on the imperial archives, a three-month stint in the British Library and working with Valerie Walkerdine, who influenced the construction of the thesis, I showed how the stereotypical signifiers are incorporated within scientific discourses and shape the fictions that the texts represent. I then drew on open-ended interviews with sixteen Maori women research scientists, who were asked to discuss their Maori woman identity in relation to being a scientist to explore the conditions by which the subject Maori woman scientist emerges. I argue in this work that the body forms a basis of a doubling—a written body and a corporeal body—that cannot be separated but continually refer to each other. The body is identified with a presence of the past and has the ability to transcend time through a web of discourses. In order to attain the status of mind, the Maori women scientists needed to flee their bodily markers of race and sex—to flee themselves. The Maori women in my work became caught within the sign of double articulation—eye/I—that manifests itself as an ambivalent desire to be brown and white through mixed and split origins and identifications.

CONCLUSION

There is much more activity in the field of culture and science education research now than fifteen years ago when I began my postgraduate study and co-wrote my first article. I have completed two postgraduate degrees, we have seen our first Maori immersion graduates through university, and we are writing our second *Putaiao* (science) curriculum. In the field there has been a number of developments, including the establishment of a journal addressing culture and science education, and a larger group of academics researching and writing in indigenous science education. However, while the field is minor in science education there is more being addressed in indigenous education and more is being published in journals addressing issues in native education.

In Maori education we have yet to reach a position where we have a critical mass of Maori science teachers to create a community of practice and interest. While there are a couple of emerging researchers in the area we are still dealing with broad issues of establishing and embedding an alternative education system. This is a feature of Maori education generally, especially Maori medium education. Whilst we have made significant changes in Maori education in the last twenty years, we still have significant issues to address. Providing a full curriculum in the medium of Maori based on Maori philosophies, values and knowledge was an extremely ambitious and unusual undertaking. This has provided us with a platform from which to carry out significant research for years to come. Such circumstances make it difficult for people outside these politics to measure the magnitude and effect of what we are doing. While some argue that we now live in a global world and the case for *place-based* politics ignores the large flow of immigrant populations, I argue that now, more than ever, we need to give our children a place to stand.

A few weeks ago one of my sisters and I visited our *urupa* (cemetery) so I could pay my respects to my uncle whose funeral I could not attend. It is situated on a small block of land just outside a small rural town that has become a fashionable escape from the pressures of the city. The town now boasts a number of boutique vineyards and popular cafés. It never used to be like that and as a teenager I considered it to be "the back of beyond" and could not wait to get out. In our usual fashion my sister and I ended up wandering among the gravestones of our ancestors (including our father) reminiscing, pausing alongside the graves of the *mokopuna* (grandchildren) buried, and wondering about our cousins who are now scattered to the four winds. We always do this when we go there. I draw great strength and comfort from knowing this will be where I will be laid to rest—in this ancestral ground—to be cared for by my descendants. This is the place where I am unquestionably welcome and where I belong despite where I might live my everyday life. For the indigenous Maori people of Aotearoa New Zealand belonging is always about the people and the land.

Elizabeth McKinley (Ngati Kahungunu/Ngai Tahu)
Faculty of Education
University of Auckland

EILEEN CARLTON PARSONS

FUNCTIONING IN TWO DISPARATE WORLDS

NO BLACK DOLL, NO DOLL AT ALL

On my fourth birthday, Mom, Dad, and I visited the local department store located just outside our rural, poor and working class community. As we approached the toy section, I tugged at the hem of my mother's dress and whispered, "Momma, am I going to get a doll? Am I?"

As Mom started to respond, my dad pulled a doll from the shelf. I immediately pushed the doll away and remarked rather loudly, "I don't want no white doll. I want a black one."

Under the disapproving gazes of white shoppers, Mom and Dad stood astonished not knowing what to do. My father asked a store clerk if the store carried any black dolls. The clerk responded, "No. We don't carry black dolls here; they usually don't sell."

I tugged at Mom's dress and expectantly asked, "Momma, am I going to get a black doll?"

Mom paused while she pondered the best way to explain to a four-year old the political and economic reasons why black dolls were unavailable. She finally replied, "No honey, they don't carry them here."

I began to cry and ask why they didn't have black dolls. After I realized my tears would not result in the store producing a black doll for me, I dried my eyes and stated my resolution, an affirmation of what is black rather than a rejection of what is white: "no black doll, no doll at all."

* * *

A little over a decade ago, I used this episode, often recounted by my mother at family gatherings, as the introduction to my dissertation. At that time, the story captured the absence of black people's voices in the science education literature. Although we are no longer completely silenced, our experiences and perspectives are marginalized in science education. I have spent a lot of time and energy over the past ten years fighting the notion that I, as an African American, am an outsider in science education, in particular, and in the academy, in general. Over the years, many senior scholars of color, especially African Americans, in the various disciplines, spoke to me about this outside status.

One scholar questioned why I was surprised by the inequality (e.g., differences in teaching load and treatment by students) and inequity (e.g., noticeable discrepancies in resources). Even though it was several years ago, I clearly recall his words and the shock I felt upon hearing them: "You're black. Expect it." The words engendered considerable angst and some anger. Why should I expect it? I've worked alongside and successfully competed with non-blacks all my life. I believe

K. Tobin, W.-M. Roth (Eds.), The Culture of Science Education, 355–366.

myself to be different, a distinction that offers an invaluable perspective, and at the same level of worth as my non-black counterparts; what I expect is to be treated accordingly. Over time, the angst and the anger subsided and my experiences in the academy substantiated the excruciating words of the more seasoned scholar of color. Eventually, I accepted the notion that my beliefs do not alter my reality of working from the margins in science education.

My research interests are only valued within certain and very small circles; I continue to work in isolation to pursue those research aims; substantive mentoring that leads to collaborative research and publishing is ever elusive; and publications of my research in science education research journals occur only when I couch the work in theory mainstream to the science education community. Marginalization makes the task of being in science education a bit more arduous and progress a little slower, but I am positive and hopeful that various efforts will inevitably alter the mainstream such that the space of marginalization is smaller in the future. In the meantime, struggles emanate from this state of marginalization.

My response as a four-year old, "no black doll, no doll at all," to the store event in the 1970s typifies my current struggles-affirming who I am in a milieu in which my identity as an African American is seldom acknowledged, welcomed, and included. W.E.B. Dubois' (1969) construct of *double consciousness* aptly characterizes my predicament. In this autobiography, I chronicle my journey. I share the challenges that necessitate the existence of a double consciousness and subsequently functioning in two disparate worlds—a world that coincides with my identity as an African American, referred to as the first world, and the world that encompasses science education, called the second world.

THE TWO WORLDS

My Core Identity: The First World

At 2:17 P.M. on August 17, 1967, a 23-year old black woman with a tenth-grade education gave birth to her sixth and final child. The anxieties of how to pay the bills on a domestic and furniture worker's wages dissipated temporarily as the 30-year old black man who dropped out of high school to help support his ten younger siblings enjoyed the long-awaited arrival of a daughter—the five previous children had been boys. Even in the dearth of material possessions and wealth and in the plenitude of social ills often associated with poverty (e.g., substance abuse, violence), the sixth and final child grew up in a stimulating, vibrant, expressive, communalistic, and spiritual environment.

With meager earnings, no one family living within the small community appropriately called "Thankful," could actually make ends meet every month, month after month. So when there were no more pennies to pinch or dollars to stretch, another family would pull from their scanty possessions to tide us over until the next payday. Money was only a part of the sharing. The offerings also included clothes, food, and homemade remedies for any ailment (utilization of the medical profession occurred only in emergencies). In the midst of poverty, the spirit of giving and the spirit of people thrived. This thriving was most apparent in the gathering of community folk.

The coming together of community residents occurred often-on the front porches of houses lacking indoor plumbing, in the old school closed after desegregation, betwixt the pews of the Baptist church, or at the street corner usually manned by the so-called community gamblers and drunkards. Sharp but playful wit, boisterous laughter, and rambunctious jostling were ever-present in the gathering of relatives, neighbors, and friends. Marked by their flare and flamboyance, the telling of happy and sad stories, and the congenial, sometimes contentious, freeflow sharing of personal experiences captivated those present; not evident by the listeners' silence but by their continual insertions, corrections, and flavorful commentary. During this time of being together, the people did not speak of the fruits of their labor, the trials of their work, or the tribulations of their mere existence; if mentioned at all, they gave thanks to God for His blessings or expressed faith that "better times were a' coming."

Genuine, open, and frank sharing, emotional expressiveness, and esteeming the needs of others over one's desires became integral parts of my worldview. I naturally adopted a lively communication style replete with colorful language, provocative and evocative undertones that featured episodes and themes rather than singular topics. As I encountered the second world, the perspective derived from my life experiences within the rural, African American poor and working class community operated as a taken-for-granted assumption.

Identity Development: Mediated Forays into the Second World

In ways still incomprehensible to me, obstacles in my life inevitably became stepping-stones. Unlike the other students of color around me, my distracting and disruptive behavior in school did not adversely impact my achievement. When a principal sought to levy my misbehavior against my academic performance, other individuals with greater authority and status intervened on my behalf. When teachers and counselors acted to keep me in standard academic tracks as a high school sophomore, a European American English teacher and special education coordinator with the backing of the African American principal made the academically gifted classes accessible to me. On one hand, enrollment in the upper echelon classes opened the door to advanced levels of English, math and science, the first key to pursuing science and science education as a career. Enrollment also provided various opportunities that led to complete funding of my postsecondary education by way of academic fellowships. The European American "Booster Club Mom" helped me to maximize these opportunities by purchasing needed clothing suitable for interviews and luggage for traveling. On the other hand, enrollment in the academically gifted classes unleashed a plethora of unexpected challenges of which low expectations were the most daunting.

When I was told that I could not win a student election like my academically gifted peers, I campaigned until I won. After I lost two elections at the predominantly white high school, students elected me as class president in my junior year

357

and student body president in my senior year. When I was told that I could not take an advanced science elective because I lacked the necessary math prerequisite, I enrolled in summer school, took the math course, made an "A," and subsequently enrolled in calculus along with the advanced science elective. When I was told that I should be pleased to receive a "C" on a paper, I worked until I won numerous essays and oratorical contests and graduated as valedictorian with a report card of only "As." The most vivid recollection regarding low expectations occurred on the night I graduated from high school. On the night of my high school graduation, a European American teacher pulled me aside. He told me that "I had my time in the spotlight" (I had won over fifty percent of the school's awards) and that I should not expect the same at The University of North Carolina at Chapel Hill. At that moment, the stubborn, resolute four-year old re-surfaced and responded, "We'll see."

After the high school graduation, my community celebrated the successes, thereby mitigating any lingering effects of the teacher's discouraging remarks. The community also pooled their resources by collecting linens, towels, and toiletries-anything they thought a youngster might need for college. The elderly within the community provided sage advice (e.g., always remember who you are and where you come from) and admonishments (e.g., don't become an educated fool). The middle aged paraphrased Bible scriptures to encourage me (e.g., in all things we are more than conquerors through Him who loves us) and my peers shared in my anxiety and excitement of starting anew.

At UNC-CH, low expectations were not an issue; perhaps, my early placement into the UNC-CH honors program served to dispel them. Who I was and what I would do were of greater concern. During my undergraduate studies at UNC-CH, I began to develop a racial and cultural consciousness, an awareness that my rural, African American community did not cultivate. Unfortunately, I did not utilize these lenses in the continual process of developing my identity and in understanding the disconnection I experienced in my selected field of study, chemistry. Instead, this racial and cultural consciousness served as an impetus for formally declaring science teaching as a major. Even though I continued to take chemistry courses (I took graduate level chemistry courses as a college senior), I viewed science teaching as a more viable avenue for addressing racial and cultural concerns. My heightened racial and cultural awareness also motivated me to become involved in predominantly African American and predominantly European American organizations on campus.

Pledging Delta Sigma Theta, a predominantly African American sorority, and participating in the black Student Movement enabled me to attain some semblance of connection, a state of being that existed in my first world. Within these groups, I found that stimulating, vibrant, expressive, communalistic, and spiritual environment reminiscent of Thankful, my home community, and a space to just "be": unquestionably, unconditionally valued.

I also became actively involved in predominantly European American organizations like student government. These groups introduced me to the intricacies of the second world; here, on a very superficial level, I became cognizant of a different set of rules, values, and norms. I did not have the language to articulate and label the ethos of the second world but I experienced them as (a) individual interests outweighed any group concerns or benefits, (b) sharing of faith in something beyond the observable and the expression of emotions were deemed irrational and inappropriate and undercut an individual's position, and (c) communication as well as any other actions were to conform to some standardized and orderly protocol. Via my active participation in the predominantly European and African American organizations, I achieved the status of student leader which resulted in the procurement of university and organizational awards and induction into several honor societies recognizing outstanding scholarship, leadership, and service. However, my campus involvement did little to prepare me for the identity crises that awaited me in graduate school.

Upon my arrival at Cornell University in Ithaca New York for graduate studies, I contacted the chapters of *Delta Sigma Theta* and the *Black Graduate and Professional Student Association* and became an active member. Slowly, the academic and psychological demands of graduate studies required more and more time. Eventually, no time was available for participating in the predominantly African American communities and I found myself functioning in the second world all of my waking hours. After two years, the kindness and guidance of Joe Novak, the chair of my graduate studies, could not alleviate the overwhelming disconnect I experienced from living solely in the second world. In this second world, statistics silenced different perspectives, experimental research designs squelched dynamic situations, and stoic, detached reporting recast virile events as sterile. In light of my stellar academic performance in graduate school, I decided that a PhD in science education was not for me. I graduated with a MSc degree in science education and returned to North Carolina (NC) to pursue studies in education administration. One year later, I returned to Cornell to finish what I had started—a PhD in curriculum and instruction/science education with minors in program evaluation and education administration. The expectation to live solely in the second world and the structures to reinforce such an expectation had not changed since my first stay at Cornell but several outlets via course work were available during my second tenure.

During my second tenure at Cornell, Deborah Trumbull, an ethnographer and science teacher educator, directed my studies. At the outset, Deborah recommended courses that deviated from the traditional, quantitative research paradigm. Taking courses in the sociology of science, divergent epistemologies, and qualitative research was the hook I needed; I began to think that perhaps there was a place for me in science education after all. The professors in these classes encouraged the expression and critique of different views and allowed me to use the communication of my community in my writing. In addition, Deborah introduced me to readings regarding the identity development of African Americans. Although my experiences did not align with the readings (e.g., the first stage of development was to hate being black) and I disagreed with much of the writings, the work encouraged me to seek out other literature.

In this search, I stumbled upon the work of Wade Boykin, an African American psychologist, and the work pertaining to the dominant culture in the United States.

The aforementioned work not only provided a framework for understanding my own development but also now serve as the conceptual underpinnings of my work as a researcher. Boykin's (1986) constructs of *verve, movement, expressive individualism, affect, communalism,* and *spirituality* captured what I lived in my first world and the notions of rugged individualism, materialistic conceptions of reality, cognition emphasis, and conformity characterized what I experienced in the second world. Lastly, realizing that she could not fully understand my experiences, Professor Trumbull introduced me to an individual who would impact my life for many years to come: Mary Monroe Atwater.

CAREER TRAJECTORY AND THE SECOND WORLD

In the Beginning

Near the end of my doctoral program, I married my hometown sweetheart of more than ten years, an individual rooted in and intricately connected to my first world. After I received the PhD at the age of 26, I returned to North Carolina to live with my husband. Motivated by and adhering to one edict of my first world, esteeming the needs of others over one's desires, I elected to return to high school science teaching armed with the belief that my expanded experiential and knowledge bases would enable me to make a difference in the lives of students. While teaching high school science, I had my first experience of working in the second world and living in the first. I contrived certain boundaries, mostly demarcated by time, for functioning in the first and second worlds. Because of a growing desire to conduct research, I eventually left high school teaching and secured a tenure-track, assistant professor position at Lenoir Rhyne, a small, liberal arts Lutheran College located in Hickory, NC.

Like most assistant professor positions at small teaching institutions, I fulfilled numerous and varied obligations associated with working with undergraduate students and taught many courses, in and outside my area of expertise. During this time, the desire to do research intensified and I began to understand the implications of my career choice to return to teaching after the completion of the PhD I called upon Novak, Trumbull, and Atwater for advice on how to undo what I had done; I needed direct guidance on how to go from a focus primarily upon teaching to one on research. The only way out was to publish and they advised me to begin with the publication of the dissertation. At this juncture, I recognized that I had spent a lot of time during my doctoral studies determining who I was in relation to the second world I was about to enter but not enough time on what I needed to do in order to succeed in that world. Professionally alone, living in my first world where an understanding of what I do is captured in "she teaches" and works at an institution where research is secondary, I faced seemingly insurmountable obstacles figuring out what PhDs in science education research did and how they did it.

While writing, I struggled to perform my duties as an assistant professor and sole science educator at Lenoir Rhyne. Cognitively, I knew I should reduce the amount of time I invested in teaching so I could write but I could not compromise my sense of commitment to do my very best in my endeavors, a by-product of spirituality esteemed in my first world. By traversing the self-imposed time boundaries I established between the first and second worlds, I managed to publish parts of the dissertation in the *Journal of Research in Science Teaching*, a top-tier journal in science education. I presented and discussed how the images of the scientists of academically gifted, African American high school females were racially and culturally embedded and the implications of such images for science instruction and career choice. During this time, I also received a small research grant of $410 from the *Institute for Excellence in College and University Teaching Small Grants Program for Research*. I used the grant to investigate my own teaching, the effectiveness of lecture versus role-playing. The project resulted in an on-line publication. Simultaneously with the grant-sponsored research project, I conducted an independent study on pre-service elementary teachers' beliefs about the nature of science. I presented the findings of the beliefs project at the annual conference of the National Association for Research in Science Teaching (NARST).

The conference presentation was my first and I did not expect the communication norms to differ from the ones to which I was familiar. Pulling from the communication repertoires within my first world, I prepared the presentation to be very interactive, creative, and entertaining in nature. Minutes prior to the presentation, the presider informed me that the presentations would occur as a group and presenters would give a five to ten-minute talk on their respective papers. In a panic, I approached Deborah Trumbull, who graciously attended the presentation, about what to do. She quickly described what should occur: a structured and linear oration commonplace to the second world. At that point, I realized that in order to learn what PhDs in science education research did and how they did it, I needed to attend NARST conferences and become involved in NARST; Mary Atwater was instrumental in these efforts.

After I attended my first NARST conference, Mary recommended me for one of the NARST committees. Serving on the Outstanding Paper committee was my first entry into NARST and provided an opportunity for me to interact and converse with science education researchers. My very first intimate encounter with a colleague was reminiscent of the low expectations I encountered during my high school years. I described my professional situation to the European American with whom I had previous contact and she informed me that it was extremely difficult almost impossible to move from a teaching to a research institution. She advised that I focus upon the scholarly activity the teaching institution valued. Somewhat despondent, I stated that my life was full of achieving the impossible. The spiritual grounding from my first world re-surfaced and Biblical scriptures replaced any thoughts of defeat: "what is impossible with man is possible with God" and "everything is possible for him who believes." Strangely motivated by the daunting challenge that lay ahead, I returned to Hickory determined to secure a tenure-track position at a research-intensive institution.

Upon the recommendation of Deborah, I applied for a Ford Foundation Postdoctoral Fellowship for minorities in science and engineering that would fund one year of full-time research. Because the funds were earmarked for the sciences and my research was in the field of science education, I did not anticipate an award. To my surprise and delight, I won the first $40,000 postdoctoral fellowship awarded to an individual in science education. The Ford Foundation Postdoctoral fellowship not only provided funds to conduct full-time research for one year, but it also offered an opportunity to interact with other scholars of color.

In addition to the $40,000, the Ford Foundation Postdoctoral Fellowship provided financial support to attend Ford Foundation conferences. Because these conferences showed me that my long, arduous, and psychologically demanding journey in the academy was common to people of color, especially African Americans, I no longer felt alone or that I was somehow deficient. From these conferences, I developed another significant support relationship with a more senior member in the academic community-a voice that would keep me from leaving the academy when I became weary of trying. In essence, the Ford Foundation Postdoctoral Fellowship facilitated my first step into research by buying the time I so desperately needed and by offering psychological and emotional support. With the backing of the Ford Foundation, I started the journey to establish a research agenda. In a trial-and-error, learning-by-doing mode, I implemented a research project that investigated the learning environments of seven elementary school teachers instructing in urban and rural areas of NC.

Venturing into Research

In order to conduct the Ford-funded research project, I took a one-year academic leave from Lenoir Rhyne College, which occurred shortly after my reappointment and the completion of my third year of employment. After the leave, with the unconditional support of my husband, and the long-distance, informal guidance of Deborah and Mary, I resigned from the tenure-track position so I would have sufficient time to analyze the project data and disseminate the study's findings.

While working on the Ford postdoctoral data, I accepted a part-time instructor's position at the University of North Carolina at Charlotte (UNC-Charlotte). In this part-time role, I had my first opportunity to work with graduate students at the master's level. I found the work with graduate students stimulating, challenging, and very satisfying. In addition to acquiring a position where research was a part of the responsibilities, the work with graduate students became another factor for pursuing employment at a research-intensive institution. After a little over a year in the part-time position at UNC-Charlotte, Joe Novak informed me of an opening at North Carolina State University (NCSU). In the tenure-track assistant professor position at NCSU, I began to realize my potential and to define the kind of work I wanted to comprise my professional achievements. Unlike many research institutions, at NCSU science education was not subsumed under a curriculum and instruction department; as such, I was part of a science education community. Being a part of this community at NCSU enabled me to make my most significant progress to date.

First, as my department chair, John Penick helped me to understand the culture and norms of science education. Via our numerous informal and formal conversations,

John offered an enlightening perspective on what science education researchers did and how they did it.

Second, John connected me with others in science education, a link not established in my previous positions. With the comfort and confidence that came with interacting with key leaders in science education, I became more assertive in my participation in NARST and the Association for Science Teacher Education (ASTE). Consequently, the ASTE membership elected me for its Board of Directors and the NARST election committee nominated me for its board. Because I was uncertain about accepting a nomination for a NARST Board seat as an untenured assistant professor, I elicited the advice of individuals with more years of experience in NARST. I was encouraged not to accept the nomination but eventually followed the recommendation of the NARST nomination committee chair. Although the NARST membership did not elect me to the Board, the mere honor of the nomination motivated me to become more involved in NARST and to learn how it operated. Prior to the nomination to the NARST Board, I actively served on numerous NARST committees and as coordinator of conference strands; after the Board seat nomination, my involvement in NARST took a significant turn. I assumed an active role on the Equity and Ethics Committee and fervently worked to help other underrepresented scholars of color to take a path more direct than my own into postsecondary science education. In this capacity, I met Okhee Lee, an individual who continues to offer sage advice as I forge ahead in this second world.

Third, with regard to research and publishing, I learned many lessons from my NCSU colleagues and reaped from their generous acts. Because of their understanding of the rules, values, and norms in science education, my NCSU colleagues protected me, to some extent, from excessive teaching and administrative duties, a necessity for most assistant professors. Consequently, during my first two years at NCSU, I published several articles from the Ford-funded project. These articles were published in national (e.g., *Journal of Research in Childhood Education*) and international (e.g., *Learning Environments Research*) journals and discussed the different forms of democracy, the mediation of white male privilege, and the manifestation of culturally relevant caring that existed within the elementary school classrooms. During this time, I revisited the work I started as a doctoral student by publishing more practitioner-oriented pieces about culture and science teaching. Subsequently, I decided, against the sentiments of some, to more fully develop a research agenda around culture, race, and African American students in science. The receipt of two prestigious research grants from the American Educational Research Association, *Office of Educational Research and Improvement,* and *Spencer Education Foundation* funded the initial establishment of this research agenda and alleviated any nagging doubts regarding the worth of such an endeavor.

Lastly, I attributed much of my growth to my work with NCSU doctoral students in science education. During my four years at NCSU, only doctoral students of color elicited my participation on their committees as chair and co-chair. Three African American females designated me as chair. These students' research interests ranged from diversity issues in informal science education to racial identity to reflection—all areas somewhat tangential to my area of expertise. In the capacity as chair, I adhered to the tenets of my first world that esteemed the needs of the group over my individual agenda, and proceeded accordingly. I viewed it as a duty and privilege to help facilitate these students' growth in their areas of interest. In order to help these students develop, I became more reflective about research in science education and what it meant to be a person of color, especially African American, in the field. I quickly learned that it was not enough to explain the written and unwritten rules (at least the ones to which I was currently privy) to the doctoral students of color, to engage them in research and writing, and to broker relationships with others in science education but it was imperative for me to prepare them for a world in which they would likely be marginalized. I considered it a responsibility to assist them in managing the contradictions between rhetoric and practice in science education, to facilitate a clear understanding and strong sense of self in the midst of these contradictions, and to help them make informed choices by considering the aforementioned.

Careful not to taint the students' experiences, I delayed sharing my views and perceptions until they encountered challenges common to individuals of color in science education. These challenges include but are not limited to (a) navigating the institutionalized and informal structures and mechanisms that diffuse interest in and pursuit of research positively centered upon people of color, (b) working in isolation with limited resources and access to such resources, (c) being invited to collaborate with others in servitude where menial labor rather than intellectual contribution is expected, and (d) fulfilling diversity-related responsibilities in addition to (rather than in lieu of) the typical obligations associated with work in science education. In assisting the doctoral students in deciding how they would address these challenges and others, it was necessary for me to be supportive and strict, brutally honest yet encouraging, and understanding but tough. On occasion, I also needed to serve as their advocate—a precarious situation for an untenured assistant professor. In the end, my efforts with the doctoral students made my work meaningful; for me, my work with the doctoral students provided greater clarity, self-understanding, focus, and determination. Motivated by the previously described internal changes, I made a professional change that removed me from the science education community at NCSU. In July 2005, I started a tenure-track assistant professor position, presently the only one in science education, at UNC-CH, a leading institution in addressing various dimensions of social justice.

MANAGING THE CONTINUOUS TENSIONS

Since I began my journey in postsecondary science education a little over ten years ago, I have made several professional moves. I began as an assistant professor in a tenure-track position at a teaching institution, worked part-time at a doctoral institution, and eventually secured a tenure-track assistant professor position at a research-intensive university. At the time of this writing, I finished my first semester at yet another research-intensive university. As I reflect upon my journey, I realize that many of the challenges were constant from one institution to the next. These challenges emerged from functioning in two distinct worlds. One world, the first world, corresponds with my identity as an African American and the other, the second world, is the world in which science education is a part. Such constancy across the varying, predominantly white institutions and the presence of the former challenges to date indicates that reconciling the two disparate worlds in which I live and work is improbable. Perhaps, the best I can hope for is managing tensions such that principle or progress, though slower than I prefer, are not sacrificed in the process.

One value that appears integral to success in the second world is rugged individualism, esteeming one's own interests above all else and acting to advance those interests. Individualism is a direct contrast to communalism and spirituality, values of the first world that partially define who I am. In mediating the tensions that arise from the values clash, the approaches I implement take several different forms. When working with colleagues, I sometimes forfeit my undertakings. In my interactions with doctoral students, I mostly alter my interests. For instances when my interests are not consonant with the concerns of others, I often pursue them alone as a way to avoid exploiting or oppressing others. In utilizing these strategies, on one hand, I retain my identity while upholding a principle of my first world but on the other hand, I compromise productivity.

A second value of the second world from which daunting anxieties emerge is conformity. Conformity, the expectation to adhere to existing norms and to perform in manners similar to others, is dialectically opposed to two aspects of the first world, expressive individualism and affect. Expressive individualism emphasizes distinctive, genuine personal expression and affect stresses openness and the sharing of emotion. To alleviate the strain of the competing values, I conform when an important principle is not jeopardized. In situations where a principle is endangered then I clearly explain the rationale behind and the plausible benefits of my non-conforming positions and corresponding actions.

Lastly, one norm in the second world that is highly problematic is communication. The mode of communication valued in the second world is written. This world's communication is linear, focuses upon one topic at a time, and makes meaning and relationships explicit. When communication is oral in the second world, speakers act as spokespersons and use language that separates them from the topic, lessens opposition, and eliminates emotion. In contrast, communication in the first world is oral. In the communication of the first world, subjects change from one moment to the next, the topics are revisited in a circular fashion, and relationships among topics are implied. When discussing topics and ideas, speakers from the first world act as passionate advocates in their delivery, creatively use language to present positions, and intentionally incite emotion. To date, I have not found an adequate approach for managing the incommensurability of the two communication modes. In order to progress in science education, I must publish; to publish, I must communicate in the manner valued by the second world. In my pursuit to communicate in a way suitable for science education, in particular, and the academy, in general, I have lost the communicative style of the first world. Whenever I am in the presence of people rooted in the first world, the majority of the significant others in my life, I am no longer able to spontaneously participate in the banter. I no longer possess the words and the facility.

The psychological toll of this double-consciousness and functioning in two different worlds is tremendous. W.E.B. Dubois (1969) speaks to this psychological turmoil in his quote "One ever feels his twoness—an American, a Negro; two warring souls, two thoughts, two irreconciled strivings; two warring ideals in one dark body, whose dogged strength alone keeps it from being torn asunder" (p. 16). At times I wonder: How long can I produce the expected quantity and quality of work under such psychological strain? Will I eventually become a statistic, another qualified scholar of color who sees opting out of this second world as the only means for managing this double-consciousness? Or, will I be pushed out by the denial of tenure? For now, the tenacious, resolute persona of that four-year old who declared "No black doll, no doll at all" is still very much alive and I plan to persevere as I have done so many times before. When replenishment is needed, I will draw from my spiritual grounding; the unconditional support of a husband, extended family, and community; and a few folk in the academy who do not hesitate to tell me like it is.

Eileen Carlton Parsons
School of Education,
The University of North Carolina at Chapel Hill

REPRODUCING SCIENCE EDUCATION

Science education constitutes a particular subculture of largely white middle-class character. As all cultures, it has its own means and practices of reproducing itself. But in a dialectical perspective of culture, reproduction also means active production that is never the same, and therefore, it means transformation. This is where we can hope to bring about changes as a consequence of which other values of systems of thought come to be acknowledged and integrated. In this commentary, we discuss the role of bodies in knowing, cultural relevance, the role of the family in the reproduction of society, issues of race, questions of race, difference, equity, and hybridity.

BODIES OF KNOWLEDGE, EMBODIED KNOWLEDGE

Largely being a servant to the natural sciences—the reproduction of which has been the major raison d'être for science education not in the least since Sputnik—has influenced the nature of the dominant epistemology not only in the beginning of science education but also to the present day. It is true, the discourse has changed such that now, in the wake of constructivism, students are allowed to *construct* knowledge, though in the end, what they are to construct is predetermined, physically embodied in the reform texts policy makers and state (provincial) curriculum departments produce. Historically, the main goal of science education was to impart the body or bodies of knowledge articulated in the textbooks constructed to meet the guidelines laid out in state (provincial) policies. But textbooks do not embody knowledge, even though the materiality of the text constitutes a body; the signs in these books becomes knowledge when real living beings mobilize them for the pressing issues at their hand. It is in their bodily actions—doing, speaking, writing—that people mobilize and enact the "bodies" of scientific knowledge.

Virtually unknown to many science educators is the work related to the embodied nature of knowing, which means that we understand and have practical comprehensions in terms of our early experiences of the material and social world. If this is so, then the very ways in which we come to learn and know are culturally specific. Having to learn something that is foreign to and incompatible with your ways of being constitutes symbolic violence.

But knowledge transcends the singularities of individual bodies and, as Elizabeth McKinley points out, their corporeality. In fact, it is possible to know only when there are multiple bodies, a plurality of bodies such that touch and proximity are fundamental to the possibility to the experience of something as other. More so, otherness is the very precondition for anything such as a Self to emerge; and it is the figure of the Other, the other's face, that enables the possibility of a thinking subject. Knowing—doing, talking, thinking—is transactional, undissociably relating the human subject with its object, and therefore the body to other bodies and to itself.

Some science educators might all too quickly dismiss the thought of knowing as embodied and the role of bodies in ways of knowing pointing out that science precisely works because it is context independent, that is, also independent from the bodies of scientists. However, like fish in the water, they tend to forget that the very system of schooling is set up such as to *discipline the body*, that the academic disciplines exist in the way they do because the bodies of the practitioners have been disciplined in the course of twenty or more years of school experience (12 years of elementary and high school, four years for a bachelors, two years for a master's degree, and three or more years for a PhD). The experience of schooling disciplines the body and regiments its way of operating, including its thinking. First, schooling requires hours of seating without movement, externally regulated periods of engagement with specific subject matter independent of emotional, epistemic, and other (bodily) needs, sudden switches between subject areas, working under supervised and controlled conditions, working independently and in silence disconnected from the more frequent social nature of human activity, and so forth. Only those who become academics generally and natural scientists and science educators specifically successfully complete this rite of passage—which essentially instills forms of rigor, working styles, and ways of relating to others. This disconnection from the otherwise normal ways of being finds its equivalent in the claimed independence of the contents of scientific knowing from the scientists as persons generally and from human bodies more specifically.

Related to the bodily and embodied nature of knowing is the idea of separate worlds, or, as others expressed it, separate realities. Eileen Carlton Parsons writes about these different worlds, engendered by the fact that we have bodies that are through and through products of their surrounding culture. Here it is necessary to introduce the important distinction that German phenomenological philosophers have introduced between *Körper* (body), a material body, and *Leib* (body), a term inadequately translated as *flesh* to denote the sensual and sensing body of a human being. Phenomenological philosophers—e.g., Edmund Husserl, Maurice Merleau-Ponty—have been concerned precisely with the interlacing of the flesh, language, and the world; sociologists—e.g., Pierre Bourdieu, Harold Garfinkel—have made the embodiment of knowledge the core of their theories. The human body qua material object is not cultured, but the flesh is, because it is endowed with senses, fully it is a cultural-historical entity. The particularity of Parson's and McKinley's experiences arises from their being in flesh, which goes beyond, as McKinley notes, mere corporeality. Flesh, because of its association with the senses, is the condition for sense, and therefore to the forms of consciousness it enables. Not surprisingly, therefore, both authors in this part of the book allude to or explicitly articulate the different form of consciousness that they experience when participating in the first, their Aboriginal or African American worlds, and the second world, constituted by the still dominant and domineering (i.e., colonizing) Western culture of the industrialized nations where they are citizens.

The question of the body poses itself when we think about the relevance of everyday experience in the science education curriculum. Are science curricula appropriate for students living in situations where they are more concerned with not going hungry than with learning something seemingly esoteric such as the number of neutrons in heavy water? Are hungry and ill-clothed bodies ready to accept the scientific bodies of knowledge? The fact is that in 1997, The Center for the Future of Children reported that 21 percent of U.S. children live in poverty, but the rate is much higher for non-whites: 47 percent of African American and 40 percent of Hispanic children live in poverty. Are science curricula appropriate for these children? How are the experiences of these children used as resources to allow them learn some form of science? The following lyrics are from the rap version of *Anything* that Sigel directs toward his mother. This question is particularly salient in the apparent gap between the experience young rappers express in their songs and what science educators attempt to achieve when they try to get students to perform according to policy documents or when they try to get students ready for high-stakes examinations. One these rappers is Beanie Sigel, who is of relevance to us because his home is Philadelphia, where we have conducted a lot of research concerning science education for African American students most of whom are living in poverty.

i know i'm ya little baby but these streets raised me crazy product of my environment/
nothin could save me thanks fah lettin me bloom, for your wisdom, for your womb/
for the roof over my head, for my shoes, for my bed/
for the most important lesson in life was when you said, strive for what you believe in,/
take those you can achieve 'em/
thanks for the days you kept me breathin, when my asthma was bad and my chest was weevin/
thanks for the look of love just as i was leavin on nights you thought i wouldn't come back/
and left you grieving thanks for holdin down the household when times was bad,/
as a man, i apologize for my dad/
when the rent was due, you would hussle like a pimp would do no wonder what life meant for you/
your a queen, you deserve the c.r.e.a.m. everything that gleems, everything that shines,/
everything thats mine

(http://www.lyricsdir.com/beanie-sigel-anything-lyrics.html)

Like so many of his raps, the lyrics contain obscenities while unfolding a strong social commentary on the ways in which, in this case, African Americans live in inner cities in the United States. The words of the rap point to a deplorable situation in which poverty robs people of their health and dignity—seemingly driving them to bleak lives with limited opportunities for mobility through a class structure that is oppressive. Sigel specifically calls attention to his asthma and alludes to his mother's wisdom. What do we know of the incidence of asthma among urban youth and to what extent are pervasive urban problems like environmental asthma included in science curricula in urban schools? Would students knowing about the possible causes of environmental asthma lead to better health? What is the distribu-

369

tion of asthma in different parts of large inner city regions and how does its incidence correlate with poverty? Are questions such as these legitimate parts of a science curriculum? How can we justify epistemologies and ontologies that support curricula based on science stripped free of contexts in which people encounter it? How can it make sense to education policy makers to advocate curricula and supporting resources that are not connected directly to the lives and aspirations of learners? If science education is to serve people in *their* lifeworlds then it is important that what they learn can be enacted throughout their lifeworlds as they pursue their goals. Hence, there needs to be a better overlap with the goals of social life—allowing those who seek to learn the opportunity to study forms of science that are central to their lives and the social mobility they seek.

Today, despite a lot of lip service to build on students' experiences to support their science learning, the negation of students' situated knowledges—not only those students of different cultural origins but also those of the white working class, and many female students—is the norm. Conceptual change theorists continue to be concerned with eliminating (mis-, alternative, naïve etc.) conceptions, which means precisely the negation of the knowledges that students have evolved as part of their everyday life outside schools. The ideas put forward by Matthews, as cited in McKinley's chapter, are abhorrent and misinformed. How can we justify science curricula that are disconnected from the lives of learners—not allowing them to use what they know, can do, and are interested in learning as a framework for science curricula? If science education is to serve all learners then it cannot be more of the same. Current models only seem to perpetuate inequities and are grounded in deficit perspectives and assumptions that learning is optimal when students are controlled. It is time for change to models that acknowledge and build on the strengths in what people know and can do.

Beanie Sigel's lyrics also make salient the role of the family in the reproduction of poverty and the struggles to make it out. He is thankful to his mother, but, as a man, apologizes to her for his dad. The family also is one of the common threads in the present as in other auto/biographies of this volume. Generations of science educators were brought up in families where one or both parents had not completed high school. In McKinley's situation, the mother spent one year in high school, the father was more frequently absent than attending; in Parson's family, her mother made it to tenth grade, her father dropped out. Although similar situations existed in the lives of other science educators—Michael Roth's mother and father never went beyond tenth and eleventh grade, in part mediated by the post-WWII situation in their native Germany—being born to parents without a high school diploma is a much more prevalent experience to children from Aboriginal peoples and African Americans. In 1999, 18.5% of those under the age of 18 were living in low-income and poverty situations. The poverty rates are almost twice in urban areas (e.g., Toronto, Montreal). A 2001 OECD report shows that Australia, Canada, Germany, and the UK have poverty rates around the 10 percent mark, whereas the poverty rate in the US is 17 percent and therefore between the less industrialized nations of Turkey (16.2%) and Mexico (21.9%).

Such numbers are important because in many instances, including those of members of the white European cultures, a situation of poverty and life in the lower social classes marks the consciousness-forming experiences. Interestingly, though, one notes the frequent presence both of a desire to know among the parents and their efforts to make their children succeed in school systems that were—and often still are—inherently biased against them.

Parent backing and, where possible, parent help in learning is especially important for those students who, unlike McKinley, Parsons, and Latin American science educators featured in this book, (initially) are unsuccessful and even repeat a year as Michael Roth had to do, though granted, this repeated fifth grade was in an academic stream. To make it, culturally others need to demonstrate academic success, which inherently means—at least in the past—success within a system that is for them culturally other. This means that such students, in part, have to be or become like the white people, something other students remark and point out as "acting white."

DIFFERENCE, EQUITY, AND HYBRIDITY

All too often, deviations from a pale skin color are used to construct boundaries and produce inequities. As gender, race is being used as a resource for keeping some human beings from accessing opportunities already available to others. In the effort to introduce social justice, however, much of the theorizing begins with ontologizing the *same*, which becomes the reference in establishing difference. To overcome inequities in science education, both in terms of learning science and in opening access to academic careers, we need a different framework for conceptualizing difference as something that exists in and for itself.

Issues of Race

Racism is clearly at play in each of the chapters, though those who are involved in the day-to-day transactions with McKinley and Parsons would no doubt strenuously deny that any of their practices are racist. It seems as if the source of racism is hegemony, schema and associated practices that are accepted as normal and sensible, and yet at the same time systematically disadvantage participants according to their race. Is it possible that the good intentions of science educators are examples of racism? Is it hegemony when Parsons is advised to follow pathways that would maintain her employment in a teaching university rather than strive to do what it takes to earn a position in a research-intensive institution? We wonder if such advice is based on deficit perspectives by which well intentioned actions are taken based on what individuals cannot do and do not do rather than on what they can do and want to do? Deficit models have saturated practices in science education and are at the core of many of the curriculum reforms and research programs associated with conceptual change. For year after year, decade after decade, science educators are exhorted to try harder, accepting paradigms that embrace deficit perspectives and teaching and learning models associated with teachers obtaining and maintaining control over students. Similarly hierarchies are created within educational systems, holding those higher in the hierarchy responsible and hence accountable for the accomplishments of those below them in the hierarchy. Rather than continue to try harder with the same game plan we suggest it is past time to alter the game plan. It is time to set aside deficit perspectives and models for social life that fail to account for the agency of all participants—not just those designated as symbolically powerful.

We are struck by McKinley's descriptions of how she is counted out in a double sense—by virtue of her race and gender. Her remarks are reminiscent of the black female principal at City High, where we did much of our research in urban high schools—referring to one of our student-researchers the principal pronounced: "She's got three strikes against her. She's black, she's female, and she's poor. In making the comment the principal was stating a truth based on her lifeworld—as a black female who had grown up in poverty she knew only too well about the social forces that held her back and similarly disadvantaged her students.

The triple quandary faced by female black students from inner city neighborhoods is no secret. Their struggles are well represented in rap music, sung by thousands of students daily in inner city high schools. Beneath the anger encapsulated in the obscenities of the lyrics, the violence and often the misogyny are example after example of systemic racism and social violence that is inflicted on racial minorities, especially those below the poverty line and often females. Yet within those same lyrics are glimmers of hope—the salience of respect, loyalty, communality, spirituality—those very qualities Parsons described when she spoke of those attributes she brings to her professional struggles to succeed.

McKinley and Parsons are successful scholars. They have good positions in science education and they know how to succeed. Yet what are the costs to produce two successes? How many others have not made it into the club? The curriculum efforts of McKinley in New Zealand are most promising because they are premised on the potential of the culture learners have and can enact. Rather than more decades of trying to reduce gaps in achievement and improved culturally sensitive ways of measuring attainment on the same canonical forms of science, we might accept the challenge of creolized sciences. What does it mean that there are creolized sciences and how might we ascertain what students have learned from programs like those being planned by scholars like McKinley?

Equity and Difference

In science education, as in other disciplines, the cultural-historical path dependence has led to confusions concerning the idea and notion of *equity*. Thus, it is not uncommon to hear that students or faculty members are treated the same in response to the kind of inequities that Parsons remembers in the narrative of her colleague who suggested that she should expect to be treated inequitably. Treating different individuals "the same," for example, by producing the same utterance, does not lead to the same action, because speech acts are the result of multiple turns. Because a speech act consists of the performance (locution), the intention (illocution), and its effect (perlocution), it is only complete in and with the performance of the second speaker. The speech act therefore is a function of the recipient as much as it is the function of the initiator, so that the same utterance initiates a different act in a different situation.

The confusion about equity in science and science education arises from an epistemology that is grounded in the idea of the same; difference is then defined negatively as deviation generally and often as deviance and deficit in particular. To deal with the issue of inequities more evenly, the culture of science education is in need of a different approach to difference, which it may find in the work of the philosophers of difference, including, for example, Gilles Deleuze (1968/1994) and Jean-Luc Nancy (2000). Thus, if we begin by taking difference as the norm, which exists in and for itself, then all sameness is the outcome and result of constructive work. From this perspective, we may not even assume that the expression "$A = A$" is true, because the ink and paper particles making the first of the two letters are different from those that make the second letter. This now allows us to specify the conditions under which the first and second figures can be taken to be the same, which are different in mathematics than in, for example, electron microscopy or chemical analysis, which will detect variations between the ink and paper of the two instances.

Once we institute such a move, then the question no longer is about equality but of equity and social justice, which may, to be realized, require different actions in each particular case rather than the same action—if something like that were possible in the strong sense of the term *same*. We begin with the assumption that even two African American students or science education scholars are different, allowing us to ask the question about what equitable treatment and opportunities might mean in each particular case. It would allow us to ask under what conditions and which actions, in their differences, might be considered (constructed) as equitable in the face of the obvious differences that exist between individuals, their cultures, and their biographies. It would then allow us to ask questions about equity with respect to different Aboriginal cultures among themselves and with the currently dominant Western thought system that underlies the sciences in their current form—though there are scientists, too, who acknowledge the value of Aboriginal and other forms of local knowledge, for example, the knowledge local (Western) fishermen have accumulated over centuries in the waters of Newfoundland, and which nowadays is used to better understand the history of cod stocks in the area.

Boundaries and Hybridity

Parsons writes about the two worlds she experiences, and moving back and forth between them. McKinley experienced the two worlds at home, where she saw how her white mother received different treatment than her Maori father in a colonial and colonizing Aotearoa-New Zealand state. Some theorists use the term *boundary crossing* to make salient the fact of the different experiences that occur when an individual moves between the first and second world. A question has to be asked, however, what does cross any existing boundary? Certainly it is not the flesh, and it is not the person who feels to have a different identity in the two worlds.

In recent years, the notion of *diaspora* has been prominent in cultural studies posed to theorize the experiences of displaced persons and migrants who find themselves in the midst of a foreign culture (second world) with their ways of feeling and thinking that have been built in their root culture (first world). Because our actions stand in a dialectical relationship with the resources available in our surrounding, the confrontation of the two worlds (external resources, internal schema) leads to a métissage or creolization of elements of the two. Boundary crossers, such as McKinley and Parsons, therefore, no longer are at home in the first or second culture; rather, cultural studies theorists such as Homi Bhabha propose that such individuals inhabit a *third space*, characterized by the hybridization of various forms of practices and resources into new forms of practices, giving new meaning to the resources in turn.

The drawback of this conceptualization is that it is built on the idea of the unicity of culture—the first and second world—rather than on the idea of difference and heterogeneity. We are no further along in theorizing inequities in science and science education, because we inherently lump all white Western scholars in the same category. But, as we know from experience, migration between societies— Australia and the US in Ken's case, and Germany and Canada, in Michael's case— lead to cross-cultural experiences and behavior possible in one society that are impossible in another (e.g., because viewed as gendered, aggressive, or inappropriate in some other ways). Complicating the matters even further, Michael's mother's tongue of his childhood now is his third language, as for more than a quarter century he has worked in an Anglo-Saxon context but has led a home life in a Francophone culture, becoming deeply familiar with French philosophy, for example. The result of the métissage in the same flesh has led to an inherent heterogeneity that also characterizes his work. It is no longer a life between two worlds cleanly separated by some boundary and a third space only to the extent that it is heterogeneous and different from itself according to the formula "third space ≠ third space," not in the least because this space continuously changes its nature and therefore like the proverbial river that one cannot step into twice.

LOOKING FORWARD IN LOOKING BACKWARD

At each moment in a cultural field, the present horizon of possibilities is shaped by the history of the field, but future developments already are contained in this horizon, which foreshadows and sets up the cultural possibilities of tomorrow. In this book, we interweave cultural historical accounts and auto/biographies to tell the story of science education of the past 45 years. In this epilogue, we take a look back over the ground covered to better understand the horizon of our field, which constitutes the very (indeterminate) ground for what science education will be tomorrow. Among the themes we address are those of inclusivity, the nurturing of newcomers that a field requires to successfully reproduce and transform itself, the function of gatekeepers and obligatory points of passage, the increasing globalization of science education, and the methods and methodologies that characterize the research typically being done and reported.

INCLUSION

Who's in and who's out? As we approach the final chapter of our journey through the cultural history of research in science education, our minds return to a keynote address delivered in 1998 by Peter Fensham in San Diego at the annual meeting of the National Association for Research in Science Teaching. He described a study that was a prelude to writing his 2004 book entitled: *Defining an Identity: The Evolution of Science Education as a Field of Research.* We were both at the back of the packed auditorium, next to each other, and felt a bit awkward asking questions. Ken did so nonetheless. As he recalls it, he asked Peter to discuss his selection criteria for inclusion of scholars in his study. Ken remembers concluding his question with the quip that some of his heroes were not included. He had in mind Mary Budd Rowe and Russell Yeany, who had been role models for him. He did not anticipate the aggression of Peter's response to his question. According to Ken: "Peter assumed that I was chiding him for not including me in his list of participants in this preliminary version of his research." Of course this was not Ken's intention but later when he discussed the issue with mutual colleagues, including David Treagust, it was clear that he too felt that the focus of Ken's query was effectively to ask: "Why did you exclude me?" So, in the spirit that readers may wonder about how we decided on issues of inclusion, we explore in this section the rationale for inclusion and, to the extent possible, how some prominent science educators came to be excluded.

Most of all, readers should be aware of the argument that we have made throughout: this is as much about the cultural history of science education as it is about the historicity of science education culture. In both, the biography of an individual is viewed as concretely realizing the cultural-historical possibilities of the times; and in their lives, works, and voices, they thereby produced and reproduced the field. The very nature of this field, the way its practitioners make sense today is mediated by the nature of the field as it was. Thus, we find that for science education, the same description holds that also holds for other sciences: The nature of the field today, that is, its typical practices, ways of thinking, writing (in journals, books), and talking (at conferences) is shaped by the contributions of its members throughout its history, so that the culture in its very essence preserves the accomplishments of its forefathers:

> [I]t is of the essence of the results of each stage not only that their ideal ontic sense in fact comes later than that of earlier results but that, since sense is grounded upon sense, the earlier sense gives something of its validity to the later one, indeed becomes part of it to a certain extent. Thus no building block within the spiritual structure is self-sufficient; and none, then, can immediately be reactivated by itself. (Husserl, 1976, p. 373)

Thus, each auto/biography included exemplifies our field in that it provides an account of some of the realized possibilities of the field as a whole.

The selections of senior science educators reflected who we knew well enough to call on to participate in the project. Our primary objective was to be as representative as possible when space limitations do not allow us to have more than 20 or 22 individuals. We wanted two science educators who were toward the end of their careers, who had enjoyed success in the international arena in terms of their own contributions to research, mentoring of others, and taking leadership in the field. Hence, Jane Butler Kahle and James Gallagher were clear choices, since each was about to commence retirement and yet they were clearly active and acknowledged leaders in their fields. Notably, both were recipients of NARST's *Distinguished Contributions to Science Education Award* (DCA). We were fortunate in that both were on our initial list and agreed to participate without hesitation. Barry Fraser was one of the first we identified. He was also a recipient of NARST's DCA, was outside of the United States, still active, and a pioneer in the field through his research on learning environments, while building a hugely successful science education center. We did invite other senior scholars and several declined for reasons of being too busy to meet the deadlines and in one case that she had moved on from science education and was committed to other pursuits.

We assumed that through the auto/biographies of those we selected the research of their peers and mentors would be acknowledged and contextualized. This proved to be the case, although one notable exception until this point in time has been Audrey Champagne, who undertook research on the cognitive structure of science knowledge and conceptual change. As a female in science education Champagne has been a role model for many science educators and assumed key leadership roles within NARST and the American Association for the Advancement of Science. Like Kahle, Gallagher, Fraser, Fensham and Treagust, she too was a recipient of NARST's DCA.

After deciding on those who were about to retire we selected colleagues from across the career ladder—all with established careers as researchers in science education. Hence, our priority was not to select the most eminent science educator from each region in the world—though we may have done this in some cases. Instead we wanted to include scholars across the career ladder and from around the world to write their auto/biographies and in so doing provide windows into the evolution and spread of science education. Some of those selected were reluctant initially, often because they felt there were others more deserving than they. However, we were persistent and explained that our goal was not to tell a history through our version of a science education hall of fame. Our selection practices reflect our concerns with issues of diversity and equity and a determination that this history of science education would not be centered on the United States. Although it is inevitable that trends in the United States would shape science education, we carefully monitored the invitations and acceptances to ensure that we had a diverse pool of authors.

The promise of writing a cultural history based on the lives of participants, using the voices of the participants, is realized in what we regard as compelling chapters that constitute the core of the book. Yet we wanted to go beyond and here we had to grapple with competing tendencies of the two editors. Michael's preference to take the auto/biographies as starting out points to ratchet up the theoretical bases for science education and Ken's preference to take the same auto/biographies as start points to tell more stories needed to come into a balance in the introductory and concluding sections of each part of the book and n this concluding section. Our struggles have produced a book that hopefully meets both goals well and in so doing meets the expectations of readers from across a broad spectrum of science education.

As we explore those who emerged in the stories of others there are two areas that we address here, while acknowledging that by identifying others we simply exacerbate the problems of those who still are not acknowledged for their efforts and place in science education. Having said that we take note of an area of research that has been prominent and widely cited in science education—involving the nature of science. Norman Lederman, now at Illinois Institute of Technology, has been highly influential in this field and his work is widely cited. Lederman has been President of NARST, a recipient of a best paper ward for a paper published in the *Journal of Research in Science Teaching*, and two of his former doctoral students, Julie Gess-Newsome and Fouad Abd-El-Khalick, are recipients of NARST's outstanding doctoral dissertation award. Abd-El-Khalick also was recipient of NARST's early career award.

Though he is mentioned in the book Joseph Krajcik has been a highly successful science educator. A graduate from the University of Iowa, Krajcik has been at the University of Michigan where he has been extensively involved in the uses of computers in science education and extensive curriculum development in urban areas such as Detroit. His work on curriculum development and technology tools for teachers has been grounded in psychological models. The citations from the *Thomson ISI-Web of Science* show his emphases with numerous coauthored articles with Phyllis Blumenfeld, Ron Marx, both from educational psychology and Elliot Soloway, an engineer. Not so apparent from the citations is the mentorship he received, first from Vincent Lunetta, with whom he pioneered research using computers, and afterwards, from his Dean at the University of Michigan, Carl Berger, himself recipient of NARST's DCA award and the 1983 NARST president.

Another area of research that has been prominent, but did not emerge through the auto/biographies in this volume involves reading, writing and science education. William Holliday, Larry Yore, James Shymansky, and Brian Hand all have been actively involved in research in this area—which has been international in scope and has connected with conceptual change researchers such as David Treagust and with Michael's work on scientific literacy and the uses of figures, photos and graphs in science texts and classrooms.

Finally, Angela Calabrese Barton, though still in the early stages of her career has been very successful in building a program of urban science education at Teachers College and for a short time at the University of Texas. She has published prolifically and already has established herself as a leader in science education. Now at the Michigan State University, she has successfully navigated many of the issues of gender that we have addressed through others' auto/biographies. She obtained tenure at a research intensive university, gave birth to two children and juggled her professional life along with her husband's career as a university engineer. She has pioneered *critical ethnography* and has involved parents and student researchers in innovative ways. She, too, has received the accolades of her peers receiving awards from NARST and AERA for her journal articles and books— including an award from Division K of AERA for her exemplary book on scientific literacy, coauthored with Michael.

NURTURING THE FIELD

To keep a field such as science education vibrant, much more than graduating newcomers is required. Traditions are established when newcomers actively are involved in renewing the field in the very process of reproducing it. Mentoring new scholars, which may include writing with them or getting grants, can shape a field substantially because it leads to stronger traditions (in the sense of "handing down") of forms of thought and forms of praxis.

Ken: It seems almost a no-brainer to argue that we need good mentors in science education. We have many mentoring programs and for some reason I have a jaundiced view of most of them. How can we filter the mentors to make sure they know what to say to students? Just this past week a colleague has been mentoring some of my current doctoral students and last night they reported to me that she is a bitter person with a distorted view of science education. Phew. That is gratifying. Perhaps my graduate students are taking my advice to look closely at who is giving the advice as well as the nature of the advice. At the same time I have spent time with two fresh assistant professors from different programs—seeking mentoring assistance. Just as some of my former doctoral students have sought postdoctoral experiences with you—I think it imperative for the health of the field that mentoring opportunities for new researchers become an established part of our field—with structures to support their expansion.

Michael: What we say when we mentor is mediated by what we stand for; my tendency, because I value research and scholarship, therefore always is to support new scholars in getting their research program going, publishing regularly, and getting grants. Already at the graduate level, beginning with their master's degrees, the students in my research team focus on scholarship and write for publication in the high-impact journals of their field—which in some cases is outside the field of science education proper. Too often, however, I see new colleagues being exploited, even female professors by female department heads, who, perhaps intending to take the path of least resistance, ask our young colleagues to do so much more in teaching and service than some of our established colleagues do.

Ken: Some newcomers want to make a difference and volunteer to do too much. Without a sense of priorities they become their own worst enemies. It is imperative that senior colleagues intervene to prevent this from happening.

Michael: It is surprising that new faculty do not leave graduate school with a better sense of how to succeed in the academy.

Ken: Absolutely. In the United States I think we have a serious problem with the structure of doctoral degrees, so that my students, for example, cannot publish in the same way yours do. Put bluntly the amount of coursework is excessive and in my mind produces graduates who are not well prepared for careers as researchers. In my present university the structure of the degree can prevent many students even from putting together a committee until end of the first two years in which they do coursework. The linearity of the coursework and research components is beyond sensibility, as is the size of the required core. The nub of the problem is in the fact that those who created the degree structure and maintain it are not researchers and do not have a record of exemplary scholarship. To be a little cynical, bean counters have appropriated the "terminal" degree. That being the case, terminal may take on a different meaning than intended. While universities like Curtin have moved into large cities like New York and offer research only (or mainly) degrees, many other universities are bogged down in structures that are defended in the name of high scholarship but have dubious claims to producing the outcomes people like me hold dear.

Michael: Normally this might be the case, too, at my university. But there is historical precedent that the supervisor can change even required courses—and in my case, I had students pursue their reading, writing, and researching interests by creating courses by special arrangement, which we organized around themes such as "research apprenticeship," "discourse analysis," or "cultural-historical activity theory." It is in the context that my students realize their research from day 1 when they enter my laboratory to the moment they successfully defend their dissertations—which in essence consists of a collection of papers suitably contextualized by introductory and concluding chapters.

Ken: In raising this issue I do not mean to imply that universities like Curtin will take over doctoral education in New York City. That will not happen. However, the take home message for our doctoral program and others like it is to look at what is done elsewhere. Ironically, so often the practices of scientists hold sway in science education and yet in this instance, with few exceptions, that does not appear to be the case. In many science departments I know of formal course work finishes after one year at which time students are attached to research groups where they do most of their intellectual work. At the very least this needs to be an option for those science education doctorates that purport to maintain the high quality of research in science education.

Michael: I do understand the resistance of some older folks in our discipline to such a model, for all too often, students do not move to the cutting edge of the field because they lack the resources for so doing. A case in point pertains to the methods of doing research when science educators are formed in science departments and without the requirement to take some course in qualitative or quantitative inquiry. From a methodological point of view, the work produced in such contexts may be rather weak and not make it through peer review. On the other hand, in such a context my own students benefit tremendously, because their learning is framed around a trajectory of becoming a successful researcher, articulated in and through successful publication of manuscripts. Thus, even my master's level students get the results of their studies—always framed in and around the research funding I obtained, and therefore also an integral part of my research program—published in our leading journals.

Ken: Creating and sustaining networks with technology appears to have opened many doors in this day and age. For example, we undertake our editorial meetings using videoconferencing using a Macintosh application called iChat—which operates across platform with the Windows operating system using AIM. With multi-party conferencing possible my research group now is beginning to use iChat more and more as a preferred way of interacting since travel can be so complicated in such a large city. Of course iChat also extends across national boundaries and we have successfully communicated using this medium with scholars in Australia, Taiwan, and New Zealand. As the capabilities of the Internet expand the possibilities for doing collaborative research and for creating international networks is expanding rapidly.

Michael: This is precisely the way in which I maintain working relationships with former graduate students (e.g., Yew Jin Lee) and postdoctoral fellows (e.g., SungWon Hwang) with whom I continue to publish more so from what I am in the process of developing rather than from what they have pursued centrally while staying in my laboratory.

GATEKEEPERS, OBLIGATORY POINTS OF PASSAGE

In science education, as in every other discipline, there are conceptual personae and central institutional figures around who it is virtually impossible to get around. Members have to cite particular authors, frame their research around certain theories, and submit their work to journals where editors and reviewers are the gatekeepers making the decisions about who gets published and who does not.

Ken: In the very early parts of this book we identified Jean Piaget as the only viable conceptual persona in science education. However, just yesterday you advised me that you have an article accepted in the *Review of Educational Research* (RER), coauthored with Yew Jin Lee, one of your former doctoral students. The article has the title: "'Vygotsky's Neglected Legacy': Cultural-Historical Activity Theory." Since I regard RER as the leading journal in which to publish a piece like this let me congratulate you on your success and ask whether or not you feel Lev Vygotsky will be our second conceptual persona? In my days at Florida State University when I was arguing so strenuously about radical constructivism and its virtues vis a vis more objectivist perspectives there was an active group arguing with me in favor of Vygotsky's social constructivism—that was two decades ago and in that time more and more people have adopted a Vygotskyan perspective—not just those who went down the path to activity theory.

Michael: Lev Vygotsky ought to be one of our conceptual personae, not just because of what he has done, but because of the developments of his (first-generation) activity theory into what are now third- and fourth-generation theories. These theories conceptualize the irreducible, mutually constitutive relation between individual and collective. Such perspectives have allowed me to theorize culture and language as the dynamic systems and subsystems that we know them to be, continually changing at the very instant that they reproduce themselves in realizing possibilities that emerged from the horizon shaped by their past.

Ken: What about Pierre Bourdieu as a future conceptual persona? In my own work he has had a profound impact and his social perspectives are also having impact in numerous research groups—including your own.

Michael: Bourdieu certainly has done a lot to move the field of practice theory ahead. Whether he is going to be a conceptual persona? Our cultural history will show. If you think for a moment about philosophy, then you know that only a few come to stand for the field in paradigmatic ways, though even the conceptual personae realize the cultural possibilities of their days. Thus, during the times of Immanuel Kant or Georg F.W. Hegel—who interacted a lot with and appropriated ideas from Friedrich Schelling—there were many other philosophers. Kant and Hegel picked up from them ideas and concepts, and yet, these two came to stand in singular ways for the philosophical dimension of humanity. Karl Marx is another philosopher who epitomizes a whole way of philosophical thinking.

Ken: Throughout this volume we have discussed the necessity to publish and obtain external support for scholarly activities. Because of the high-stakes nature of these activities it seems imperative that we look closely at the persons who end up becoming the gatekeepers. In the case of journals it seems apparent that editors and associate editors have enormous responsibilities and power. The high-stakes part comes with the high-impact requirement that many countries and institutions within countries place on publishing in this journal rather than that. How do the editors of those journals get appointed and what are the conditions for their retention? What are the qualifications that fit someone to assume such a position? Similar queries can be raised about those who become gatekeepers on the resources to fund research—within government and private foundations. Of all the field shapers government and private foundations appear to have the greatest potential to transform science education or reproduce it in this form rather than that. Hence the presence of gatekeepers and their preservation is a focus for considerable thought, discussion and other potential forms of action.

Michael: Science education centers play an important role in the production and reproduction of the field as a whole and of particular ways of praxis more specifically. I am thinking about the way in which the Centre of Science and Mathematics Education at Curtin University of Technology has shaped the recent evolution of our field through its innovative doctoral program. With it, forms of doing research have been reproduced, including Barry Fraser's ways of studying learning environments, David Treagust's conceptual change approach to studying knowing and learning in science, Peter Taylor's ways of studying personal growth, or Léonie Rennie's methods for studying informal learning.

Ken: Centers come and centers go. There are exceptions to that of course and we have shown through the chapters of this volume that Leeds, King's, Kiel, Curtin, Monash, Michigan State, and Georgia are among the key centers for science education in the four decades or so that we have explored. Otherwise, one or two key faculty move and a center seemingly evaporates as another starts to emerge. Hence, Norm Lederman leaves Oregon State University and the science education program there wanes whereas his new institution, Illinois Institute of Technology, appears on the science education map. Similarly, as Angela Calabrese Barton leaves Teachers College it is anticipated that the large and successful urban education program there will diminish in its international significance. On a larger scale the exodus of science educators from Purdue due to retirements and faculty moves (e.g., Jane Kahle to Miami; Dudley Herron to Morehead State; and later Sandra Abell to Missouri), diluted what had been a potent critical mass of science educators, perhaps about to be revitalized with the recent hiring of John Staver from Kansas State. As science educators move around from one university to another and from one center to another center, programs that tend to focus on a set of scholars, can appear and disappear almost overnight. As the current crop of researchers approach retirement it will be interesting to see if some of these staples identified above disappear overnight as leaders in our field emerge in institutions that are just beginning to make their marks in science education.

GLOBALIZATION

As the United States produced doctoral graduates from overseas, who subsequently returned to their home countries, it is no surprise that gradually doctoral education became internationalized. Then, with the advent of new centers offering doctoral degrees in international sites, exporting doctoral education became a possibility. What soon became apparent was that universities like Curtin were marketing convenience. While many universities in the United States, like the past two I have been at, have high rejection rates, Curtin encourages applications, accepts students into the program, and offers the flexibility of convenient residence requirements and the flexibility of mainly research requirements (i.e., little coursework to complete). As we pointed out earlier some students may flounder in such circumstances and isolation can be a major problem for those who enroll for a doctorate from an institution in another country—unless they relocate to that country and attend on campus.

The emphasis on producing publication that is now an international given occurs in a context of English language hegemony. The high impact journals publish only in English. A challenge for our field is to provide structures for publication in the native language of the researcher. Whether or not this will be possible remains to be seen—but perhaps with the advances in technology the time is not too distant when different forms of representing scholarship will replace the familiar textual forms of journal writing. The move from paper to electronic form may have opened floodgates, heralding not only longer papers (as Dillon predicted), but also uses of photographs and movies. Already the science education journal we co-edit, *Cultural Studies of Science Education*, has had several articles that move in the direction of visual ethnography. With new technologies will arise new hegemonies—but perhaps these will not be English centered as is presently the case.

The high stakes tests within cities, states, and countries have been moved to the international domain and for very many years now being number 1 is not just confined to sports' teams. Being number 1 in mathematics and science was an imperative in the mid 1980s and to some extent it still is a priority and gets enshrined in accountability policies. Within science education careers have been built on developing tests to measure science achievement in an international context and the results from those texts shape political agendas and mobilize the private sector that has a stake in producing the best scientists in the world. TIMSS, TIMSS+, PISA: what is to come and where will it end? On the one hand, technologies offer us unprecedented opportunities to collaborate globally for the benefit of humankind and yet national priorities still project the goal to be number 1 and compete with others in what just may be a zero sum game. Clearly not every country can be number 1 and it might just be the case that international perspectives can breed new forms of collaboration that lead to a form of science education that is transformative in a global sense and produces more productive forms of social life for all—no matter where they commence their lives.

METHODS AND METHODOLOGY IN SCIENCE EDUCATION

An important distinction not often made in our field is that between method and methodology. By methodology, literally the science of method, we mean the principles that determine how methods are deployed and interpreted in science education. We show in this book how changes have occurred concomitant with acceptance of a greater range of theories that underpin research in science education. For example, Ken's very early research in science education was a conceptual maze. In his first study he explored the relationships between teacher wait time and the science achievement of students. In his thinking about teaching and learning he used a Piagetian theory for learning and constructed research questions and hypotheses accordingly. However, he used a quasi experiment because at the time this is what was necessary for peers to regard the study as scientific. Because of similar reasons, Michael did a quantitative study of the development of proportional reasoning in adulthood. Our understanding of research and research design introduced the use of quantitative measures, random assignment of participants to treatment groups and a representation of social life in classrooms in terms of independent, dependent variables and covariates. The positivism underpinning our research design and associated research methods were largely tacit. We accepted the tenets of good design and research practices based on what we were taught in the early 1970s and had virtually nobody to talk to about what we were doing. For example, Russell Docking taught Ken about research methods and several years later he introduced Ken to Barry McGaw who taught him more about such issues, internal and external validity, and especially how to apply sophisticated multivariate analyses, including path and factor analyses.

Although there were pioneers in education who already were using ethnography when we began our doctorates we did not know about them and our research methods continued to embrace positivism. We did not seriously contemplate alternatives until we were led to interpretive methods and ethnography by the twin frustrations of the reductionism of depicting social life in terms of variables and the myopia of a priori hypothesizing. Thus, in the mid 1980s Ken began to realize the salience of seeking coherence in terms of the theories he used as a basis for his research focus (and the depiction therein of social life) and the methods he employed to do his research. When Michael returned to teaching high school students, he realized that the deviations from the norm that are treated in statistical approaches as error are those deviations science teachers have to understand to address the particular and singular needs individual students have. Fortunately for us the field was willing to accept our realizations and associated practices. The journal editors of the day, Russell Yeany (*Journal of Research in Science Teaching* [JRST]), Leo Klopfer (*Science Education*), and Richard Kempa (*International Journal of Science Education*) all were willing to consider for review papers that incorporated different theoretical frameworks and employed diverse methods. That is, the methodology of science education began to shift discernibly in the mid 1980s.

In the 20 years since the mid 1980s we cannot claim that it has been smooth sailing. As the number of journals has grown the editors have in some ways used practices that tended to reinforce reproduction of conservative approaches to science education. We do not mean to imply that the editors are less scholarly or lack vision—just that structures were used to make it more difficult to publish research away from the traditional mainstream without struggles with reviewers.

As the manuscript flow increased, editors of JRST felt a need to have a more sophisticated division of labor than the editor and assistant editor type of structure that had characterized the journal through its first two decades. To avoid offending particular editors and associate editors we will not get too specific with our remarks here—suffice to note than some of the decisions made sometimes made us wonder about the nature of peer review. In what senses did peers review our manuscripts? The process was uneven—some reviewers and associate editors seeming to be more in field than others. Perhaps the task of editing a journal with a large manuscript flow was becoming unmanageable on the one hand and the process of getting associate editors produced a lack of coherence in the criteria applied for acceptance and rejection, on the other hand. Our point here though is not to gnash our teeth and flail our arms but to point out that the change trajectory was not smoothly upward—but followed a fine structure that included regression to traditional rationale for what constituted scientific research. Possibly this tendency was fuelled in part by scientists making a transition into science education and being assigned editorial responsibilities from the outset of their journey. In contrast to Penny Gilmer doing a second doctorate before assuming the mantle of science educator.

The new journal we have created, *Cultural Studies of Science Education*, is explicit in relation to what we will and will not accept for possible publication. We are clear in our orientation toward sociocultural and cultural-historical lenses on social life and its study and representation. As we enact our roles as gatekeepers we remain cognizant of the perspectives of those on the outside trying to get their work into the journal. Rather than creating new forms of hegemony we must be on the lookout for ways to remain current and at the forefront of theory and associated methodologies that can inform scholarship in science education.

LOOKING AHEAD

The roads ahead are already prepared by what has come before. Yet the possibilities are infinite. The field of science education includes many fine scholars and more will be produced in the years ahead to collectively expand scholarly practice in science education. In both the optimistic and pessimistic senses, an observer situated in the early 1960s could not have imagined the present state of scholarship in science education. So, as we look ahead to the next 50 years, we expect even greater changes in the faces of science education than have occurred in the past 50 years. Preparing scholars for tomorrow's science education obviously needs more than reproduction of the status quo—today's trail blazers need bold visions and even bolder actions if scholarship in science education is to have a place in tomorrow's world. We join them on this mission and wish them success.

REFERENCES

Aikenhead, G. S. (1996). Science education: Border crossing into the subculture of science. *Studies in Science Education, 26*, 1–52.

Arzi, H. J. (1998). Enhancing science education through laboratory environments: More than walls, benches and widgets. In B. J. Fraser & K. G. Tobin (Eds.), *International handbook of science education* (pp. 595–608). Dordrecht, Netherlands: Kluwer Academic Publishers.

Arzi, H. J., & White, R.T. (2006). *Change in teachers' knowledge of subject matter: A 17-year longitudinal study.* Manuscript submitted for publication.

Arzi, H. J., Ben–Zvi, R., & Ganiel, U. (1985). Proactive and retroactive facilitation of long–term retention by curriculum continuity. *American Educational Research Journal, 22*, 369–388.

Arzi, H. J., Ben–Zvi, R., & Ganiel, U. (1986). Forgetting versus savings: The many facets of long–term retention. *Science Education, 70*, 171–188.

Atkin, J. M., & Black, P. (2003). *Inside science education reform: A history of curricular and policy change.* New York: Teachers College Press.

Atkin, J. M., & Karplus, R. (1962). Discovery or invention? *Science Teacher, 29*(5), 45.

Bakhtin, M. (1981). *The dialogic imagination.* Austin: University of Texas.

Becker, J., & Varelas, M. (2001). Piaget's early theory of the role of language in intellectual development: A response to DeVries' Piaget's social theory. *Educational Researcher, 30*(6), 22–23.

Boykin, A. (1986). The triple quandary and the schooling of Afro-American children. In U. Neisser (Ed.), *The school achievement of minority children: New perspectives* (pp. 57–92). Hillsdale, NJ: Lawrence Erlbaum Associates.

Bransford, J. D., Brown, A. L., & Cocking, R. R. (1999). *How people learn: Brain, mind, experience, and school.* Washington, DC: National Academy Press.

Bruffee, K. A. (1993). *Collaborative learning: Higher education, interdependence, and the authority of knowledge.* Baltimore: Johns Hopkins University Press.

Clark, D., & Jorde, D. (2004). Helping students revise disruptive experimentally supported ideas about thermodynamics. *Journal of Research in Science Teaching, 41*, 1–23.

Cobb, P., Confrey, J., diSessa, A., Lehrer, R., & Schauble, L. (2003). Design experiments in education research. *Educational Researcher, 32*(1), 9–13.

DeBoer, G. (1991). *A history of ideas in science education: implications for practice.* New York: Teachers College Press.

Deleuze, G. (1994). *Repetition and difference* (P. Patton, Trans.). New York: Columbia University Press. (First published in 1968)

Deleuze, G., & Guattari, F. (1994). *What is philosophy?* New York: Columbia University Press. (First published in 1991)

Derrida, J. (1998). *Monolingualism of the Other; or, The prosthesis of origin.* Stanford, CA: Stanford University Press.

Driver, R., & Easley, J. (1978). Pupils and paradigms: A review of literature related to concept development in adolescent science students. *Studies in Science Education, 5*, 61–84.

Driver, R. (1983). *The pupil as scientist.* Milton Keynes, UK: Open University Press.

Driver, R., Guesne, E., & Tiberghien, A. (1985). *Children's ideas in science.* Milton Keynes, UK: Open University Press.

Driver, R., & Oldham, V. (1986). A constructivist approach to curriculum development in science. *Studies in Science Education, 13*, 105–122.

DuBois, W.E.B. (1969). *The souls of Black folk.* New York: New American Library.

Duit, R., Gropengießer, H., & Kattmann, U. (2005). Towards science education research that is relevant for improving practice: The model of educational reconstruction. In H. E. Fischer (Ed.), *Developing standards in research on science education* (pp. 1–9). London: Taylor & Francis.

Duit, R., Roth, W.-M., Komorek, M., & Wilbers, J. (1998). Conceptual change cum discourse analysis: Towards an integrative perspective on learning in science. *International Journal of Science Education, 20,* 1059–1073.

Duit, R., & Treagust, D. (2003). Conceptual change: A powerful framework for improving science teaching and learning. *International Journal of Science Education, 25,* 671–688.

Edwards, D., & Potter, J. (1992). *Discursive psychology.* London: Sage.

Engeström, Y., & Miettinen, R. (1999). Introduction. In Y. Engeström, R. Miettinen, & R.-L. Punamäki (Eds.), *Perspectives on activity theory* (pp. 1–16). Cambridge, UK: Cambridge University Press.

Erickson, F. (1986). Qualitative research on teaching. In M. C. Wittrock (Ed.), *Handbook for research on teaching* 3rd ed. (pp. 119–161). New York: Macmillan.

Falk, J. H., & Dierking, L.D. (1998). Free-choice learning: An alternative term to informal learning? *Informal Learning Environments Research Newsletter.* May/June 1998 Washington, DC: American Educational Research Association.

Fanon, F. (1967). *Black skin, white masks.* New York: Grove.

Fensham, P. J. (2004). *Defining an identity: The evolution of science education as a field of research.* Dordrecht, Netherlands: Kluwer Academic Publishers.

Fuss, D. (1995). *Identification papers.* New York: Routledge.

Gallagher J., Contreras, A., Dawson, G., & Gallard, A. (1985). *Science education in the Americas: A report of the Inter–American conference on science education.* East Lansing: Michigan State University.

Gallagher, J. J. (1965). *Children's explanations of scientific phenomena: initial appraisal and modification after instruction.* Unpublished Doctoral Dissertation, Harvard University.

Gallagher, J. J. (1971). A broader base for science teaching. *Science Education, 55,* 329–338.

Gallagher, J. J. (1989). Research on secondary school science teachers' practices, knowledge and beliefs: A basis for restructuring. In M. L. Matyas, K. Tobin, & B. Fraser (Eds.), *Looking into windows: Qualitative research in science education* (pp. 43–57). Washington: American Association for the Advancement of Science.

Gallagher, J. J. (Ed.). (1991). *Interpretive research in science education.* Monograph Series of National Association for Research in Science Teaching.

Gallagher, J. J. (2007). *Teaching science for understanding: A practical guide.* Columbus: Merrill.

Gallagher, J. J., & Cline, D. A. (1988). *US–Japan seminar on science education.* Washington, DC: National Science Teachers Association.

Gallagher, J. J., & Parker J. (1996). Using assessment to improve teaching and learning in science. Michigan State University. Developed under NSF Sponsorship.

Gallagher, J., & Tobin, K. (1987). Teacher management and student engagement in high school science. *Science Education, 71,* 535–555.

Gilmer, P. J. (2004). *Transforming university biochemistry teaching through action research: Utilizing collaborative learning and technology.* Unpublished doctoral thesis, Curtin University of Technology.

Giroux, H. A. (2003). Public time and educated hope: Educational leadership and the war against youth. Available at http://www.units.muohio.edu/eduleadership/anthology/OA/OA03001.html. (Accessed June 4, 2006)

Grayling, A. C. (1997). Intellectual or academic. *Prospect,* 15. Available at http://www.prospect–magazine.co.uk/article_details.php?id=4656 (accessed on June 4, 2006).

Heidegger, M. (1977). *Sein und Zeit.* Tübingen, Germany: Max Niemeyer.

Henriksen, E. K., & Jorde, D. (2001). High schools students' understanding of radiation and the environment: Can museums play a role? *Science Education, 85,* 189–206.

Hewson, P. W., & Lemberger, J. (2000). Status as the hallmark of conceptual learning. In R. Millar, J. Leach, & J. Osborne (Eds.), *Improving science education: The contribution of research* (pp. 110–125). Buckingham, UK: Open University Press.

Hewson, P. W., Beeth, M. E., & Thorley, N. R. (1998). Teaching for conceptual change. In K. G. Tobin & B. J. Fraser (Eds.), *International handbook of science education* (pp. 199–218). Dordrecht, Netherlands: Kluwer Academic Publishers.

Hewson, P. W., Curtis, M. D., Schneckloth, S. E., & Damonse, B. (2005). Building education research capacity: Collaboration between the United States and South Africa. *Journal of International Cooperation in Education, 8*, 61–80.

Hewson, P. W., Tabachnick, B. R., Zeichner, K. M., Blomker, K., Meyer, H., Lemberger, J., Marion, R., Park, H., & Toolin, R. (1999). Educating prospective teachers of biology: Introduction and research methods. *Science Education, 83*, 247–273.

Hewson, P. W., & Hewson, M. G. (1988). An appropriate conception of teaching science: A view from studies of science learning. *Science Education, 72*, 597–614.

Holland, D., & Lave, J. (2001). History in person: An introduction. In D. Holland & J. Lave (Eds.), *History in person: Enduring struggles, contentious practice, intimate identities* (pp. 3–33). Santa Fe, NM: School of American Research Press.

Holliday, W. G. (2003). Influential research in science teaching: 1963–Present [Special Issue]. *Journal of Research in Science Teaching, 40(Supplement)*, v–x.

Hurd, P., & Gallagher, J. J. (1968). *New directions in science teaching*. San Francisco: Wadsworth.

Husserl, E. (1976). *Husserliana VI: Die Krisis der europäischen Wissenschaften und die transzendentale Phänomenologie. Eine Einleitung in die phänomenologische Philosophie*. The Hague: Kluwer Academic Publishers.

Inhelder, B., & Piaget, J. (1958). *The growth of logical thinking from childhood to adolescence*. New York: Basic.

Jorde, D., & Bungum, B. (Eds.). (2003). *Naturfagdidaktikk—perspektiver, forskning, utvikling*. Oslo, Norway: Gyldendal akademisk.

Jorde, D., & Lea, A. (1996). Sharing science: Primary science for both teachers and pupils. In L. Parker, L. Rennie, & B. Fraser (Eds.), *Gender, science and mathematics*. Dordrecht, The Netherlands: Kluwer Academic Publishers.

Kaestle, C. F. (1993). The awful reputation of education research. *Educational Researcher, 22*(1), 23, 26–31.

Kahle, J. B. (Ed.). (1998). *Journal of Women and Minorities in Science and Engineering, 4*(2 & 3), 91–320.

Kahle, J. B. (2004). Will girls be left behind? Gender differences and accountability [Guest editorial]. *Journal of Research in Science Teaching 41*, 961–969.

Kahle, J. B., & Lakes, M. K. (1983). The myth of equality in science classrooms. *Journal of Research in Science Teaching, 20*, 131–140.

Kahle, J. B., & Meece, J. L. (1994). Research on gender issues in the classroom. In D. Gabel (Ed.), *Handbook of research in science teaching and learning* (pp. 1559–1610). Washington, DC: National Science Teachers' Association.

Kahle, J. B., Meece, J., & Scantlebury, K. (2000). Urban African–American middle school science students: Does standards–based teaching make a difference? *Journal of Research in Science Teaching, 37*, 1019–1041.

Kahle, J. B., Parker, L. H., Rennie, L. J., & Riley, D. (1993). Gender differences in science education: Building a model. *Educational Psychologist, 28*, 379–404.

Kingsland, S. E. (1995). *Modeling nature: Episodes in the history of population ecology* (2nd ed.). Chicago: University of Chicago Press.

Knain, E. (1999). *Naturfagets tause stemme. Diskursanalyse av Natur- og miljøfag i et allmenndannelsesperspektiv.* [The silent voice of school science.] Doctoral dissertation, University of Oslo.

Knorr-Cetina, K. D. (1981). *The manufacture of knowledge: An essay on the constructivist and contextual nature of science*. Oxford: Pergamon Press.

REFERENCES

Kolstø, S. D. (2001). *Science education for citizenship. Thoughtful decision-making about Science–Related social issues*. Doctoral dissertation, University of Oslo.

Kuhn, T. S. (1970). *The structure of scientific revolutions* (2nd ed.). Chicago: University of Chicago Press.

Kyle, W. C., Abell, S. K., Roth, W.-M., & Gallagher, J. J. (1992). Toward a mature discipline of science education. *Journal of Research in Science Teaching, 29*, 1015–1018.

Latour, B., & Woolgar, S. (1979). *Laboratory life: The social construction of scientific facts*. Beverly Hills, CA: Sage.

Lave, J. (1988). *Cognition in practice: Mind, mathematics and culture in everyday life*. Cambridge, UK: Cambridge University Press.

Lave, J., & Wenger, E. (1991). *Situated learning: Legitimate peripheral participation*. Cambridge, UK: Cambridge University Press.

Layton, D. (1973). *Science for the people. The origins of schools' science curriculum in England*. London: George Allen & Unwin.

Lee, O. (1999a). Equity implications based on the conceptions of science achievement in major reform documents. *Review of Educational Research, 69*, 83–115.

Lee, O. (1999b). Science knowledge, worldviews, and information sources in social and cultural contexts: Making sense after a natural disaster. *American Educational Research Journal, 36*, 187–219.

Lee, O. (2002). Science inquiry for elementary students from diverse backgrounds. In W. G. Secada (Ed.), *Review of research in education Vol. 26* (pp. 23–69). Washington, DC: American Educational Research Association.

Lee, O. (2005). Science education and English language learners: Synthesis and Research Agenda. *Review of Educational Research, 75*, 491–530.

Lee, O., & Anderson, C. W. (1993). Task engagement and conceptual change in middle school science classrooms. *American Educational Research Journal, 30*, 585–610.

Lee, O., & Luykx, A. (2005). Dilemmas in scaling up innovations in science instruction with nonmainstream elementary students. *American Educational Research Journal, 42*, 411–438.

Lee, O., & Luykx, A. (2006). *Science education and student diversity: Synthesis and research agenda*. New York: Cambridge University Press.

Lee, S., & Roth, W.-M. (2003). Of traversals and hybrid spaces: Science in the community. *Mind, Culture, & Activity, 10*, 120–142.

Lemke, J. L. (1990). *Talking science: Language, learning and values*. Norwood, NJ: Ablex.

Matthews, M. (1995). *Challenging New Zealand science education*. Palmerston North: Dunmore Press.

McKinley, E. (1996). Towards an Indigenous science curriculum. *Research in Science Education, 26*, 155–167.

McKinley, E. (2003). Brown bodies in white coats: Maori women scientists and identity. *Journal of Occupational Science, 9*(3), 109–116.

McKinley, E. (2005a) Locating the global: culture, language and science education for indigenous students. *International Journal of Science Education, 27*, 227–241.

McKinley, E. (2005b) Brown bodies, white coats: Postcolonialism, Maori women and science. *Discourse, 26*, 481–496.

McKinley, E., Waiti, P. & Bell, B. (1992) Language, culture and science education. *International Journal of Science Education, 14*, 579–595

Nancy, J.-L. (2000). *Being singular plural*. Stanford, CA: Stanford University Press.

Nancy, J.-L. (2002). *Hegel: The restlessness of the negative*. Minneapolis: University of Minnesota Press.

National Research Council. (2006). *America's lab report: Investigations in high school science*. Washington, DC: The National Academies Press.

Ødegård, M. (2001). *The drama of science education. How public understanding of biotechnology and drama as a learning activity may enhance a critical and inclusive education*. Doctoral dissertation, University of Oslo.

Ogborn, J. (2005). 40 years of curriculum development. In K. Boersma, M. Goedhart, O. De Jong, & H. Eijkelhof (Eds.), *Research and the quality of science education* (pp. 57–65). Dordrecht, Netherlands: Springer.

Palmer, P. J. (1998). *The courage to teach: Exploring the inner landscape of a teacher's life.* San Francisco: Jossey-Bass.

Parker L. H., Rennie, L. J., & Fraser B. J. (eds.). (1996). *Gender, science and mathematics. Shortening the shadow.* Dordrecht, The Netherlands: Kluwer Academic Publishers.

Piaget, J., & Inhelder, B. (1958). *The growth of logical thinking, from childhood to adolescence* London: Routledge & Kegan Paul

Posner, G. J., Strike, K. A., Hewson, P. W., & Gertzog, W. A. (1982). Accommodation of a scientific conception: Towards a theory of conceptual change. *Science Education, 66,* 211–227.

Potter, J., & Wetherell, M. (1987). *Discourse and social psychology: Beyond attitudes and behaviour.* London: Sage.

Prenzel, M., Duit, R., Euler, M., Geiser, H., Hoffmann, L., Lehrke, M., Müller, C., Rimmele, R., Seidel, T. & Widodo, A. (2002). Zum Zusammenspiel von Unterrichtsskripts und Lernprozessen. Erste Ergebnisse einer Videostudie Physik. In R. Brechel (Hrsg.), *Zur Didaktik der Physik und Chemie. Probleme und Perspektiven. Vorträge auf der Tagung für Didaktik der Physik/Chemie* (S. 313–315). Alsbach/Bergstraße: Leuchtturm.

Reyes, L. Y., & Molina, A. (2005). *Scientific literacy: Beliefs, roles, goals and context for a better world.* Paper presented at the VII International Congress in Research on the didactics of science. Granada, Spain.

Reyes, L., Salcedo, L. E., & Perafán, G. A. (1999). *Acciones y creencias. Tomo I. Tesoro Oculto del Educador.* Bogotá, Colombia. Universidad Pedagógica Nacional.

Rickinson, M., Dillon, J., Teamey, K., Morris, M., Choi, M. Y., Sanders, D., & Benefield, P. (2004). *A review of research on outdoor learning,* Preston Montford, Shropshire: Field Studies Council.

Rorty, R. (1989). *Contingency, irony, and solidarity.* Cambridge: Cambridge University Press.

Rossiter, M. W. (1982). *Women scientists in America: Struggles and strategies to 1940.* Baltimore, MD: Johns Hopkins University Press.

Roth, W.-M. (1995). *Authentic school science: Knowing and learning in open–inquiry science laboratories.* Dordrecht, The Netherlands: Kluwer Academic Publishing.

Roth, W.-M. (Ed.). (2005a). *Auto/biography and auto/ethnography: Praxis of research method.* Rotterdam: SensePublishers.

Roth, W.-M. (2005b). Publish or stay behind and perhaps perish: Stability of publication practices in (some) social sciences. *Soziale Systeme, 11,* 129–150.

Roth, W.-M., & Barton, A. C. (2004). *Rethinking scientific literacy.* New York: Routledge.

Roth, W.-M., & Roychoudhury, A. (1992). The social construction of scientific concepts or The concept map as conscription device and tool for social thinking in high school science. *Science Education, 76,* 531–557.

Roth, W.-M., & Tobin, K. G. (2002). *At the elbow of another: Learning to teach by coteaching.* New York: Peter Lang.

Said, E. (1978). *Orientalism.* London, Penguin.

Scantlebury, K. (2005). A snake in the nest or in a snake's nest: Peer review for a female science educator. In W.-M. Roth (Ed.), *Auto/biography and auto/ethnography: Praxis of research method* (pp. 331–338). Rotterdam: SensePublishers.

Schmidt, W., Jorde, D. et al. (1996). *Characterizing pedagogical flow. An investigation of mathematics and science teaching in six countries.* Dordrecht, The Netherlands: Kluwer Academic Publishers.

Schreiner, C., & Sjøberg, S. (2006). Science education and youth's identity construction—two incompatible projects? In D. Corrigan, J. Dillon, & R. Gunstone, (Eds.), *The re-emergence of values in the science curriculum.* Rotterdam: Sense Publications.

Schreiner, C. (2006). *Exploring a ROSE–garden. Norwegian youth's orientations towards science— seen as signs of late modernities.* Doctoral dissertation, University of Oslo.

391

REFERENCES

Sewell, W. H. (1992). A theory of structure: Duality, agency, and transformation. *American Journal of Sociology, 98,* 1–29.

Sewell, W. H. (1999). The concept(s) of culture. In V. E. Bonnell & L. Hunt (Eds.), *Beyond the cultural turn* (pp. 35–61). Berkeley: University of California Press.

Sinnes, A. T. (2005). *Approaches to gender equity in science education: Two initiatives in sub-Saharan Africa seen through a lens derived from feminist critique of science.* Doctoral dissertation, University of Oslo.

Sjøberg, S. (2002). Science for the children? Report from the Science and Scientists project. *Acta Didactica.* University of Oslo.

Sjøberg, S. (2004). *Naturfag som allmenndannelse.* Oslo: Gyldendal Akademisk.

Sjøberg, S., & Imsen, G. (1987). Gender and science education. In P. Fensham (Ed.), *Development and dilemmas in science education.* London: Falmer.

Taylor, P. C., Gilmer, P. J., & Tobin, K. (Eds.). (2002). *Transforming undergraduate science education: Social constructivist perspectives.* New York: Peter Lang.

Tobin, K. (1993). *The practice of constructivism in science education.* Washington, DC: AAA Press.

Tobin, K. (2000). Becoming an urban science educator. *Research in Science Education, 30,* 89–106.

Tobin, K., Espinet, M., Byrd, S. E., & Adams, D. (1988). Alternative perspectives of effective science teaching. *Science Education, 72,* 433–451.

Tobin, K., & J. Gallagher (1987a). What happens in high school science classrooms? *Journal of Curriculum Studies, 19,* 549–560.

Tobin, K., & J. Gallagher (1987b). The role of target students in the science classroom. *Journal of Research in Science Teaching, 24,* 61–76

UNDP (2005). *Human development report 2005.* New York: United Nations Development Programme.

Varelas, M., Becker, J., Luster, B., & Wenzel, S. (2002). When genres meet: Inquiry into a 6th grade urban science class. *Journal of Research in Science Teaching, 39,* 579–605.

Varelas, M., & Benhart, J. (2004). Welcome to rock day. *Science and Children, 40*(1), 40–45.

Varelas, M., Luster, B., & Wenzel, S. (1999). Meaning making in a community of learners: Struggles and possibilities in an urban science class. *Research in Science Education, 29,* 227–245.

Varelas, M., Pappas, C., Barry, A., & O'Neill, A. (2001). Examining language to capture scientific understandings: The case of the water cycle. *Science and Children, 38*(7), 26–29.

Varelas, M., & Pineda, E. (1999). Intermingling and bumpiness: Exploring meaning making in the discourse of a science classroom. *Research in Science Education, 29,* 25–49.

von Aufschnaiter, S., & Welzel, M. (1999). Individual learning processes—A research programme with focus on the complexity of situated cognition. In *Research in Science Education in Europe* (pp. 209–215). Dordrecht, The Netherlands: Kluwer Academic Publishers.

Welzel M. et al. (1998). Teachers' objectives for labwork. Research tool and cross country results. In *Labwork in science education.* URL: http://www.physik.uni–bremen.de/physics.education/aufschnaiter

Welzel, M. (1997). Student centred instruction and learning processes in physics. *Research in Science Education, 27,* 383–394.

Welzel, M., & Roth, W.-M. (1998). Do interviews really assess students' knowledge? *International Journal of Science Education, 20,* 25–44.

Welzel, M., & Stadler, H. (Eds.). (2005). "Nimm doch mal die Kamera!" Zur Nutzung von Videos in der Lehrerbildung—Beispiele und Empfehlungen aus den Naturwissenschaften. Münster: Waxmann.

White, M. (1999). *Isaac Newton: The last sorcerer.* Toronto: HarperCollins.

Wilbers, J., & Duit, R. (2005). Post-festum and heuristic analogies. In P. J. Aubusson, A. G. Harrison, & S. M. Ritchie (Eds.), *Metaphors and analogy in science education* (pp. 37–49). Dordrecht, The Netherlands: Springer.

NAME INDEX

SUBJECT INDEX

A

Accountability, 33, 58, 254, 256, 257, 383
Achievement: gap, 30, 256; retention, 27, 28, 29, 32, 111, 155, 232, 295, 341, 382; science, 29, 30, 32, 48, 49, 50, 222, 257, 383, 384; student, 76, 212, 256, 257
Action, 4, 66, 70, 96, 102, 104, 106, 128, 133, 143, 144, 148, 190, 195, 200, 207, 212, 227, 264, 266, 321, 326, 372, 373, 382
Action research, 128, 133, 143, 144, 190, 195, 200, 227, 266, 321
Activity theory, 66, 134, 144, 303, 379, 381
Administration, 13, 17, 42, 65, 76, 77, 169, 182, 189, 285, 294, 297, 359
Agency, 57, 69, 134, 142, 201, 318, 326, 332, 351, 372
Alternative paradigms, 99
Anderson, Charles (Andy), 252
Antioch College, 13
Apartheid, 130, 339
Appleby College, 63, 64, 65
Argumentation, 245, 257, 272, 273
Arizona State University, 256, 329
Artifact, 21
Assertions, 64, 82, 335
Assessment, 19, 23, 33, 44, 76, 122, 173, 254, 256, 257, 306, 315, 326, 328, 344; formative, 18, 293; high-stakes, 256, 257; multiple-choice items, 38, 285, 330
Attitudes, 22, 30, 32, 34, 79, 93, 102, 105, 116, 199, 272, 316
Attractors: strange, 78
Ausubelian theory, 81, 113
Authentic assessment, 30
Authentic science, 64
Auto/biography, 3, 6, 69, 73, 105, 133, 147, 152, 154, 197, 220, 249, 285, 288, 323, 324, 341, 356, 376
Auto/ethnography, 56
Awards, 29, 40, 49, 52, 53, 55, 56, 58, 86, 98, 105, 153, 231, 279, 296, 312, 319, 324, 330, 331, 346, 358, 359, 362, 377, 378; Distinguished Contributions to Science Education Through Research, 153, 329, 331, 376, 378; outstanding paper, 40, 44, 128, 361

B

Bar Ilan University, 291
Behaviorism, 79, 108, 111, 126, 152, 225, 226, 243, 268
Belief, 2, 3, 22, 27, 43, 54, 102, 126, 129, 130, 133, 173, 175, 194, 197, 199, 200, 202, 204, 257, 259, 292, 293, 296, 297, 339, 345, 349, 356, 360, 361
Bildung, 108, 110, 115, 119, 151
Bloom's taxonomy, 123, 291
Bologna Declaration, 286
Brain worker, 311, 317
British Council, 97, 210, 296, 318
Brock University, 69
Bryn Mawr College, 135, 141

C

California Institute of Technology, 14
Capitalism, 68
Chaos theory, 113
Citation, 126, 156, 280, 281, 329, 381
Class: working, 20, 346, 355, 357, 370
Classroom: interactions, 34, 100, 292; research, 52, 149, 190, 214, 254, 330, 331, 334; studies, 106
Cogenerative dialogue, 57
Cognition, 64, 83, 331
Cognitive apprentice, 64, 67
Cognitive structure, 127, 376
Colgate University, 13
Collaboration, 11, 18, 19, 38, 39, 40, 43, 45, 52, 53, 55, 56, 57, 66, 69, 74, 76, 80, 81, 87, 92, 122, 126, 128, 129, 133, 134, 135, 136, 137, 138, 139, 143, 144, 145, 150, 151, 152, 153, 154, 179, 181, 182, 188, 193, 194, 195, 200, 221, 227, 246, 248, 249, 254, 265, 266, 269, 272, 298, 320, 327, 334, 356, 380, 383; international, 87, 101, 193, 194, 195, 272, 309
Collective, 3, 4, 5, 6, 58, 59, 68, 73, 79, 85, 154, 158, 200, 201, 205, 226, 324, 325, 342, 343, 381
College science teaching, 140
Colonizing, 341, 348, 368, 373
Commodification, 73, 85, 231

ABOUT THE AUTHORS

Ken Tobin is Presidential Professor at the Graduate Center of the City University of New York. In 2004 he was recognized by the National Science Foundation as a *Distinguished Teaching Scholar* and by the Association for the Education of Teachers of Science as *Outstanding Science Teacher Educator of the Year.* Prior to commencing a career as a teacher educator, Ken taught high school science and mathematics in Australia and was involved in curriculum design. His research focuses on the teaching and learning of science in urban schools, which involve mainly African American students living in conditions of poverty. A parallel program of research focuses on coteaching as a way of learning to teach in urban high schools. Recently Ken edited a Handbook about *Teaching and Learning Science* (Praeger), *Doing Educational Research* (with Joe Kincheloe), and *Improving Urban Science Education* (with Rowhea Elmesky and Gale Seiler). Earlier this year Ken, with Michael, co-authored a third book on coteaching and the uses of cogenerative dialogues—*Teaching to Learn.* With Michael, Ken is founding co-editor of *Cultural Studies of Science Education.*

Wolff-Michael Roth is Lansdowne Professor of Applied Cognitive Science at the University of Victoria, British Columbia, Canada. For most of the 1980–1992 period, he taught science, mathematics, and computer science at the middle and high school levels. From 1992 on, already working at the university, he taught science in British Columbia elementary schools at the fourth- through seventh-grade levels always associated with research on knowing and learning. More recently, he has conducted several ethnographic studies of scientific research, a variety of workplaces, and environmental activist movements. His research focuses on cultural-historical, linguistic, and embodied aspects of scientific and mathematical cognition and communication from elementary school to professional practice, including, among others, studies of scientists, technicians, and environmentalists at their work sites. His recent books include *Rethinking Scientific Literacy* (2004, with A. C. Barton), *Talking Science: Language and Learning in Science Classrooms* (2005), *Doing Qualitative Research: Praxis of Method* (2005), and *Learning Science: Singular Plural Perspectives* (2006).

NEW DIRECTIONS IN MATHEMATICS AND SCIENCE EDUCATION

Volume 1
Learning Science:
A Singular Plural Perspective
W.-M. Roth, *University of Victoria, Canada*
Paperback ISBN 90-77874-25-9 Hardback ISBN 90-77874-26-7

Volume 2
Theorems in School:
From History, Epistemology and Cognition to Classroom Practice
P. Boero, *Universita di Genova, Italy* (Ed.)
Paperback ISBN 90-77874-21-6 Hardback ISBN 90-77874-22-4

Volume 3
The Culture of Science:
Historical and Biographical Perspectives
K. Tobin, *The Graduate Center, City University of New York*, USA & W.-M. Roth, *University of Victoria, Canada* (Eds.)
Paperback ISBN 90-77874-33-X Hardback ISBN 90-77874-35-6

Volume 4
Understanding Teacher Expertise in Primary Science:
A Sociocultural Approach
A. Traianou, *Goldsmiths College, University of London, UK*
Paperback ISBN 90-77874-88-7 Hardback ISBN 90-77874-89-5

By the same authors:

Doing Educational Research: *A Handbook*
Kenneth Tobin, *The Graduate Center, CUNY, USA* and **Joe Kincheloe,** *McGill University, Montreal, Canada* **(eds.)**

Doing Educational Research explores a variety of important issues and methods in educational research. Contributors include some of the most important voices in educational research. In the handbook these scholars provide detailed insights into one dimension of the research process that engages both students as well as experienced researchers with key concepts and recent innovations in the domain. The editors and authors believe that there is a need for a handbook on educational research that is both practical as it introduces beginning scholars to the field and innovative as it pushes the boundaries of the conversation about educational research at this historical juncture.

In this collection the authors explore a variety of topics from methodologies such as ethnography, action research, hermeneutics, historiography, psychoanalysis, literary criticism to issues such as social theory, epistemology, and paradigms. The book addresses complex topics in an accessible and readable manner. The book will be very useful as a text in educational research at the graduate and the undergraduate level.

September 2006, 480 pp
paperback: ISBN:90-77874-48-8
hardback: ISBN:90-77874-01-1
SERIES: BOLD VISIONS IN EDUCATIONAL RESEARCH 1

Auto/Biography and Auto/Ethnography: *Praxis of Research Method*
W. -M. Roth, *University of Victoria, Canada* **(ed.)**

In a number of academic disciplines, auto/biography and auto/ethnography have become central means of critiquing of the ways in which research represents individuals and their cultures. The contributors to this volume explore, by means of examples, auto/biography and auto/ethnography as means for critical analysis and as tool kit for the different stakeholders in education.

The book was written to be used by upper undergraduate and graduate students taking courses in research design andd professors, who want to have a reference on design and methodology.

July 2005, 448 pp
paperback: ISBN:90-77874-04-6
hardback: ISBN:90-77874-49-6
SERIES: BOLD VISIONS IN EDUCATIONAL RESEARCH 2

Doing Qualitative Research: *Praxis of Method*
W. -M. Roth, *University of Victoria, Canada*

The author takes readers on a journey of a large number of issues in designing actual studies of knowing and learning in the classroom, exploring actual data, and putting readers face to face with problems that he actually or possibly encountered, and what he has done or possibly could have done. The reader subsequently sees the results of data collection in the different analyses provided. The book is organized around six major themes (sections), in the course of which it develops the practical problems an educational researcher might face in a large variety of settings.

The book was written to be used by upper undergraduate and graduate students taking courses in research design and professors who want to have a reference on design and methodology.

August 2005, 508 pp
paperback: ISBN:90-77874-05-4
hardback: ISBN:90-77874-51-8
SERIES: BOLD VISIONS IN EDUCATIONAL RESEARCH 3

Learning Science: *A Singular Plural Perspective*
W.-M. Roth, *University of Victoria, Canada*

How do you *intend* (to learn, know, see) something that you do not yet know? Given the theory-laden nature of perception, how do you *perceive* something in a science demonstration that requires knowing the very theory that you are to learn? In this book, the author provides answers to these and other (intractable) problems of learning in science. He uses both first-person, phenomenological methods, critically analyzing his own experiences of learning in unfamiliar situations *and* third-person, ethnographic methods, critically analyzing the learning of students involved in hands-on investigations concerning motion and static electricity.

This book, which employs the cognitive phenomenological method described in the recently published *Doing Qualitative Research: Praxis of Method* (See page 1 of this brochure), has been written for all those who are interested in learning science: undergraduate students preparing for a career in science teaching, graduate students interested in the problems of teaching and learning of science, and faculty members researching and teaching in science education.

March 2006, 372 pp
paperback: ISBN:90-77874-25-9
hardback: ISBN:90-77874-26-7
SERIES: NEW DIRECTIONS IN MATH AND SCIENCE EDUCATION 1

Teaching to Learn: *A View from the Field*
Kenneth Tobin, *The Graduate Center, CUNY, USA* and **W.-M. Roth**, *University of Victoria, Canada*

A recurrent trope in education is the gap that exists between theory, taught at the university, and praxis, what teachers do in classrooms. How might one bridge this inevitable gap if new teachers are asked to learn (to talk) about teaching rather than to teach? In response to this challenging question, the two authors of this book have developed coteaching and cogenerative dialoguing, two forms of praxis that allow very different stakeholders to teach and subsequently to reflect together about their teaching. The authors have developed these forms of praxis not by theorizing and then implementing them, but by working at the elbow of new and experienced teachers, students, supervisors, and department heads. Tobin and Roth describe the many ways coteaching and cogenerative dialogues are used to improve learning environments—dramatically improving teaching and learning across cultural borders defined by race, ethnicity, gender, and language. Teaching to Learn is written for science educators and teacher educators along the professional continuum: new and practicing teachers, graduate students, professors, researchers, curriculum developers, evaluation consultants, science supervisors, school administrators, and policy makers. Thick ethnographic descriptions and specific suggestions provide readers access to resources to get started and continue their journeys along a variety of professional trajectories.

July 2006, 282 pp
paperback: ISBN:90-77874-81-X
hardback: ISBN:90-77874-91-7
SERIES: NEW DIRECTIONS IN MATH. AND SCIENCE EDUCATION 4

Out in 2007

Science, Learning, and Identity: *Sociocultural and Cultural-historical Perspectives*
W. -M. Roth, *University of Victoria, Canada and* **Kenneth Tobin,** *The Graduate Center, CUNY, USA* (eds.)

For more information on these and our other titles go to
WWW.SENSEPUBLISHERS.COM

Printed in the United Kingdom
by Lightning Source UK Ltd.
121359UK00001B/52/A